Cosmology, History, and Theology

Contributors

Hannes O. Alfvén

Asim O. Barut

Peter G. Bergmann

Allen D. Breck

Robert S. Cohen

Patrick A. Heelan

Philip J. Hefner

Jean Heidmann

Michael A. Hoskin

Stanley L. Jaki

Philip J. Lawrence

William H. McCrea

Jacques Merleau-Ponty

Harles W. Misner

Kristian P. Moesgaard

John D. North

Arthur R. Peacocke

David W. Peat

Arno A. Penzias

Karl Philberth

Kenji Tomita

Jean-Pierre Vigier

Gerald J. Whitrow

Wolfgang Yourgrau

Cosmology, History, and Theology

Edited by

Wolfgang Yourgrau
and Allen D. Breck

University of Denver
Denver, Colorado

WITHDRAWN

PLENUM PRESS · NEW YORK AND LONDON

Library of Congress Cataloging in Publication Data

Main entry under title:

Cosmology, history, and theology.

"Based on the third international colloquium held at the University of Denver,
November 5-8, 1974."
Includes bibliographical references and index.
1. Cosmology—Congresses. 2. Astronomy—History—Congresses. 3. Theology—
Congresses. I. Yourgrau, Wolfgang. II. Breck, Allen duPont. III. Denver. University.
QB981.C82   523.1   76-54269
ISBN 0-306-30940-8

Based on the Third International Colloquium held at the University of
Denver, November 5—8, 1974

© 1977 Plenum Press, New York
A Division of Plenum Publishing Corporation
227 West 17th Street, New York, N.Y. 10011

Printed in the United States of America

This volume is dedicated to the memory of
a great cosmogonist
Abbé Lemaître

FOREWORD

It is difficult to doubt that we suffer at present from the manifold aspects of an economic crisis which affects all walks of life. Well, men in almost every epoch in history have maintained that they were going through a crisis which was supposed to be always more grave than any preceding critical phase. Very often those crises were not of an economic nature, but concerned either health, the political structure, the opportunity of acquiring knowledge, and so on.

I think that we would consider today that some of those claims that were made in various historical epochs were often exaggerated if viewed from a historical point of view. However, it seems undeniable that we at present are in the middle of a universal economic crisis which has affected almost every facet of our daily life.

And yet, the fact that despite these adverse conditions it is still possible to gather scholars from all corners of the world to deal with often sheer theoretical and sometimes abstract pursuits is a refutation of any facile pessimism—it is reassuring to all who wonder where political and social events are taking us. Our salvation may well come from those acts of the mind so characteristic of the pure scientist and scholar.

The appearance of this volume, which is devoted to cosmology, history of science, and theology, confirms my firm belief that man's craving for knowledge and understanding is an innate and permanent characteristic, and that there will always be men dedicated to overcoming ignorance and extending and deepening knowledge in the various domains of scholarly study.

This International Colloquium demonstrates clearly that, in spite of the economic crisis and world-wide social unrest, man is capable of devoting the best part of his mind to issues that are non-practical, and not even social or economic—issues such as cosmology, history, and theology. Why have we chosen these subjects for this Colloquium? The answer is very simple: That we continue to study the nature of our Universe and the history of scientific ideas and achievements, as well as the everlasting questions raised in theology, is proof that we are capable of confronting these great topics even while the search to improve socioeconomic conditions goes on. This Colloquium is very convincing evidence that islands of pure search for knowledge are the firmest ground for optimism and hope for the future of man's intellectual survival and growth in the midst of the sometimes depressing circumstances surrounding the more prosaic fields of human existence.

We are very proud that our little family here forms such an island. I am particularly happy that the scholarly endeavors exemplified by the contributions to this Colloquium in the form of lectures, discussions, and dialogues took place at the University of Denver. And I wish to conclude my remarks by thanking all of those involved, in the name of the University at large and myself, for having confirmed and strengthened my belief that nothing whatsoever can arrest our intrinsic faith in the advance of human knowledge. And the proper forum to communicate such advance is, in our age, still the University.

University of Denver   Maurice B. Mitchell
Denver, Colorado   Chancellor

PREFACE

The University of Denver has held two highly successful international conferences, both of which were a somewhat new experiment in American academic life, within the past few years. Participating were scholars in the fields of physics, biology, history, and philosophy, from Europe, the Western Hemisphere, and Africa. These first two colloquia were made possible by grants from the Martin-Marietta Corporation Foundation.

The results of these conferences were published in two substantial volumes, which received world-wide distribution and testified to the importance and significance of these conferences. The first volume, *Physics, Logic, and History*, was published in hard cover by Plenum Press (New York and London) in 1970. It was dedicated to George Gamow, one of our chief participants. Results of Colloquium II (again published by Plenum Press) appeared in 1972 and the volume was dedicated to another participant, Albert Szent-Györgyi. This volume was entitled *Biology, History, and Natural Philosophy*. The paperback edition appeared in 1974.

We are now engaged in a third major international colloquium. This time we intend to address ourselves to the important problems which appear under the heading, *Cosmology, History, and Theology*. We are assembling this year some twenty American and international cosmologists, historians of science, and theologians. The international character of the conference is of vital relevance to our third effort. By considering the enormous implications which current research brings to all of us, and not only to the academic community, these international scholars will approach as closely as possible the solution of the issues and problems which lie ahead.

We believe that workers in each of the three disciplines—cosmology, history, and theology—have much to say to contemporary man about the nature of the world in which he lives. But we also believe that scholars should not devote themselves to their disciplines in isolation. Hence our desire to bring them together, to speak to one another, and to write for the widest possible audience.

The Board of Higher Education of the United Methodist Church has contributed funds for this Third Colloquium in honor of Chester M. Alter, who was Chancellor of the University of Denver from 1953 to 1966. He is a vigorous proponent of the physical sciences, a liberal education, and the integrative approach to all learning symbolized by this Colloquium.

Further financial support of Colloquium III comes from the firm of Nelson, Haley, Patterson, and Quirk, Inc., of Greeley, Colorado, a subsidiary of C-E Tec, Waltham, Massachusetts.

The Editors

PARTICIPANTS

HANNES O. ALFVÉN, Applied Physics and Astrophysics, the Royal Institute of Technology, Stockholm and the University of California, San Diego. General cosmology, origin and evolutionary history of the solar system, plasma physics; Nobel prize in physics 1970 for fundamental work in magnetohydrodynamics.

ASIM O. BARUT, Physics, University of Colorado. Elementary particles, gravitation, mathematical physics, statistical mechanics, quantum field theory.

PETER G. BERGMANN, Physics, Syracuse University. General researches in theoretical physics, particularly in special and general relativity theories, gravitation, irreversible processes and statistical mechanics, underwater sound inquiries, mathematical physics; assistant to Albert Einstein, 1936–1941.

ALLEN D. BRECK, History, University of Denver. Cochairman of the Colloquium; medieval history, historiography, philosophy of history, Copernican studies, Wyclyf studies.

ROBERT S. COHEN, Physics and Philosophy, Boston University. Editor of *Boston Studies in the Philosophy of Science*; history and philosophy of science, plasma physics, relations of history and physics.

JÜRGEN EHLERS, Astrophysics, Max-Planck-Institut für Physik und Astrophysik, Munich. Mathematical foundations of Einstein's theory of gravitation, description of matter in GRT and applications of GRT to astrophysics and cosmology, axiomatics in physics, relationship between physical theories.

PATRICK A. HEELAN, S.J., Philosophy, State University of New York at Stony Brook. Elastic wave propagation in high-energy physics and in inhomogeneous media, philosophy of quantum mechanics, quantum logic, geometry of pictorial space, philosophy of perception, the hermeneutics of experimental science.

PHILIP J. HEFNER, Systematic Theology, Lutheran School of Theology, Chicago. Nineteenth-century German theology, the interrelationship between science and theology, contemporary philosophy; editor of *Zygon, Journal of Religion and Science*.

JEAN HEIDMANN, Astronomy, Paris Observatory in Meudon. Cosmology, galactic structure, radioastronomy; editor of *Astronomy and Astrophysics*.

MICHAEL A. HOSKIN, History and Philosophy of Science, University of Cambridge. Founder-editor of *History of Astronomy* (journal), editor of *General History of Astronomy*, Coeditor of *Journal of the History of Science*; sidereal astronomy, especially William Herschel.

STANLEY L. JAKI, Benedictine priest; History of Science, Seton Hall University. History of Astronomy, philosophical and theological presuppositions in the development of cosmology, history of physics, lack of finality in physical theories, their interaction with other disciplines of human culture; Gifford lecturer at the University of Edinburgh, 1974–1975 and 1975-1976.

PHILIP J. LAWRENCE, History of Science, Harvard University. Histories of cosmology, biology, geology, and theology in the 19th century and their interactions; theory of uniformitarianism.

EDWARD A. LINDELL, Dean of the Faculty of Arts and Sciences, University of Denver.

WILLIAM H. McCREA, FRS, Emeritus, Astronomy, University of Sussex. Astrophysics, relativity theories, cosmology, rate of accretions of matter by stars, mechanism for radio-galaxies and quasars, the Universe, its structure and properties; coeditor of *Cambridge Monographs on Mathematical Physics* (Cambridge U.P.).

JACQUES MERLEAU-PONTY, Epistemology, University of Paris, Nanterre. Cosmology (especially of the eighteenth and nineteenth centuries), physical time, causality, structure and development of modern sciences.

CHARLES W. MISNER, Physics and Astronomy, University of Maryland. Gravitational collapse, conservation laws and the boundary of a boundary, relativity and topology, cosmology and gravitational radiation.

MAURICE B. MITCHELL, Chancellor, University of Denver.

KRISTIAN P. MOESGAARD, History of Science, University of Aarhus. Mathematical astronomy before Newton, Copernican theory of precession, Copernican influence on Tycho Brahe, comparison between Arab scholars, Ptolemy, and Copernicus.

JOHN D. NORTH, History of Modern Cosmology and the Exact Sciences, Museum of the History of Science, University of Oxford. Editor-in-chief of *Archives internationales d'histoire des sciences*; medieval intellectual life, especially mathematics and astronomy, origins of calculus of tensors, interrelationship of geometry and logic in the nineteenth century.

ARTHUR R. PEACOCKE, Dean of Clare College, University of Cambridge. Theology: science and theology, thoughts about the Eucharist. Biophysical chemistry: molecular basis of heredity, physical chemistry of biological macromolecules, e.g., osmotic pressure of biological macromolecules.

DAVID W. PEAT, Institute of Astronomy, University of Cambridge. Astrophysics, especially stellar spectroscopy, evolution and chemical history of stars and the galaxy, and related fields of nucleosynthesis; theology of nature and its cosmological aspects.

ARNO A. PENZIAS, Radioastronomy Physics, Bell Laboratories, Crawford Hill Laboratory. Microwave physics, satellite communications, atmospheric physics, interstellar matter cosmology, astronomical uses of TV.

KARL PHILBERTH, fam. O.S.B., Puchheim, West Germany. Glaciology, Cold Regions Research and Engineering Laboratory, Hanover, New Hampshire. The relevance of general relativity theory and cosmology to religious problems.

KENJI TOMITA, Physics, Research Institute for Theoretical Physics, Hiroshima University. Cosmology, general relativity, gravitational collapse, galaxy formation, chaotic universe.

JEAN-PIERRE VIGIER, Theoretical Physics, Institut Henri Poincaré, Paris. Cofounder with Louis de Broglie and David Bohm of a new elementary-particle model; proponent of nonzero photon rest mass; atomic physics, general relativity theory, cosmology, theory of levels in physical reality.

GERALD J. WHITROW, History and Applications of Mathematics, Imperial College of Science and Technology, University of London. Cosmology, cosmogony, general relativity theory, history of astronomy, of cosmology, and of mathematics; aspects of the study of time.

WOLFGANG YOURGRAU, History of Science, University of Denver. Cochairman of the Colloquium. Quantum theory, gravitation, general relativity, classical and nonclassical thermophysics, variational principles, cosmology; cofounder–editor of *Foundations of Physics*; assistant to Erwin Schrödinger, 1930-1933.

CONTENTS

Chapter I

Cosmology: Myth or Science?

Hannes Alfvén

University of California, San Diego
and
Royal Institute of Technology, Stockholm

EARLY COSMOLOGY

Cosmology began when man began to ask: What is beyond the horizon and what happened before the earliest event I can remember? The method of finding out was to ask those who had traveled very far, and they reported what they had seen and also what people they had met far away had told them about still more remote regions. Similarly, grandfather told about *his* young days and what his grandfather had told him and so on. But the information became increasingly uncertain the more remote the regions and the times.

The increasing demand for knowledge about very remote regions and very early times was met by people who claimed they could give accurate information about the most distant regions and the earliest times. When asked how they could know all this they often answered that they had direct contact with the gods, and got revelations about the structure of the whole universe and how it was created. And some of these prophets were believed by large groups of people, and myths about the creation and structure of the universe were incorporated as essential parts of religious traditions.

In early mythologies the whole world was usually considered to be eternal. When the gods "created" the world, this meant essentially that they brought order into an initial chaos. In the Mediterranean and Middle East regions the creation was held to have taken place several thousand years ago, whereas in India the time scale is much more grandiose. In some cases time was counted in *Kalpas*, or days of Brahma; one Kalpa is 4 or 5 billion years.

1

COSMOLOGY AND SCIENCE[1]

Very early it was obvious that astronomy was important for cosmology. The rise of science and philosophy, especially in Egypt and Greece, of course, influenced the views of how the universe was structured. Above all the Pythagorean way of thinking was decisive. The discovery that music could be understood in terms of simple mathematical relations as well as the development of geometry, introduced a new era in philosophy and science. These discoveries necessarily had an enormous impact on Platonic and Aristotelian thinking.

THE PTOLEMAIC SYSTEM

The Ptolemaic system was a result of this outlook. One may say that its basic principle is that because the world was created by the gods, there must be a sublime order in its basic structure—even if many regrettable local disorders were obvious. According to the Pythagoreans, the most "perfect" of all geometrical figures is the circle, and the most "perfect" of all solid bodies is the sphere. *Ergo* the Earth must be a circular disk or a sphere, surrounded by a number of crystal spheres, on which the planets and the stars were located. Further, the most perfect motion was uniform motion. *Ergo* the crystal spheres must rotate with uniform velocity.

THE CREATION *EX NIHILO*

The rise of the monotheistic religions meant that one of the gods became more important than the others; he became God. He also became more important than the material world. He alone was eternal: The whole world was a secondary structure created by Him. In the Bible the creation takes six days. To a certain extent it still has the character of bringing order into a preexisting chaos, but soon creation was thought of as the production of the world *ex nihilo*. God is powerful enough to create the whole world by just pronouncing some magic words, or by his will power. In the Aristotelian philosophy the material world was "ungenerated and indestructible." It was not until medieval times that Aristotle's views were accommodated to the idea of creation *ex nihilo*, essentially by St Thomas, who remodeled the Aristotelian philosophy in accordance with the requirements of ecclesiastical doctrine.[1]

---

[1] Detailed accounts of the historical development of scientific thinking are found in a number of papers presented at this conference. See also Singer.[1]

COMPARISON WITH OBSERVATION

In one respect this Ptolemaic cosmology seemed to be confirmed by observation: The outermost crystal sphere, the one on which the stars were fixed, did apparently move with a constant speed. This was just what could be expected because this sphere was the outermost one, closest to God, and hence most divine. Unfortunately the theory did not agree so well with observational results when applied to the planets, including the Sun and Moon. The Sun and the Moon sometimes moved more to the north, sometimes to the south, and a planet like Jupiter sometimes reversed its motion in relation to the stars.

It was obvious that something was wrong. But the basic principles—uniform motion and perfect geometrical figures—were sacrosanct and could not be given up even if they were in conflict with observations. Instead, a very ingenious idea was forwarded: The planets were not directly fixed on the crystal spheres, but each was on a small circle—an epicycle—which moved with a constant velocity with its center fixed on the crystal sphere. For a time this theory looked promising, but better observations soon demonstrated that it was not accurate. Increasingly complicated additions to the system were made, and one can very well understand what was meant by the famous astronomer, King Alphonse X of Castile, when he said: "Had I been present at the creation, I could have rendered profound advice." But at the same time as the system became more complicated, it also became more sacrosanct.

MYTHICAL VERSUS SCIENTIFIC APPROACH

The Ptolemaic system was initially a quite attractive theory but, during the centuries, it developed into an almost sacred and rigid structure incapable of incorporating new discoveries. The reason for this was that basically the approach was not scientific but mythological.[2] The basic ideas were the perfect geometrical figures and uniform motion. The idea of building a world system on such general principles represented great progress, because earlier it had been generally believed that events in the world were governed by the will or the whimsies of the gods. The Ptolemaic system did not necessarily question that the celestial system was created by God, but it claimed that He must have acted according to certain philosophical or mathematical principles which it was possible to analyze and understand.

---

[2] It is a semantic question whether a model initially deriving from "divine inspiration" should be called a myth even if it includes philosophical and mathematical elements. Some would no doubt prefer to call it, for example, "*a priori* metaphysics."

The Pythagorean philosophy had a logical beauty which well could be called "divine." By pure abstract thinking the theoreticians claimed to have discovered the principles according to which God acted when he created the world. And when these principles were found, it was held that the world *must* be structured according to them. Observations of physical reality were not really necessary. The system was based on divine inspiration. If Galileo claimed that he saw celestial bodies or sunspots which should *a priori* not exist, it was his telescope and not the theoretical system which was wrong.

THE COPERNICAN SYSTEM

Under the impact of more accurate observations the Ptolemaic system was replaced by the Copernican system. The real importance of this was not that a geocentric system was replaced by a heliocentric system, but that the latter could develop in such a way that it was able to absorb new empirical material, supplied by Tycho Brahe and many others. In the hands of Galileo, Kepler, and Newton it became a cosmology which was not primarily based on some philosophical or mathematical principle. Instead it was an empirical synthesis, a summary of all astronomical observations ever made. It led to the discovery of new basic laws of nature, which agreed very well with the observed motions of the celestial bodies and indeed were simpler and even more beautiful than the old laws. But it is important to note that these laws are not everlasting. When it became obvious that Newtonian mechanics was not applicable to atoms, it was replaced in this field by quantum mechanics.

The difference between myth and science is the difference between divine inspiration or "unaided reason" (as Bertrand Russell puts it) on one hand and theories developed in observational contact with the real world on the other. Newton said, "Hypotheses non fingo."

VICTORIOUS SCIENCE

It took more than two centuries for the victory of science over myth in the field of celestial mechanics to spread to the realm of biology.

In our century the scientific approach has embraced other areas which earlier were alien to it, such as the origin of life and the functioning of the human brain.

THE NEW MYTHS

However, this does not mean a complete and definite victory of common sense and science over myth. In reality we witness today an antiscientific attitude and a revival of myth. This tendency may have several causes, but in a way the most interesting and also most dangerous threat comes from science itself. And in a true dialectic sense it is the triumph of science which has released the forces which now once again seem to make myths more powerful than science.

One of the most beautiful results of science was the *special* theory of relativity. It was essentially based on the Michelson–Morley experiment and on Maxwell's theory of electromagnetism, which in an elegant way described all the results of the study of electric, magnetic, and optical phenomena. Already when expressed in an ordinary three-dimensional Cartesian coordinate system the special theory of relativity is a beautiful theory, but its mathematic beauty is definitely increased somewhat if it is expressed in four-dimensional space.

This fact was given an enormous importance. It was claimed that "Einstein has discovered that space is four dimensional," a statement which obviously is nonsense. But it had great publicity value. After one or two decades of propaganda the four-dimensional world became immensely popular, especially when one learned that the fourth dimension was not time, but time multiplied by $\sqrt{-1}$.

To most people this was impossible to understand. Indeed it requires considerable mathematical insight to understand the deep meaning of it—and still profounder insight to realize that it is mostly mathematical jargon with not much valuable meaning for our views of physical reality.

Many people probably felt relieved by being told that the true nature of the physical world could not be understood except by Einstein and a few other geniuses. They had tried hard to understand science, but now it was evident that science was something to believe in, but nothing possible to understand. Paradoxically enough, Einstein may have been hailed by the general public not because he was a great thinker, but because he saved everybody from the duty of having to think.

Soon the bestsellers among the popular science books became those that presented scientific results as insults to common sense. The more abstruse the better! Contrary to Bertrand Russell, science became increasingly presented as the negation of common sense. The limit between science and pseudoscience was erased. To most people it was increasingly difficult to find any difference between science and science fiction.

THE THEORY OF RELATIVITY

But let us return to the theory of relativity and its direct impact on scientists. The four-dimensional presentation of the *special* theory of relativity was rather innocent. This theory was and is used every day in laboratories for calculating the behavior of high-energy particles, etc. As experimental physicists have a strong feeling that their laboratories are three dimensional, and firmly located in a three-dimensional world, the four-dimensional formulation is taken for what it is: A nice little decoration comparable to a cartoon or a calendar pin-up on the wall.

On the other hand, in the *general* theory of relativity, the four-dimensional formulation is somewhat more important. This theory is also more dangerous because it came into the hands of mathematicians and cosmologists, who had very little contact with empirical reality. Furthermore, they applied it to regions which are very distant, and counting dimensions far away is not very easy. Many of these scientists had never visited a laboratory or looked through a telescope, and even if they had, it was below their dignity to get their hands dirty. They looked down on the experimental physicist whose only job was to confirm the high-brow conclusions they had reached, and those who were not able to confirm them were thought to be incompetent. Observing astronomers came under heavy pressure from theoreticians.

GENERAL RELATIVITY AND THE UNIVERSE

The general theory of relativity opened an extremely fascinating possibility. Similar to the Earth's surface, which is without borders but is still finite, one can in a four-dimensional space have a "hypersphere" without any limits and still with a finite volume. This idea was certainly worthwhile to investigate.

Einstein's equations allowed for a type of solutions which suggested a Universe in a state of expansion. Some of the solutions had a "singular point," meaning that at a certain time the whole Universe consisted of one single point only. From this singular point the Universe began to expand, so that later all parts of it rush away from each other with velocities which are proportional to the distance between them. These types of mathematical solutions seemed to be applicable to the "expanding Universe" which Hubble's famous empirical law describes. The way was now open for a grand new cosmology.

The originator of this was Abbé Lamaître, who called the Universe when it was at the singular point "l'Atome Primitive." Its great propagandist was Gamow, but it was Hoyle who humorously named the beginning of the expansion the "*Big Bang*."[3] Neither Lemaître nor Gamow went to the extreme in postulating that the whole Universe ever was a mathematical *point*. The "initial state" was supposed to be a concentration of "all mass in the Universe" in a very small sphere. This mass is heated to a temperature of several billion degrees. When this "atomic bomb" explodes, its parts are thrown out with relative velocities which are sometimes close to the velocity of light.

This model, which at least from certain points of view was fascinating, was believed to explain the main evolution and the present structure of the Universe.

---

[3] A detailed account of the Big-Bang theory is given by Heidmann.[2]

The following consequences were claimed to derive from it:

1. In less $\frac{1}{2}$ hr after the explosion the elements we find now were formed by nuclear reactions in the very hot and very dense matter.

2. At an early time a heat radiation was produced which, on further expansion, cooled down and should be now observed as a blackbody radiation with a temperature of $50°K$. With revised values for galactic distances this temperature was adjusted down to $20°$. (Through a number of *ad hoc* assumptions it can of course be reduced still more.)

3. At a later stage the expanding matter condensed to form the galaxies we observe today.

4. The average density in the Universe should be at least 10^{-29} g/cm^3. (This value should be corrected for the new determination of galactic distances.)

5. A fifth conclusion, which is seldom drawn explicitly, is that the state at the singular point necessarily presupposes a divine creation!

To Abbé Lemaître this was very attractive, because it gave a justification to the creation *ex nihilo*, which St Thomas had helped establish as a credo. To many other scientists it was more of an embarrassment because God is very seldom mentioned in ordinary scientific literature. Hence the problem of how the singular state was produced is usually not raised. There have been a number of attempts to explain how the singular state could be reached from an early state similar to the present state in the Universe, but none seems to be successful.

BIG BANG AND OBSERVATIONS

Only by working out the consequences of a model is it possible to check whether it gives a fair description of the real world or not. Hence it was quite legitimate to devote much work to the evaluation of the Big-Bang model. After nearly half a century of work, the time seems ripe for drawing conclusions about the validity of the model. These are discouraging.[4] The model obviously cannot explain a number of the phenomena it claimed to account for, and the observations seem to disagree with the theoretical predictions.

1. It seems impossible to account for the formation of the elements by the Big-Bang process. Perhaps the observed abundance of helium could have been formed by it, but for the other composite elements the observed abundances are wrong by orders of magnitude. Hence the measured cosmic abundances do not give the expected support to the Big-Bang theory (or rather conjecture), but neither do they contradict it.

2. An isotropic microwave radiation has been detected, which the Big-Bang propagandists have baptised to "the $3°K$ blackbody radiation" in spite of the fact that it still is impossible to decide whether it is a blackbody or a greybody

---

[4] A review of objections to the Big-Bang theory has been given by Burbidge.[3]

radiation, and whether it is a temperature radiation. Looking at the observational material without Big-Bang goggles, the greybody interpretation of the observational data seems somewhat more likely, and this would disprove the Big-Bang cosmology. However, it cannot be *excluded* at present that it is a 3° black-body radiation. If this is so, the Big-Bang cosmologists claim that this is the isotropic radiation which they expected in spite of the fact that they supposed a radiation with about seven times the temperature, and hence several thousand times higher energy density. (Of course, it is easy to introduce additional effects to explain the discrepancy.)

In reality they claim that the existence of this very "cold" radiation proves that the temperature of the universe once was *10 billion degrees* (sic!), an extrapolation over more than nine orders of magnitude. This extrapolation requires among many other things, that we know the state of the universe at all epochs after the Big Bang with such a certainty that we can exclude that microwave radiation was produced later. It should be noted that there are also isotropic x and γ radiations which need other explanations, and that there are a number of celestial objects (quasars, etc.) which release enormous energies which the Big-Bang cosmologists do not claim to understand.

3. The Universe as we see it is obviously *not* homogeneous, as the Big-Bang model requires, but consists of a multitude of galaxies. These must have been formed at some evolutionary stage of the Universe, but so far no acceptable theory of the formation of galaxies has been derived from the Big-Bang model.

4. Even if it can be claimed that galaxies are "local" phenomena which need not necessarily be included in a large-scale cosmology, it is more difficult to neglect the existence of large, sometimes very large, clusters of galaxies. Most embarrassing is that the distant quasars ($z = \Delta\lambda/\lambda > 1.5$) are located exclusively in two regions, one near the galactic north pole and the other in the south galactic hemisphere. Hence the large-scale isotropy of the universe, which is a cornerstone in the Big-Bang cosmology, is contradicted by observations.

5. The mean density in the Universe is, according to observations, 10^{-31}, a factor of almost 100 too low. Intense efforts have been made to find the 99% which is missing ("black footballs" and "black holes" have been suggested) but there is no support—rather counterindications—of their existence. (Revision of galactic distances changes both the theoretical and the observational value but does not decrease the discrepancy.)

6. The validity of Hubble's law cannot be counted as support of the theory, because, of an infinity of possible mathematical solutions, one was chosen which fitted Hubble's law. Recent results, however, make it doubtful to what extent Hubble's law really is valid. An increasing number of flagrant deviations from the linearity of Hubble's law have been reported.[5] This is serious to the Big-Bang model.

---

[5]The variations of the Hubble parameter are summarized by Vigier.[4] See also Fig. 4 in Heidmann's paper.[2]

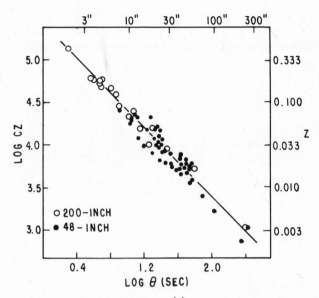

Fig. 1. Measurements of the Hubble parameter.[7] The diagram gives the redshift as a function of apparent diameter θ of the galaxies. It can be interpreted as giving the velocity as function of distance (inversely proportional to θ).

There are regions of galaxies with redshifts systematically different by 20% (Vera Rubin). Furthermore, there are pairs of galaxies with vastly different redshifts, which seems to show that there are other galaxy populations not obeying Hubble's law. This, too, is fatal to the Big-Bang model.

7. The Hubble diagrams no doubt show that the metagalaxy (Universe) is expanding, which means that at earlier times the galaxies were closer together. But it is not legitimate to conclude that observations suggest that once all the galaxies were formed by one big lump of matter. With the uncertainties in the measurements one can only conclude that the metagalaxy once was less than, say, 4×10^9 light-years = 4×10^{27} cm.[6] But this is many orders of magnitude from the Big-Bang radius, which, according to some authors, is even $\ll 1$ cm! See Figs. 1 and 2.

Hence to the best of our knowledge the Big-Bang cosmology is not in agreement with the Universe we observe. It can be brought into apparent agreement only by a number of *ad hoc* assumptions. After all, it seems to be much closer to a myth, in certain basic respects of the same kind as the Ptolemaic system, which also needed an increasing number of *ad hoc* assumptions—epicycles. It is

---

[6] For detailed arguments why the minimum size of the metagalaxy is not likely to have been smaller than one billion light years, see Alfvén.[5,6] Reference 6 also contains literature references.

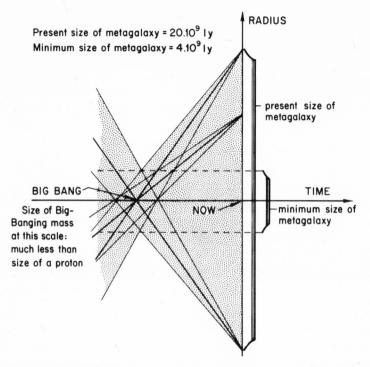

Fig. 2. If the straight line in Fig. 1 is used, a coalescence of all the galaxies at one point is obtained which is interpreted as support of the Big-Bang model. If instead the observational points are used directly for a linear extrapolation, the size of the metagalaxy ("Universe") is found to vary as shown by the shaded area. A minimum size of some billion light years is indicated. Only if the spread of the dots in Fig. 1 is almost exclusively due to observational errors is a much smaller minimum size obtained.

a myth which is decorated with sophisticated mathematical formulas. These make it more respectable, but not necessarily more true.

THE PTOLEMAIC COSMOLOGY AND THE BIG BANG

Our observational knowledge of the Universe is enormously greater now than during ancient times. However, the human mind probably works in essentially the same way today as it did some thousands of years ago, and the basic character of our attempts to widen our field of knowledge may also be similar today to what they were in past periods. Cosmology represents a pioneer field: One tries to study regions where facts and guesses necessarily mix. From this point of view it is of interest to compare the Ptolemaic cosmology and the Big-Bang theory. Such a comparison may also be of some interest as a contribution to the sociology of science.

Both the Ptolemaic and the Big-Bang cosmologies started from unquestionably correct and extremely beautiful philosophical–mathematical results. No one can study Pythagorean science comprising the mathematical theory of music and the theory of regular polyhedra without being immensely impressed. The same holds for the theory of relativity.

In both cases the beautiful structures established by mathematical thinking seemed to call for further development, and a cosmological application was natural. In both cases the steps in this direction were rather arbitrary, and the justification was perhaps in reality largely "divine inspiration." The Ptolemaic system, with the complicated system of crystal spheres generating the "harmony of the spheres," could not be produced from any prior state of the world. It could only be achieved by gods who were skilled workmen or artists. But observational facts demonstrated that in no way with any reasonable display of skill could they have achieved a model which was in agreement with reality.

Similarly, the Big-Bang theory represents only one of an infinite number of possible solutions. If we assume the observed average density of the Universe, the result is an essentially "flat" space. This means that *according to observations, the "universe" we perceive could be treated essentially with ordinary three-dimensional geometry* (of course using the *special* theory of relativity, here as in the laboratory). But accepting the observational results means that the general theory of relativity loses its place as the foundation of cosmology. It would still be of interest as a small correction in many cases, and it would be applicable to "black holes" (if these exist!) But the fascinating view of a limitless and still finite universe would be lost and with it much of the unquestionable philosophical–mathematical beauty of general relativity. In a way this is regrettable.

On the other hand, what the Big-Bang cosmologists tell us is that once upon a time the whole earth, the sun and the planets, the hundred billion stars in our galaxy, and, moreover, all the hundred billion galaxies which can be observed—all this enormous universe was compressed into one small ball. There are different views about the size of the ball, but some even claim that it was smaller than the head of a pin! Few claim explicity that this exploding super-atomic bomb was created by God; most avoid making an explicit statement. And all pretend that they know what happened during the first few *seconds*—or even microseconds—after creation.

If the Big-Bang cosmology is beautiful to the mind of mathematicians, it is abstruse to most people if not presented in a camouflaged way. No science fiction writer would dare to make his readers believe in a story in such a striking contrast with common sense. But when hundreds or thousands of cosmologists dress this story in sophisticated equations, and against the truth claim that this nonsense is supported by all that has been observed in the giant telescopes—who dares to doubt? If this is regarded as science, there is a conflict between science and common sense. The cosmological doctrine of today is an antiintellectual factor, which possibly is rather an important state of affairs.

THE COSMOLOGICAL ESTABLISHMENT

When the Ptolemaic system was threatened by an increasing number of observations which were adverse to it, it reacted in an authoritarian way. It was already a powerful establishment which had become sacrosanct and it did not tolerate any objections. Not even the mentioning of the existence of heretic views was tolerated. Copernicus complains that it was almost impossible to find any philosophical book which contained objections to the Ptolemaic system—until he finally found a reference to the old heliocentric system of Aristarchus. It is interesting to note that the present cosmologist reacts in a somewhat similar way. "Modern cosmology" nowadays is claimed to mean "relativistic cosmology," and in textbooks with the title "Modern Cosmology" it is often difficult to find any unbiased presentation of the objections, including the fact that, according to the best observations, space is flat, and that the general theory of relativity is essentially irrelevant to modern cosmology. Similarly, at international conferences on "Cosmology" it is very difficult to get even 10 minutes to question the importance of general relativity to cosmology. The prevailing attitude is that all the objections to Big-Bang cosmology are swept under the rug. And this is the fate also of the Creator, who is indispensable for manufacturing the Big-Banging atomic bomb.

The way the cosmological discussion has been conducted during the last decades will make many people believe that any criticism of the Big-Bang cosmology necessarily is a support of the "steady-state" or "continuous-creation" cosmology. Such support can certainly not be deduced from the present analysis. As is now generally realized, the continuous-creation cosmology is in a decisive way irreconcilable with observations. However, much of the objections against the Big Bang which the advocates of the steady-state theory have presented are correct and it is not fair to neglect these, as the Big Bang scientists usually do.

SCIENCE VS. MYTH

Since the Big-Bang hypothesis is unacceptable, the question arises of what other hypothesis we should place in its stead. The answer is simple and straightforward: *none!*

The Big-Bang conjecture is a myth, a wonderful myth maybe, which deserves a place of honor in the columbarium which already contains the Indian myth of a cyclic Universe, the Chinese cosmic egg, the Biblical myth of creation in six days, the Ptolemaic cosmological myth, and many others. It will always be admired for its beauty and it will always have a number of believers, just as the millennia-old myths. But nothing is gained if we try to place another myth in the place which the Big-Bang myth occupies now, not even if this new myth is adorned with still more beautiful mathematical formulas.

The scientific approach to cosmology is necessarily drastically different from the mythological approach. First of all, it must be absolutely clear that if a scientist makes a guess about the state of the Universe some billion years ago, the chance that this guess is realistic is negligible. If he takes this guess as the starting point for a theory, this is unlikely to be a scientific theory but very likely will be a myth.

The reason why so many attempts have been made to conjecture what the state was several billion years ago is probably the general belief that long ago the state of the Universe must have been much simpler, much more regular than today, indeed so simple that it could be represented by a mathematical model which could be derived from some fundamental principles through very ingenious thinking. Except for some vague and unconvincing reference to the second law of thermodynamics, no reasonable scientific motivation for this belief seems to have been given. This belief probably emanates from the old myths of creation. God established a perfect order and "harmony" and it should be possible to discover which principles He followed when He did so. He was certainly intelligent enough to understand the general theory of relativity, and if He did, why shouldn't He create the Universe according to its wonderful principles?

Accepting that it is just as likely that in the past the state of the Universe was essentially as complicated as it is today, how should we approach cosmology? Obviously in the same way as we have conjectured the first man did before any prophet had invented a myth. We should try to clarify the present state of our close surroundings, and from that proceed to more distant regions and to successively earlier epochs.

To try to write a grand cosmic drama leads necessarily to myth. To try to let knowledge substitute for ignorance in increasingly larger regions of space and time is science.

But we must always keep in mind that the further we proceed from here and now, the more hypothetical, the more speculative will our description of the cosmos necessarily be. Of course, we should try to reduce speculation as far as possible. But to eliminate it altogether is impossible—and perhaps not even desirable. Of decisive importance is that the approach be empirical, not mythological. The difference between science and myth is the difference between critical thinking and the belief in prophets; it is the difference between "De omnibus est dubitandum"[7] and "Credo quia absurdum."[8]

[7]"Everything should be questioned" (Descartes).
[8]"I believe because it is absurd" (Tertullian).

REFERENCES

1. C. Singer, *A Short History of Scientific Ideas* (Oxford, 1959).
2. Jean Heidmann, this volume, Chapter V.
3. G. Burbidge, *Nature* **233,** 36–40 (1971).
4. J.-P. Vigier, this volume, Chapter XI.
5. H. Alfvén, *Worlds-Antiworlds* (Freeman, San Francisco, 1968); *Sci. Am.* **1967** (April), 106.
6. H. Alfvén, *Physics Today* **24** (2), 28–33 (1971).

Chapter II

Elementary Particles, Universes, and Singularity Surfaces

Asim O. Barut

University of Colorado

At the turn of the century physicists began to be dissatisfied with the dualism of a theory admitting two kinds of fundamental physical reality: on the one hand the field and on the other hand the material particles. It is only natural that attempts were made to represent the material particles as structures in the field, that is as spaces where the fields are exceptionally concentrated. Any such representation of particles on the basis of the field theory would have been a great achievement, but in spite of all efforts of science it has not been accomplished. It must even be admitted that this dualism is today sharper and more troublesome than it was ten years ago. This fact is connected with the latest impetus of developments in quantum theory where the theory of the continuum (field theory) and the essentially discontinuous interpretation of the elementary structures and process are fighting for supremacy.

A. Einstein, New York Times, February 3, 1929

FIELDS AND MATTER

Nearly half a century after this assessment by Einstein of the state of physical theory we still have today a dualistic theory with two kinds of physical realities: matter on the one hand, and electromagnetic field and gravitation on the other hand. Although in modern quantum field theory both entities are described mathematically by fields, they are *distinct* fields, a field $A_\mu(x)$ for the electromagnetic field, for example, and fields $\Psi_i(x)$ for the electrons, protons, etc.—or a curved space and fields $\Psi_i(x)$, to describe the interaction of matter and gravitational field. The action (of the world) in the first case is a sum of three terms, one for the matter field (or fields), one for the radiation field, and a local coupling or interaction term involving the products of both kinds of fields.

This separation, together with the local coupling (i.e., point particles), seems to be at the root of the infinite renormalization terms of both classical and quantum theories, aside from the fact that a unitary theory is logically superior to a dualistic theory of which Einstein speaks.[1]

There is a further difficulty in the separation of the electromagnetic field $A_\mu(x)$ and matter field $\Psi(x)$ when both of these fields are quantized. Whereas the field $A_\mu(x)$ has an interpretation as a classical field, the field $\Psi(x)$ is a probability amplitude, and one quantizes further this probability amplitude (that is why this further quantization is sometimes called "second quantization"). One is tacitly assuming that the classical field equations describing a strong beam of electrons, for example, are those governing the probability amplitudes $\Psi(x)$, in the same manner as $A_\mu(x)$ describes the behavior of a classical beam of light. If we had correct classical field equations for both the electromagnetic field and matter there should logically be, of course, only one quantization, and the asymmetric physical interpretation of both fields would not arise. It is not easy to verify experimentally the classical field equations for an electron beam.[2] Because of their charge the electrons in a strong beam of electrons repel each other, and one cannot study the matter field independently of the electromagnetic field. This simple experimental behavior is a direct impetus toward a unified theory of charged particles and the electromagnetic field. We also know from nuclear physics that it is impossible *in principle* to "switch off" the electromagnetic interactions to study nuclear forces, although it can be done approximately in practice.

The unitarian point of view of field and matter does not mean that they are on the same footing. In fact, if matter is the singularity of the field, we expect the laws governing the motion and interaction of the singularities to look different than those of fields in singularity-free regions. But they must be related to each other. In quantum field theory, assuming that there are *classical* matter and radiation fields to be quantized, the difference is dramatically shown by the quantization with commutators of the dynamical degrees of freedom (normal modes) of the radiation field[3] and by the stranger quantization with anticommutators of the degrees of freedom of the matter field.[4] In a unitary theory of matter and field this distinction, which leads to the experimentally verified different statistics for photons and electrons and to the exclusion principle, has to be a *derivable* result. This will be a test and a triumph of any unitary theory of matter and field.

A field and its singularity seem to provide the clue to explaining the fundamental difference, from the point of view of measurement theory, between particles and fields. In Dirac's words,[5]

"If we wish to make an observation on a system of interacting particles, the only effective method of procedure is to subject it to a field of electromagnetic radiation and see

---

[1] Compare the discussion of "dualistic" and "unitarian standpoints" in the Introduction to Ref. 1.

how they react. Thus the role of the field is to provide a means of making observations. The very nature of an observation requires an interplay between the field and the particles. We cannot therefore suppose the field to be a dynamical system on the same footing as the particles and thus something to be observed in the same way as the particles. The field should appear in the theory as something more elementary and fundamental."

SINGULARITIES OF FIELDS

For definiteness we shall consider an electromagnetic field which is singular at a point, along a line, or surface, or in a region. A singular field $F_{\mu\nu}^M$ satisfies the free Maxwell equations everywhere except at the world lines of the singularities, where its divergence is equal to a singular current density j_μ. If we subtract the singular part $\Lambda_{\mu\nu}$ from $F_{\mu\nu}^M$, we obtain a regular field $F_{\mu\nu}$ everywhere. The Lagrangian density is quadratic in the fields: $L = (1/16\pi)\, F_{\mu\nu}F^{\mu\nu}$. If we re-express $F_{\mu\nu} = F_{\mu\nu}^M - \Lambda_{\mu\nu}$, we obtain three terms: the Lagrangian density of the Maxwell field $(1/16\pi)F_{\mu\nu}^M F^{\mu\nu M}$ and the terms $(1/16\pi)\,\Lambda_{\mu\nu}\Lambda^{\mu\nu}$ and $-(1/8\pi) F_{\mu\nu}^M \Lambda^{\mu\nu}$, which we can recast into the Lagrangian density for matter and for the interaction between matter and field, $j_\mu A^\mu$, respectively.[6] In the case of a point singularity, for example (i.e., a single world line), the matter Lagrangian turns out to have the usual form $\int m\, ds$ when we regularize the action of the world-line singularity by giving a certain thickness to the world line. Thus, an infinite singular field action on the world line (a distribution) can be represented by a finite action proportional to the line element ds at the expense of introducing the concept of mass or inertia. Similarly, a field singular along a line or surface (of dimension r) gives rise, after regularization along the lines indicated above, to a finite "matter" action proportional to the invariant surface area of the form $\omega \int \sqrt{g}\, d^r\sigma$, at the expense of introducing the concept of *surface tension* ω (or line tension in the case of a string). In other words mass m (or surface tension ω) is the *measure* or *weight* of the singularity which, when multiplied with the geometrical line (or surface) element, yields the same physical action as the singular field:

$$m \int ds\ \left(\text{or}\ \omega \int \sqrt{g}\, d^r\sigma\right) = (1/16\pi) \int d^4x\, \Lambda_{\mu\nu}\Lambda^{\mu\nu}$$

Thus a way is shown to "represent the material particles as structures in the field, that is as spaces where the fields are exceptionally concentrated."

Next we ask, more specifically, of what stuff the singularities are made, and about their nature. For this purpose we consider now an extended singularity, for a point singularity is a limiting case and we do not see there the complete structure. Furthermore, an extended structure will permit us to construct a finite quantum electrodynamics, because for a point singularity the interaction term $j_\mu A^\mu$ is still infinite at the position of the singularity. The fundamental particles have intrinsic properties besides charge and mass, like spin, magnetic moments, form factors, and even properties like isospin, hypercharge, etc. Because a point cannot have any properties, the structure of these particles must be extended. An extended structure need not be composed of "constituents"; an extended singularity of a field is of this type.

EXTENDED MODELS OF SINGULARITIES

A remarkable example of an extensible model of a charged particle, due to Lees[7] and Dirac,[8] is a spherical shell held together by a Coulomb energy and a surface tension whose interplay result in a stable configuration around which the shell can perform oscillations. This model can be formulated in a relativistically invariant form. The space inside the shell does not have any contribution to the action, and hence need not exist. The oscillations of the boundary of the shell can be quantized, giving the ground state and excited states of the quantized singularity. The mass m is determined by the charge e of the shell and the surface tension constant ω (conversely, ω may be fixed by e and m). Thus, even though the model is extended, the particle has only the two parameters e and m, exactly like a point particle—no new size parameters are introduced.

The oscillations of the electromagnetic field outside the shell can also be quantized[9] so that the interaction between field and matter comes about through the motion of the boundary of the shell. Thus, we have here a *finite* model of interacting matter and radiation.

In order to see that it is also a unitary model in the sense discussed in the first section, we must simulate the surface tension by a singular electromagnetic field. Indeed, the boundary conditions on the surface are such that the electric field is perpendicular to the surface, like a conductor, and we can have a magnetic field parallel to the surface. By the method outlined in the preceding section, the surface tension can be obtained by a singular magnetic flux flowing from north pole to the south pole on the surface of the shell.[9] In the case of the one-dimensional singularity, a string, the flux flows along the string from one end to the other, and in the case of a point singularity the two end points coincide so that we do not see the flux line at all.

A magnetic flux flowing from a point A to another point B along any line, surface, or volume together with an electric charge implies a nonvanishing total angular momentum proportional to the flux. This is seen by performing the integral $\int \mathbf{r} \times (\mathbf{E} \times \mathbf{B}) \, dV$ over the whole space.[10] Hence if the flux is quantized, so is the total angular momentum, or spin, and *vice versa*. Consequently, we are able not only to reduce the mass, but also the other mechanical attribute of the particle, the spin, to an electromagnetic origin.

These results show that the fundamental idea of a unitary theory of matter and radiation can today be implemented by quantitative physical analysis and models.

COSMOLOGICAL MODEL

The model with a singular boundary is a universal physical idea applicable to all situations when two different physical realities meet each other, like radiation and matter. I wish now to discuss an extreme case of this principle by considering a finite cosmological model of the universe which will be an inverse of the elementary particle model of the preceding section.

It is well known that because gravitation is always attractive, general relativity does not give, in the static case, a stable universe and the so-called cosmological term is necessary for stability.[2] Physically the cosmological term indicates a repulsion of very distant matter to compensate the local attraction of gravity. The stability problem is thus very similar to that of the charged particle discussed above. What is this distant repulsion?[3]

The gravitational action combined with the cosmological term has the form

$$A = \int R\sqrt{-g}\,d^4\sigma + \lambda \int \sqrt{-g}\,d^4\sigma$$

We can interpret the second term as a "surface tension," with coefficient λ, of a four-dimensional surface embedded in five dimensions; and, if the first term is written as

$$\int_{\sigma_5 = \text{const}} B\,d^4\sigma\,d\sigma_5$$

i.e., the value of a field action in the five-dimensional space evaluated on the four-dimensional surface σ_5 = const, we have the precise analogy with our extended elementary particle model. We have consequently a stable static solution, around which the Universe can make oscillations. One could go even further, following the analogy, and quantize these oscillations, and represent the surface term as the singularity of the field.

We can also apply the same idea to a Newtonian cosmological model. It is known that even in Newtonian cosmology[14] an additional term is necessary to have a static, stable universe. Consider a spherical shell. For the action we take that of the field of a uniform mass distribution inside the shell and a surface tension term again simply proportional to the invariant surface area: $A = A_{\text{field}} + A_{\text{surface}}$. There is no contribution form outside the shell, i.e., the space outside does not exist. We take as dynamical variables the gravitational field ϕ and the radius of the surface $R(t)$, which is assumed to be a function of t only. (We use a coordinate condition $z_0 = t$ for all points of the surface, and spherical coordinates.) The action principle yields the usual field equations inside the surface

---

[2] Of course, nonstatic expanding models do not need this cosmological term.[12]

[3] The cosmological term seems again to reemerge from the oblivion into which it was sent by expanding time-dependent cosmological models, this time via observational cosmology.[13]

and an equation of motion for $R(t)$ of the same type as in our elementary particle model, again with a stable solution, oscillations, etc. And we can speculate about the singular surface term.

CONSEQUENCES

We have examined the duality between field and matter, and their possible unification by the representation of matter as extended singularities of fields. We have considered a specific, rather universal, model of *form* or *structure*, which could be applied to phenomena all the way from elementary particles to models of cosmology.

In the electromagnetic model discussed above, the equilibrium radius a is related to the charge e and the surface tension parameter ω by $a^3 = e^2/4\omega$. Before quantization the total classical energy is the sum of the Coulomb energy plus the surface energy: $E = \omega a^2 + (e^2/2a) = 3\omega a^2 = \frac{3}{4} e^2/a$. If we equate this energy to the mass m of the particle, we can obtain[8] a relation between radius a, mass m, and the coupling constant e^2:

$$\tfrac{3}{4} e^2/a = m$$

Now in the cosmological model, e^2 is replaced by GM^2 and we obtain the relation (in units $c = 1$)

$$GM \sim R$$

(up to a numerical factor). This relation, which is well satisfied by the estimated mass and radius of the Universe, is empirically known and is now theoretically derived from our model. It is interesting that the stable radius that one gets is of the same order of magnitude as the Schwarzschild radius $2GM$.

Furthermore, we find that frequency of small oscillations around the equilibrium to be the Hubble constant, $\nu \sim c/R \sim 10^{-17} \sec^{-1}$.

REFERENCES

1. M. Born and L. Infeld, *Proc. Roy. Soc. (London) A* **144**, 425 (1934).
2. H. Hepp and H. Jensen, *Sitzungsber. Heidelberger Akad. Wiss.* **4**, 89–122 (1971).
3. P. A. M. Dirac, *Proc. Roy. Soc. (London) A* **114**, 243–65 (1927).
4. P. Jordan and E. Wigner, *Z. Physik* **47**, 631 (1928).
5. P. A. M. Dirac, *Proc. Roy. Soc. (London) A* **136**, 453–64 (1932).
6. A. O. Barut, in *Proc. First International Symposium on Math. Physics, Warsaw 1974*, ed. by K. Maurin and R. Raczka (D. Reidel, 1975); in *Quantum Theory and the Structure of Time and Space*, ed. by L. Castell, M. Driescher, and C. F. von Weizsäcker (C. Hauser Verlag, München, 1975).
7. A. Lees, *Phil. Mag.* **28**, 385–95 (1939).
8. P. A. M. Dirac, *Proc. Roy. Soc. (London) A* **268**, 57–67 (1962).

9. A. O. Barut, to be published.

10. A. O. Barut and G. Bornzin, *Nucl. Phys. B* **81,** 477 (1974); A. O. Barut and H. Schneider, *J. Math. Phys.* **17,** 1115 (1976).

11. A. Einstein, *Sitzungsber. Preuss. Akad. Wiss.* **1917,** p. 142.

12. A. A. Friedman, *Z. Phys.* **10,** 377 (1922); **21,** 326 (1924); H. P. Robertson, *Proc. Nat. Acad. Sci.* **15,** 822 (1929); *Astrophys. J.* **83,** 187, 257 (1935); A. G. Walker, *Lond. Math. Soc.* **42,** 90 (1936).

13. J. E. Gunn and B. M. Tinsley, *Nature* **257,** 454 (1975).

14. E. A. Milne and W. H. McCrea, *Quart. J. Math. Oxford, Ser.* **5,** 73 (1934).

Chapter III

General Relativity and Our View of the Physical Universe

Peter G. Bergmann

Syracuse University

INTRODUCTION

The task that I have set myself is to review some of the aspects of my own chosen field of research, general relativity, that attracted me to that theory in the first place, that is to say, its contributions to our general view of the physical universe.

REVIEW OF SPECIAL AND GENERAL RELATIVITY

There are two distinct theories in physics that bear the name of relativity, the special (or restricted) theory and the general theory. They represent two distinct stages in Einstein's approach to space and time. The special theory of relativity has become an integral part of the whole of physics, whereas the general theory of relativity has remained, until quite recently, a domain of afficionados. In what follows, I shall deal with the two theories together, as the conceptual developments of the special theory also form part of the general theory.

The first major aspect of relativity is that it deprives time of the absolute character that it had had in Newtonian physics. According to prerelativistic physics, the whole of our experience in space and time can be segmented into consecutive slices of "now" and "later." If two events lie in the same slice, they are said to be simultaneous. We call time absolute to express the thought that the three possible relationships that one event may have with respect to another, "earlier," "simultaneous," and "later," are independent of the conditions under

[1] Research supported in part by the National Science Foundation Grant GP-43759X.

23

which the determination is being made. In particular the state of motion of the observer and of his instruments does not matter.

Relativity denies this absoluteness, or universality of time. It contends that for one observer the event "*A*" may precede "*B*," whereas the reverse may be true for another observer. That this possibility does not form part of our every-day experience is because we usually do not have observers who are moving rel-ative to each other at "relativistic" velocities, velocities amounting to at least several ten thousand kilometers per second.

Minkowski discovered that the relativistic notions of space and time lend themselves to a unified four-dimensional description, and that the four-dimen-sional approach is indeed the most appropriate for relativistic physics. Tran-sitions from one frame of reference to another, moving relative to the first one, resemble in some respect rotations in ordinary three-dimensional space. They differ radically in that under these transitions ("Lorentz transformations") the possible trajectories of light rays remain unchanged; that is to say, in this formal-ism there is a whole family of straight lines (null lines) that are turned only into each other, whereas in ordinary rotations every straight line can be turned into the position of any other line.

Whereas special relativity does away with the notion of absolute, or univer-sal time, general relativity abolishes the privileged role of the inertial frames of reference. Briefly, this comes about as follows. In a field of gravity, according to Newton, all freely falling bodies follow the same trajectories; they are acceler-ated the same amount. If their motions are referred to a frame of reference that is itself in free fall, such as a space vehicle when the power is turned off, objects inside behave exactly as if no gravitational field were present. Locally, at least, the gravitational field has been "transformed away." That there is any field pres-ent at all becomes clear only if observations are extended over a domain of suf-ficient size, in which the direction of a plumb line varies noticeably.

The underlying principle, known as the principle of equivalence, has been formulated in a variety of ways. One statement, certainly correct as far as it goes, is to the effect that any body's inertial mass, the ratio between acting force and resulting acceleration, equals its gravitational mass, which determines how it responds to an ambient gravitational field. An alternative formulation of the same principle is that "locally" physical processes when referred to a free-falling frame of references are indistinguishable from like processes referred to an inertial frame of reference in the absence of gravitational forces. This latter formulation is open to objections, but in spite of its flaws it conveys some of the substance of the principle. Perhaps the main consequence that I will be con-cerned with is the fact that though the presence of a gravitational field can be ascertained by extending the domain of experimentation, once a gravitational field is present, its effects cannot be disentangled from purely inertial effects. The presence of a gravitational field renders the determination of an inertial frame of reference an impossible task. Whereas the special theory of relativity pictures the spacetime manifold as a quasi-Euclidean "flat" space with straight lines, planes, and so forth, general relativity deforms that nice straight manifold into something that, in four dimensions to be sure, resembles the skin of an orange, with a large-scale topography that cosmologists have to worry about, and with innumerable local irregularities, called forth by the presence of stars and galaxies.

If the geometry of spacetime is not foreordained, but determined by small-scale and large-scale concentrations of masses, we must conclude that the whole web of spacetime cannot be taken for granted; this reservation applies equally to associated ideas such as cause and effect, categories that are based in part on certain anthropomorphic notions of the nature of space and time. Philosophers and others not in close contact with theoretical physicists may be surprised to hear that considerations of the principle of causality loom large in our work. That these worries are usually buried in technical jargon does not make them less real. Incidentally, the need for technical jargon arises not out of a need for ritual secrecy, but so as to distinguish various generalizations and modifications of the classical notions of causality that are characteristic of certain types of physical theories and hypotheses.

To summarize: The theory of relativity does away with the notions of absolute time; it does not permit the introduction of the concept of absolute rest, nor of absolute motion; it does away with the notion of absolute acceleration (which forms part of Newtonian physics). Relativity even breaks down the idea of universal interaction in certain circumstances. For instance, it is possible for two particles, a finite distance apart from each other, to travel on perfectly reasonable trajectories, yet be incapable of interacting with each other in any way. We then speak of "horizons." For instance, if two particles are approximately one light year apart from each other, and if they are accelerating away from each other each at the rate of acceleration equal to the gravitational acceleration on earth, then each is behind a horizon for the other. Never will they "see" each other, never interchange information or signals, directly or through intermediaries. Moreover their mutual distance from each other will never change, in spite of the acceleration. This one example may serve to indicate that in relativistic physics our normal intuitive feelings as to what might happen can lead us badly astray.

NONFLAT SPACETIMES

In this section I shall be concerned with spacetimes that deviate in some qualitative away from the smooth and flat spacetime of the special theory of relativity, the so-called Minkowski universe.

I have mentioned that any modification of the classical ideas about space and time may force on us changes in the concept of causality. On a small scale, if of two events "*A*" and "*B*" either may precede the other, depending on the state of motion of the observer, then neither can have a substantial effect on the other. Technically, we say that the two events are in a "spacelike" relationship to each other, and hence cannot have a causal relation. On a large scale, however, there are spacetimes imaginable in which time is closed, in the sense that proceeding into the future, you eventually find yourself confronted with your own ancestors, and finally with yourself. One calls a universe (= spacetime) causal if there are spacelike three-dimensional regions (these are comparable to "all space at one time") through which every possible history of a particle (timelike or lightlike trajectory) passes exactly once. There is nothing in the general theory that tells us that models of the Universe must be causal in this sense. Moreover, there is no evidence that forces us to the conclusion that our Universe, the one in which we actually live, is causal. Perhaps, as astronomers succeed in looking further and further into the history of our Universe, we shall ultimately get hold of evidence that bears on that question.

Another unresolved question is whether our Universe is "compact" or not. Roughly speaking, one calls a manifold compact if it is finite, such as the surface of a sphere. The example of the sphere shows that a compact manifold need not have a boundary. If you travel about a compact manifold, you might never come to a point where you cannot go on. Rather, you might find yourself returning to a region that you have already traversed previously. Our Universe might be compact in space but not in time, as was one of Einstein's early models (the so-called Einstein universe). Or it might be compact altogether, or infinitely extended in all respects. Still other combinations are imaginable. Current cosmological observations may soon lead to some decision as to the spatial compactness of the universe; a decision on timewise compactness may be off much farther into the future.

To continue, manifolds are said to be "multiply connected" if there are alternative paths from one point to another that cannot be deformed continuously into each other. For instance, a circle is multiply connected (you can go around it clockwise or counterclockwise), whereas the surface of a sphere is simply connected. If our Universe were multiply connected, we should get a glimpse of that fact eventually, though right now there is nothing to suggest that it resembles a pretzel in this respect. In any case, multiple connectedness is a conceptual possibility that one should not discount entirely.

Finally, let me mention one possibility that is right now very much in the focus of theoretical research. A manifold may contain regions in which the normal properties of smoothness are somehow violated. Such regions are known as singular regions, or as singularities. There are many examples of manifolds with singularities, but it is very difficult to define singularities comprehensively, or to classify them exhaustively in terms of their characteristics. That the possibility of singularities of spacetime is not an idle speculation is brought home by two examples. One is the possibility of "black holes," very large stars that have collapsed under the influence of their own gravity to a stage where their interior can no longer be described with the means of standard differential geometry. The other is the possibility that the "Big Bang," the early stage of our whole Universe, may be represented by a singularity.

A singularity represents a region in space and time in which the ordinary laws of nature do not hold. But a singular region is not one in which "everything goes." It is probably not the place for "miracles"; or at least the miracles are rigidly restricted. As long as the singularities are surrounded by ordinary spacetime regions, these surroundings force even on singularities certain rules that cannot be violated; at least that is how we understand the situation at present.

One can get rid of singularities at least in one's description of nature by cutting them out, by ignoring all singular regions in inventorying a model universe. Unfortunately, the cut will "bleed": a manifold from which parts have been removed will be "incomplete" in a visible manner. And much of the work on singularities is presented in the guise of articles on imcomplete manifolds. Instead of classifying singularities, one can attempt to classify the boundaries of incomplete manifolds; this is right now the preferred method of dealing with singularities.

What makes the research on singularities interesting is the existence of "no-go" theorems that have been developed in the past decade. These theorems assert that with certain, apparently very reasonable conditions imposed on a space-time model the occurrence of a singularity, or of a boundary, is unavoidable. All the no-go theorems known at present assume that there are no negative masses; and this might be an unwarranted restriction. The fact is, though, that no one has as yet discovered a particle or other physical object whose mass is not positive.

Before coming to the final part of this presentation, let me again summarize: By doing away with the quasi-Euclidicity of spacetime, the general theory of relativity opens the door to a number of possibilities that were not entertained previously. These include noncausal universes, universes with various degrees of compactness, multiply connected universes, and universes with various types of singularities.

CRITIQUE OF PRESENT CONCEPTS

Whenever there has occurred a major revision of then-current concepts, that conceptual revolution has given rise to critical evaluation, both retrospective and forward-looking. The retrospective critic asks: Is this trip really necessary? The future-directed critic asks: Why stop here? The history of the physical sciences has usually shown that the retrospective critique was sterile, whereas the prospective questioning is justified.

As for the principles of relativity and of equivalence, I should remark that the former seems to be contradicted by our cosmological experience. Whereas in the laboratory experience, say with elementary particles, the laws of relativ-

istic physics are empirically confirmed again and again, to an almost unbeliev-
able degree of accuracy, the average state of motion of matter at large in any
confined region of the universe is pretty well defined. That is to say, within a
galaxy, stars move relative to each other rather sluggishly, at speeds of, say, one
ten-thousandth of the speed of light, and so do neighboring galaxies relative to
each other. In recent years there seems to have occurred a number of contrary
observations on certain radio stars, but the possibility of observational or interpretative error has not yet been excluded; this appears an open question.

As for the principle of equivalence, I have mentioned that its various formulations appear to be flawed, either in that they are too weak or not sufficiently
hedged about. However, the resulting theory, the general theory of relativity,
appears to be internally consistent, and completely confirmed by observational
and experimental evidence to date. Like all physical theories, it will undoubtedly
be superseded by a more sophisticated theory eventually, but right now we can
only speculate when and how this will come about.

But if we permit ourselves this kind of speculation, there are a few obvious
possibilities. First, among the so-called exact sciences cosmology is a discipline
sui generis, in that it is concerned with a single object, our Universe. Cosmological laws, so-called, cannot possibly incorporate reproducibility, the prime characteristic of laws of nature postulated in other disciplines. Thus we cannot subject the principle of relativity to a test within the environment of another
universe, in which matter does not have a well-defined local state of motion.

Aside from these global considerations, the general theory of relativity,
which was created almost half a century ago, still preserves some aspects of that
period. Though the rigidity of the Minkowski universe has been abolished, the
building blocks of that universe, the world points, have been retained and so has
been their connectedness in the small, the concept of manifold. By now mathematicians have constructed a sufficient variety of deviant structures that we
know one thing for certain: The manifold is not the only kind of substrate in
which one can construct a geometry, nor is a metric structure the only geometry
imaginable for a manifold. Indeed, the principal impediment to rapid progress
along these lines is not, to my mind, the dearth but the overabundance of alternatives, and I, for one, find myself stymied by the problem confronting
Buridan's ass.

From all that I have said, I hope that I have left two impressions. For one,
general relativity has shaken many of our anthropomorphic notions of space, including time, causality, and other general concepts underlying our basic approach to the physical universe; this aspect of general relativity has probably not
yet been assimilated in its totality. Further, general relativity has begun a revolution of physical concepts but has not completed it. We do not know where the
exploration of those problems will lead us next; but we can feel assured that
the road ahead is long and arduous; none of us will live to see the end of that
journey.

Chapter IV

Quantum Relativity and the Cosmic Observer

Patrick A. Heelan

State University of New York at Stony Brook

By a critique of the opposing characteristics of general relativity and quantum theory, I want to assess the possibility and form of a quantum relativity, and thereby—through the "observer problem"—to arrive at a view of the scientist as an incarnated or embodied knower, not beyond nature, but naturalized in the material universe. My point of view is that of a philosopher and a physicist, and the manner is more heuristic than technical, although the technical material could be provided if necessary. In my treatment of the quantum theory, I am presupposing that some form of the Copenhagen Interpretation is valid.[1]

First of all, I want to treat briefly of the principal areas of conceptual conflict between quantum mechanics and general relativity. Quantum mechanics is nonlocal, in the sense that any reconstruction of it in terms of a hidden-variable theory has to be nonlocal and noncausal. General relativity, on the other hand, is a local and causal theory. There will then be a problem in linking or joining the two. Second, quantum mechanics obeys a quantum logic, while general relativity obeys a classical logic of propositions. Third, quantum mechanics is said to need classical physics as a necessary presupposition, while general relativity is not tied to models of the universe with asymptotic classical regions. Finally, since quantum fluctuations of the spacetime metric would exist in a quantum relativity, the question of "cosmic observers" has to be raised.

First, quantum mechanics has been attacked from its earliest presentation as a nonlocal theory and therefore incompatible with relativistic physics. Einstein, Podolsky, and Rosen were the first to do so, followed by von Neumann, Bohm, Jauch, Piron, and Bell.[1] Bell's theorem,[4] for instance, says that a local hidden-variable theory cannot reproduce all the statistical predictions of quantum

---

[1] See Bub[2] and Jammer,[3] Chapter 6.

mechanics.   However, although hidden-variable reconstructions of quantum mechanics are incompatible with relativity, still it can be shown that under quantum mechanics in its standard form, no nonrelativistic signals need in fact be assumed.

Let me pass rapidly to the second point: Quantum mechanics obeys a non-classical logic of propositions, a so-called quantum logic. The prevalent school of quantum logicians, represented, say, by Jauch, Piron, Finkelstein, and others, conclude that the logical connectives "and," "or," "if–then," etc. have different semantical meanings in quantum mechanics than they have in ordinary dis-course, classical or general relativistic.[2]   One cannot simply conjoin quantum mechanical propositions using the operators of classical logic and hope always to make sense: What results from the combination of true or false atomic proposi-tions may be neither true nor false. For the prevalent schools of quantum logicians, a physical theory is a triplet comprising a set of empirical propositions E, a phase or state space H (for example, the Hilbert state space of quantum mechanics), and an abstract satisfaction function S, which associates with every empirical proposition a subset or a subspace of the phase space. The relation-ships between the sets of H onto which the individual propositions of E are mapped constitute what is called the logic or the kinematic logic or the pre-theoretic logic of the system. For classical theories, the logic of the system is Boolean. Up to this point in the account, truth and falsity are not involved, since no assertion has yet been made about the real world. Assertions about the real world arise only "in a model," that is in a mapping of real and actual states of affairs onto the elements of H. The semantical model of a science just de-scribed is not merely a theory about the structure of quantum mechanics, but a general theory about the structure of all physical theories. Quantum logicians say that quantum mechanics has an anomalous, that is, nonBoolean, set struc-ture in the phase space, leading to an anomalous logic of propositions, for exam-ple, to a complemented, nondistributive lattice of propositions or a quantum logic. What is assumed in this view of a physical theory is that corresponding to every element in the Hilbert space, which elements in the case of quantum me-chanics are the wave function, there is a unique *real* state of affairs in one-to-one correspondence with it. All the semantical complexity of the system is assumed by the abstract phase space and none by the mapping process whereby the ele-ments of H are mapped onto the real world. Now this is not compatible with the Copenhagen Interpretation, which requires that there be a double mapping— a mapping first of the measurement or observation context and then a mapping of an individual wave function into an observed event. Out in the real world, there are two kinds of things: a measuring context, which restricts the domain of possible events, and the observed event, which is physically conditioned by its context. Instead of just one simple one-to-one correspondence between states

[2] See Bub[2] and Jammer,[3] Chapter 8.

of the world and the elements of the Hilbert space, there is in the Copenhagen Interpretation a complex mapping of context and event. A consequence of this, for instance, is that it may not be meaningful to talk about the quantum wave function of the Universe, since there is no outside physical context in which the Universe as a totality can be observed. I want to return later to why I think it important to hold on to the Copenhagen view.

Let me pass on to the third point, the presupposition that the Copenhagen Interpretation makes that the domain of the measuring instrument is the domain of classical physics. Now a measuring instrument is conceptually a complicated thing: It mediates between the object and the person of the scientific inquirer. It has two faces. It has a face turned toward the object, which is a purely physical face, one where a controlled physical interaction takes place. It also has a face turned toward the subject, which is not a purely physical face: Its relation to the subject is hermeneutical, that is, it generates at the interface coded information about the object. The measuring instrument consequently is a peculiar kind of thing; it is both a physical object and a generator of ciphers. It prints, as it were, the text of a page from the book of nature, coded in such a way that a properly trained subject can read it. It should be remembered that the book of nature is not nature pure and simple, but what nature prints for man to read through nature's interaction with instruments. Now, the Copenhagen view of quantum mechanics is compatible with this analysis of the measuring instrument, but this is not the case with regard to the prevalent school of quantum logicians. The Copenhagen view claims that one does not simply have an account of the object devoid of its relationships to the rest of nature, but that quantum mechanics relates an object to sets of measuring instruments that constitute the variety of reference contexts for the description of the object—with the constraint, however, that some pairs of reference contexts are mutually incompatible. Observations or events in this view are contextualized by the measuring context: The object interacts with the measuring instrument and prints a coded message decipherable by the trained human subject.

This I see as an essential character of the Copenhagen Interpretation. I do not want to say that the Copenhagen Interpretation is necessarily true! What I say is that it is an interesting point of view and more interesting than its opposite. It is a view that, as Niels Bohr saw, has applications in the contextuality of ordinary language, of the psychological and social sciences, as well as in the history of the changing frameworks that constitute the history of science. Because of its power as a heuristic notion in so many sciences, it is to be preserved in physics at least until it is shown to be inapplicable. It may well turn out that it has no place in physics, but that it properly belongs exclusively to the social sciences and the humanities, but that remains to be seen.

It is often said that there are three kinds of physics, depending on the size of the object: quantum physics for the micro-domain, general relativity for the astronomically large macro-domain, and classical physics for the human-sized, mesoscopic or middle-sized domain. Now, such a division is, I believe, not very useful. There are quantum phenomena of general relativistic scale, like black holes and neutron stars, and of mesoscopic scale, like superconductivity. The three kinds of physics are not differentiated by the size of the domains to which they refer, but by other characteristics. Now, in my modified version of the Copenhagen Interpretation, a necessary presupposition of quantum physics is not classical physics—as the orthodox view goes—but the existence of sufficiently efficient information processors of mesoscopic scale. These information processors do not have to be classical: Their interaction with objects could be quantum mechanical or general relativistic, but what they must do is provide signals that can be distinguished from one another and can serve the purpose of a text for a human subject to read. In that sense, they must be efficient information processors of mesoscopic scale. But one may answer: Is not that rather parochial and anthropocentric? Couldn't we think of Hoyle's Black Clouds as scientific inquirers capable of knowing what we know about nature? Well, yes: They would know nature, but whether they would express this knowledge in the same form as we do is not certain. The science we know is an activity of intellectual creatures of the size that we are, with bodies, and capable of hermeneutical activity.

Fourth we ask: What would it mean in general relativity to say that a fundamental (cosmic) observer is in a superposition state of geometries? Three answers have been given: (i) the modified Copenhagen view to which I shall return, (ii) Wigner's view,[5] which introduces consciousness as the factor in the physical process that reduces the wave packet—in this case, which reduces the superposition state of metric coefficients to those of a particular geometry; and (iii) the view of Everett, Wheeler, Graham, and de Witt,[3] which says that the world which we experience is only one branch of a highly multiple actuality that also includes all branches of a physical system represented in superposition states. Those not actually observed are not suppressed, as in the orthodox Copenhagen view, but merely become inaccessible to the observer following a measurement. Reality, then, is an infinite multiplicity of branching universes of which we have access only to one—the one that we ourselves experience.

These three views will be criticized in reverse order.

Everett, Wheeler, Graham, and de Witt argue that since quantum mechanics has been enmeshed for fifty years in apparently insoluble controversies about the meaning of superposition, one ought to adopt the simplest solution that cuts the Gordian knot of complexity: This, they say, is to give a straightforward realistic interpretation to everything that is symbolized in the formalism. Such an interpretation, they contend, is consistent and not contradicted by experience, although they grant it is strange and even implausible. Their solution is defended on the grounds that a theory ought to dictate its own interpretation.

---

[3] See Jammer,[3] pp. 507–21.

Observations, they say, do not create theoretical categories, but what it is that is observed when observations take place is determined by the theory prior to all observations. Hence, it is the theory that dictates its own interpretation and we ought simply to affirm as real the elements and the structure of the theory, despite our inability to verify all aspects of the conclusions empirically.

Now, with regard to the simplification effected by Everett, Wheeler, Graham, and de Witt, it is my view that such a simplification removes from quantum mechanics what is significantly new, exciting, revolutionary, and pregnant with novelty and growth. With regard to the second principle, namely, that the theory ought to dictate what the world is like, this of course is a sound principle, but it needs interpretation and skillful application. It must be remembered that Heisenberg defended quantum mechanics on the basis of this principle, that Einstein opposed Heisenberg on the grounds of the same principle, and that Everett, Wheeler, Graham, and de Witt oppose both in the light of the same principle. How could one and the same principle produce such different points of view? One would have supposed that Everett, Wheeler, Graham and de Witt had asked themselves this question. They seem, however, not to be aware of the history, use, and interpretations of the principle, nor do they say what they mean by the principle nor defend their use of it in the context of what others have said.

Passing on to Wigner's view, which physicalizes consciousness, it is a viewpoint that—however implausible—he feels follows from quantum mechanics. Noting that the reduction of the wave packet is a physical change in the real world resulting from the suppression of those correlations that were present before the reduction, we ask: How in Wigner's view does consciousness bring about the change? In the measuring process, he says, the object affects the instrument, this in turn affects the retina of the observer, which affects the optic nerve, which affects the cortex, and so on, until at some locus, consciousness envelopes the process and the process becomes self-conscious. At some point then in the brain of the scientist, consciousness intervenes to provide a self-transparency with regard to its own physical state. This is expressed in the reduction of the wave packet.

First of all, let me say that this employs a very peculiar view of consciousness. Consciousness is not primarily self-consciousness in that sense. Consciousness is primarily a consciousness of meanings—of objectivities, as Husserl would say, of idealities and objects, not of neurological causes. It is a consciousness of shared descriptive norms, not of the causes that produce the brain states that accompany consciousness.

Wigner's view is also vitiated by an absence both of the sense of the contextuality of quantum mechanics and of the sense of the hermeneutic complexity of the process of scientific observation. In other words, if there is, as he believes, a one-to-one relationship between a mathematical model, e.g., the elements of the Hilbert space and the real world which it pictures, then the reduction of the wave packet must correspond to a change in the real world and this, it would seem, is produced by the activity of conscious observation. Now, the conception that creating a scientific theory is to construct a mathematical model that pictures the real world is, of course, a very common one and was operative throughout the history of science from Galileo and Newton, for instance, to Einstein, Wheeler, and Wigner. The notion that nature is mathematical and can be represented by a mathematical model is derived from—or certainly associated strongly with—the view that a perfect knower who knows the world would know it in and through a mathematical theory. The ideal of what it is to know scientifically is to share the mathematical knowledge of a perfect and supreme knower—God. In some unproblematic way it was supposed that, knowing the formalism of the theory, one knows the real world. What was not apparent, and is still hidden from many, is that mathematical formalisms in physics may need complex interpretations to connect them with the real world.

In the case of quantum mechanics, the Copenhagen Interpretation says that the mathematical formalism of quantum mechanics is applied to the world in two stages: to the measuring context (which gives the basis of eigenstates in terms of which the wave function is expressed), and to the act of observation which selects, of the range of possible outcomes, the one that is actually realized.

There is then a hermeneutic applied at two levels. The model of the mathematician God, as the perfect scientific knower, still controls implicitly the search for scientific knowledge by providing an ideal of science. This ideal, however, is a remnant of an outdated theological system and a remnant of an outdated epistemological system.

To return to the modified Copenhagen view, how would a superposition of two different spacetime metrics be interpreted in quantum relativity? I would suppose that, as viewed by some set of efficient information processors of mesoscopic scale (cosmic observers or observers in and of the cosmos), the fundamental particles of the universe would be found embedded now in one, now in another of a variety of different metrics governed by a schedule of probabilities. What are these efficient information processors of mesoscopic scale—cosmic observers—and what are the fundamental particles of the universe? Are they identical with one another? In relativistic cosmology, the fundamental particles are generally taken to be objects of astronomical scale, like galaxies which carry local atomic or other clocks on the basis of which the local geometry is estimated. They are not mesoscopic information processors, but they are large enough to contain such processors, for example, telescopes, computers, photographic recorders, etc., as well as interpreters belonging to a scientific community.

All of these instruments are human artifacts, which suggests that quantum physics is a physics for men—embodied, intellectual, hermeneutical knowers—and not physics from a divine standpoint. Human-sized instrumentation is known to exist only in our own galaxy, but cosmological thinking would tend to deny that our galaxy is special in any significant way. It postulates the presence, or at least the possible presence, of men with their scientific artifacts on all fundamental particles of the universe. Man is potentially everywhere, not in the disembodied way of the divine mind or of the supreme mathematician, presumed to be omnipresent in space by the physicists of the 17th and 18th centuries, but in the way required by the modified Copenhagen Interpretation, that is, through the necessary use and hermeneutical interpretation of measuring instruments. This kind of presence is a very material, incarnated, embodied presence of a cosmic community of scientific inquirers and cosmic quantum observers who are both mathematicians and physical information processors. Quantum relativity then supposes that the fundamental particles of the Universe contain, in principle, human observers and mesoscopic recorders or information processors, that is, forms of embodied hermeneutical rationality.

Implied by all this is a notion of man as a naturalized citizen of the Universe, a material part of it, and using a physical body essentially in the acquisition and expression of scientific knowledge. What is man's body insofar as he is a scientific knower? His body, being on the side of the knowing subject vis-à-vis the object known, is never part of what is objectified; it does not constitute part of what any particular physical inquiry is directly concerned with. The objects of physics then are objects for embodied subjects: The body is that which mediates between the subject and the object without, however, being part of what is objectified by the inquiry. Consequently, the instrument as a mediator of scientific knowledge becomes part of the body of man when its signals are read for what they tell about the object, for instance, its spin, charge, mass, etc., but not when its signals are described, for instance, as a black or red spot. Should the observer look at the instrument as an object, then it ceases to be an information processor: The processing of information in that case takes place in the neurological system, which provides the signals which mediate between the object and the subject, and the interpretation of which constitutes knowledge.

In summary, the body of man is whatever mediates knowledge but is not objectified by the knower, observer, or scientist during the process of observation. In this epistemological sense, the measuring instrument can be part of the human body, an extension of the natural organism beyond the surface of the skin. The epistemological body is not limited to its biological boundaries: It can extend beyond it, as we have seen; it could also contract to some part less than the biological whole, for instance, when one objectifies one's own hand as an object. What emerges from this analysis is that the human body, as part of an incarnate knower, is not a thing of definite dimensions: It is flexible, it can extend its corporate presence beyond the natural organism, or it can retire into one part of that organism. It can also share the same instrument with a variety of observers, implying that the potentialities of embodiment of a human observer do not exclude the possibility of a shared embodiment within the same piece of instrumentation. One can think of a variety of such examples, e.g., telephones, the printed page, data processing equipment, etc. The telephone I use is both a part of my body and of that of the person to whom I am speaking. Further, we have a new view of man as experimental scientist, subject to a certain indeterminacy with regard to the scope of his embodied (or subjective) presence in the world. This indeterminacy manifests itself in the embodied subject's power to extend itself beyond the natural organism to envelope many different networks of communication, for instance, in the laboratory, or extending over the face of the earth, or reaching out into space to communicate with objects of astronomical scale. In all of these cases, instrumental devices write texts that the subject then reads. Consequently, the Universe divides up into subject and object: An inquiring subject embodied in one part of the Universe and his object, which is the rest of the Universe. It is clear from this analysis that neither the subject nor the object of the empirical study can be coincident with the totality of the cosmos.

What we are trying to do is to give an account of the inquiring scientist as subject. Such an account will inevitably be indirect, unclear, and tinged with the unfamiliar and the extraordinary. Science speaks of objects, never of subjects, and a science of inquiring subjects will never look like a conventional scientific theory or be couched in terms that are straightforward and familiar.[6]

Some interesting and unusual consequences emerge from the theory of the embodied scientific subject. Our capacity for truth is quite clearly a function of a trained, structured material body with bones and muscles and a neurological system that is, moreover, capable of being extended into a domain of artifacts that play the role of mesoscopic information processors providing printouts about different segments of the world around us. The body in that sense—through a scientific technology—does provide new capacities for knowledge and new capacities for truth. Since scientific technology is a dimension of cultural history, man as an embodied knower is part of the evolutionary process that links him in a special way with the development of mesoscopic information processors.

The evolution I have described is characterized by a flexible boundary between subject and object. The boundary may well extend beyond the limits of the natural organism—it may be deep in space—or it may contract to inhabit only a part of the biological body. The flexibility of the boundary throws a novel light on scientific objectivity or on the objectivity of nature: Both are seen to be relative to an embodied inquirer whose subjectivity makes objectivity possible, but whose subjectivity necessarily at the same time inhabits and cuts off from an objective gaze some part of the universe.

Finally, a comment about values. Man is creative of cognitive structures through the structuring of the instrumental environment around him. These cognitive structures become transformed into value structures in a way I will now explain. Our physical theories tell us about the interaction between different parts of matter and, therefore, about the instrument–object interface. The instrument itself can be treated as an object, explainable by physical theories. How it is modified in the course of an interaction with its object belongs to a kind of knowledge I shall call "thing-to-thing" knowledge. The subject infers from the interaction how the instrumental channels will be modulated and can then infer from the instrumental response to the state of the object. However, to the extent that the instrumentation becomes a familiar part of the environment, the modulated signals produced are not observed as intermediate objects from which inferences are made, but are read directly like a text. The instrument then becomes part of the body of the knower and reveals to man the scientific object as something directly related to man and for man. The scientific object now becomes a value for man, in this respect like the everyday objects we experience. A chair is known for the function it plays in accommodating a certain anatomical posture; it is an object revealed by and in relation to a system of values and a life-style. Physics can thus pass from an original stage of revealing thing-to-thing relations to being appropriated by human subjects for human subjects. Physics (science in general) becomes transmuted in the laboratory, and then through technology into a system of values, or, in the words of Habermas,[7] of human interests. In that sense, in which science is a creator of values, I see an unavoidable and necessary consequence of the peculiar character of the scientific knower, who is not as such in the 17th and 18th century image of the divine, but is truly a naturalized part of a material cosmos.[8] He is a piece of nature, but a piece that happens to be a subject in the cosmos, not an object.

REFERENCES

1. P. Heelan, Complementarity, Context-Dependence, and Quantum Logic, *Found. Phys.* **1**, 95–110 (1970).
2. J. Bub, *Interpretations of Quantum Mechanics* (Reidel, Dordrecht, 1974).
3. Max Jammer, *The Philosophy of Quantum Mechanics* (John Wiley, New York, 1974).
4. J. S. Bell, On the Einstein Podolsky Rosen Paradox, *Physics* **1**, 195–200 (1964).
5. E. Wigner, *Symmetries and Reflections* (Indiana University Press, Bloomington, Indiana, 1967), pp. 171–84.
6. P. Heelan, Hermeneutics of Experimental Science in the Context of the Life-World, *Philosophia Mathematica* **9**, 101–44 (1972).
7. J. Habermas, *Knowledge and Human Interests* (Beacon Press, Boston, 1971).
8. P. Heelan, Nature and Its Transformations, *Theological Studies* **33**, 486–502 (1972).

Chapter V

The Expansion of the Universe in the Frame of Conventional General Relativity

Jean Heidmann

Paris Observatory, Meudon

THE STAGE

I shall skip straight through the historical part in two leaps:

964 A.D.: Abd-al-rahman al Sufi is the first to record on a map the dim light which, a millenium later, will open up to us the extragalactic universe.

1667: An engraving appearing in a paper by Bouillaud relates what was then known about that same Andromeda nebula after the introduction of the first telescopes (Fig. 1).

1923: The extragalactic world is reached: Hubble recognizes the true nature of Messier 31.

Shortly afterward came the discovery of the redshifts of other galaxies, which was interpreted with the Hubble law as an expansion of the Universe, according to

$$v = H_0 D \tag{1}$$

where v is the recession velocity of a galaxy and D its distance.

This law may be expressed as an expansion of the coordinate scale, if we write

$$D = r R(t) \tag{2}$$

where r is a dimensionless radial coordinate, constant for each galaxy, and $R(t)$ a scale length, a function of some time t. Then H becomes the actual value of the Hubble parameter,

$$H(t) = \dot{R}(t)/R(t) \tag{3}$$

39

Fig. 1. In this 1667 engraving from a paper by Bouillaud, NGC 224 is represented by the small elliptical cloud in front of the large fish's mouth under Andromeda's arm.

What can the expansion tell us about the Universe? And what can the Universe tell us about the expansion? In order to get more insight into these questions, we use the conventional world models of general relativity theory.

The most general form of the Einstein equations relating the geometry of space-time to its matter-energy content makes use of a geometrical tensor

$$C^{\alpha\beta} + \Lambda g^{\alpha\beta}$$

where $C^{\alpha\beta}$ is the curvature tensor (scalar and Ricci), $g^{\alpha\beta}$ the metric tensor, and Λ an arbitrary constant, the cosmological constant.

We add three very realistic simplifying assumptions: (i) The content of the Universe is a perfect fluid whose particles are the galaxies; (ii) the fluid is isotropic as seen from the Earth; (iii) its pressure is negligible. We obtain thus solutions of the Einstein equations which are the zero-pressure models, in which space-time is generated by a curved space, with the same radius of curvature at each of its points (constant-curvature space; constant across space, but not versus time).

But first, are these assumptions justified?

Number 1, yes; as we know it, most of the matter-energy content of the Universe is due to galaxies: 3×10^{-31} g/cm^3 as against $<10^{-32}$ g/cm^3 for molecular hydrogen, $<10^{-31}$ g/cm^3 for atomic hydrogen, and $\lesssim 10^{-33}$ g/cm^3 for photons, cosmic rays, and neutrinos. Ionized hydrogen and black holes are still badly known: $\lesssim 10^{-30}$ g/cm^3.

A density ρ and a pressure p are enough to describe the content of the Universe.

The second assumption, yes, to within a factor of two at most from galaxy counts and from radiosource counts; to within 4% from diffuse x- and γ-ray background; and to better than 0.1% for the 3°K blackbody radiation.

And for the third, yes, as the kinetic energy of galaxies is much smaller than their rest mass energy, their random velocities being evaluated at less than $0.001c$.

Then, in the zero-pressure models, the radius of curvature of the constant-curvature space is a function $R(t)$ of a particular time t, the cosmological time:

$$R(t) = cH_0^{-1} f(t; q_0, \lambda_0) \tag{4}$$

where q_0 and λ_0 are the actual values of the deceleration parameter

$$q(t) = -\ddot{R}(t)R(t)/\dot{R}^2(t) \tag{5}$$

and of the reduced cosmological constant

$$\lambda(t) = \Lambda/3\,H(t)^2 \tag{6}$$

In these models H_0 enters simply as the scale length cH_0^{-1}, while the various shapes of the $R(t)$ function depend on q_0 and λ_0. They are displayed in Fig. 2.

Space may be spherical, Euclidean, or hyperbolic and it may vary with time with a maximum (oscillating case), in a monotonic way with a singularity, or it

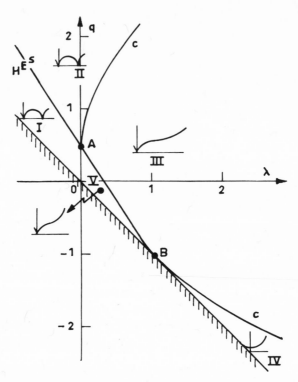

Fig. 2. Various shapes of the $R(t)$ function according to the values of the deceleration parameter q and of the reduced cosmological constant λ. The hatched area corresponds to forbidden negative densities, the E line to Euclidean space, the S region to spherical space, and the H region to hyperbolic space.

may exhibit a nonzero minimum. The simplest case is the Einstein–de Sitter model with $q_0 = 1/2$, $\lambda_0 = 0$, and density

$$\rho_{0,\text{EdS}} = 3H_0^2/8\pi G \qquad (7)$$

Note that there exist hyperbolic oscillating universes and spherical montonic ones. It should be recalled, too, that there are closed hyperbolic spaces (nearly always forgotten!).

Equation (7) shows a relation between the expansion rate H_0 and the density of the Universe.

Another relation exists with the age of the Universe. For a model with a singularity, this age is

$$t_0 = \int_0^{R_0} dR/\dot{R}(t) = H_0^{-1} h(q_0, \lambda_0) \qquad (8)$$

where H_0 enters just as a multiplying factor for a function h which has been mapped on the (q_0, λ_0) plane.

We have set the stage: The main actor for us is the expansion, with his main face H_0, and his deviants, q_0 and λ_0; but he is involved with Rhozero, Teezero, and Space-geometry.

THE HUBBLE CONSTANT

Two different sets of data determine the Hubble constant: velocities and distances.

The recession velocities are perturbed by the peculiar motions of galaxies; they may be as high as 300 km/sec.

They are also perturbed by the peculiar kinematics of the local metagalaxy.

According to de Vaucouleurs, our local group of galaxies has a peculiar velocity around 500 km/sec with respect to the Local Super Cluster. This is quite a large velocity, yet uncertain, and close to the value given by the very small anisotropy of the blackbody 3° radiation.

Farther away the Hubble diagram worked out by Sandage and Tammann for the brightest elliptical galaxies in 82 clusters (Fig. 3) provides some evidence: In the range 1000–4000 km/sec, i.e., from approximately 15 to 50 Mpc, there is no deviation from linear expansion larger than about 200 km/sec. In particular,

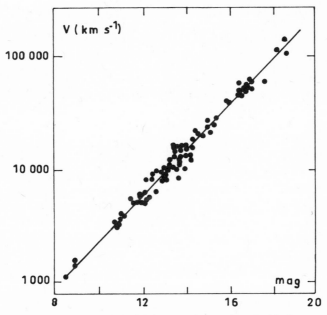

Fig. 3. The Hubble diagram for the brightest galaxy in 82 clusters, giving the logarithm of the redshift versus the corrected apparent magnitude. The straight line is the best one with slope 0.2.

the point for the Virgo cluster nearly lies on the best straight line, indicating no peculiar differential expansion between it and our galaxy.

Still farther the Rubins and Ford found a large difference of radial velocities for giant *Sc* galaxies presumed to be at the same distance from us, in two regions of the sky. Whether this is due to large flows, or to anomalous redshifts, or to some other cause, is not yet clear.

The kinematics of the local metagalaxy, up to distances of the order of 30 Mpc, is not yet completely worked out. In this respect, important progress may be expected from the discovery of nearly 1000 dwarf galaxies by Tully in this portion of space and their study in the 21-cm line of neutral hydrogen.

As to the question of distances, the best indicators are the classical Cepheids of Population I. Sandage and Tammann introduced a third parameter in the period–luminosity relation and worked out a period–luminosity–color relation which is unique for our Galaxy, the Magellanic Clouds, and Messier 31.

The relation was applied to seven nearby galaxies, most of them in the Local Group, yielding primary distances reliable at the 15–20% level. These distances are corroborated by other primary distance indicators, as was shown by van den Bergh. However, these galaxies are too close to allow the determination of H_0; they serve as calibrators for secondary-distance indicators which are used farther in space.

These secondary-distance indicators are mainly based on four methods:

(i) Use of the diameters of giant $H\,\text{II}$ regions. Initiated by Sersic and recalibrated by van den Bergh on primary distances, this method derives the distances from the apparent diameters of these regions and gives

$$H_0 = 99^{+16}_{-13} \text{ km sec}^{-1} \text{ Mpc}^{-1} \tag{9}$$

(ii) Use of the luminosity classes introduced by van den Bergh. From morphological criteria it is possible to obtain the absolute magnitudes of late spiral and of irregular galaxies. When calibrated on primary distances, the apparent magnitudes give the distances, and van den Bergh obtains

$$H_0 = 93^{+18}_{-15} \text{ km sec}^{-1} \text{ Mpc}^{-1} \tag{10}$$

(iii) Use of the luminosity–diameter relation due to Heidmann. This relation, which was found to hold between the luminosities L of spirals and their photometric diameter A, reduced to face-on view, has the form $L \propto A^{2.6}$; since the exponent is different from 2.0, the simultaneous measurement of the apparent magnitude and of the apparent diameter of a spiral galaxy provides its distance. Applied to the galaxies in the Virgo cluster and calibrated on primary distances, it gives

$$H_0 = 82^{+18}_{-14} \text{ km sec}^{-1} \text{ Mpc}^{-1} \tag{11}$$

(iv) Use of the groups of galaxies. In a statistical study, de Vaucouleurs showed that most galaxies occur in groups; resorting then to the mean diameter or the mean magnitude of the five first-ranked galaxies in each group, calibrated on primary distances, he finds

$$H_0 = 110\text{-}130 \text{ km sec}^{-1} \text{ Mpc}^{-1} \tag{12}$$

Sandage and Tammann go farther out, beyond the Local Super Cluster, by use of tertiary-distance indicators: They observed 50 giant *Sc* field galaxies. Their absolute magnitudes are used to obtain their distances, and these absolute magnitudes were established from closer galaxies whose distances were derived from their giant *H*II region diameters. They have shown that these diameters increase when the luminosity of the galaxy to which they belong increases, thus improving and expanding use of these diameters. Though details on all the successive steps are not yet published, Sandage and Tammann arrive at

$$H_0 = (55 \pm 7) \text{ km sec}^{-1} \text{ Mpc}^{-1} \tag{13}$$

Other secondary-distance indicators, statistically less reliable, are available and have been reviewed by Tammann and by van den Bergh. They calculate

$$H_0 = 70\text{-}100 \text{ km sec}^{-1} \text{ Mpc}^{-1} \tag{14}$$

Figure 4 provides a summary of the H_0 determinations. There is a general agreement except between de Vaucouleurs' and Sandage and Tammann's values. Clearly more work should be done to make possible the mapping of the local velocity field. Also, other methods should be used outside the Local Super Cluster. My group of researchers is engaged in measuring H_0 on distant *Sc*I galaxies using 21-cm-line distance criteria which we worked out in previous studies

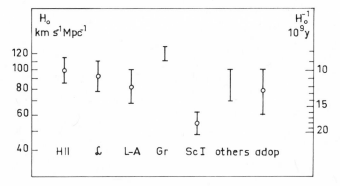

Fig. 4. The determination of the Hubble constant H_0 according to giant *H*II regions, luminosity classes ($\mathcal{L}$), the luminosity–diameter (*L-A*) relation, groups (Gr), giant *Sc* galaxies, and other methods. The adopted value is shown; the right-hand scale gives H_0^{-1}.

As a compromise, a value

$$H_0 = (80 \pm 20) \text{ km sec}^{-1} \text{ Mpc}^{-1} \tag{15}$$

may now be adopted. Then $H_0^{-1} = 10\text{--}17$ billion years.

THE DECELERATION PARAMETER AND THE COSMOLOGICAL CONSTANT

These finer parameters are still more difficult to determine. From the theoretical point of view there is no *a priori* reason for λ_0 to be zero; its value has to be decided from observations.

From the observational point of view a first indication is given by the density of the universe,

$$\rho_0 = 2(q_0 + \lambda_0) \rho_{0,\text{EdS}} \tag{16}$$

where $\sigma_0 = q_0 + \lambda_0$ is the density parameter. With $\rho_0 = 3 \times 10^{-31}$ g/cm³ for the actually observed matter, with no hidden mass, and with the value (15) for H_0, one gets

$$\sigma_0 \simeq 0.01 \tag{17}$$

Second, from the observed Hubble diagram and the theoretical curves for various q_0 values, Sandage obtains $q_0 = 0.3\text{--}1.4$ (for $\lambda_0 = 0$). From other arguments based on the fact that the kinetic energy of the expansion should be larger than the potential energy fluctuations due to the clustering of galaxies, he determines that q_0 must be very small, nearer to 0.1. At the present time we may adopt the value

$$q_0 = 1 \pm 1 \tag{18}$$

pertaining to $\lambda_0 = 0$. For $\lambda_0 \neq 0$, theoretical curves calculated by Tomita and Hayashi for the Hubble diagram indicate only that $\lambda_0 < 30$.

Observations undertaken on clusters with redshifts up to about $z = 0.6$ will perhaps bring more information.

THE 3° BACKGROUND RADIATION

As already mentioned, the large-scale isotropy of the 3° background radiation is now established to within 0.1%; though important and still uncertain (at such a level), corrections have to be applied to the observations to subtract the contribution arising from the flat-component emission of our own Galaxy.

At small angular scale, Pariisky has shown that, down to 2 arcmin, there is no variation larger than 0.003%. This excludes the possibility that the origin of this radiation is from a large number of individual sources scattered over the sky, unless their area density is larger than 10^{14} per steradian, a value which is too large to be realistic.

These isotropies are then in favor of a large-scale cosmic origin.

The spectrum has been well observed from about 1 mm to 1 m for some years. Crucial measurements in the submillimeter region are very difficult because of the Earth's atmosphere, but recent observations by Beckmann and Robson show the dip of the spectrum at 0.7 mm. These observations correspond to a blackbody spectrum at temperature $(2.7 \pm 0.1)°$K.

Finally, observations of various excited rotational states of interstellar molecules such as CN, functioning as thermometers, show that this $2.7°$ radiation pervades space not only near the Earth but also at various places in our Galaxy.

This radiation is then most probably universal and its simplest interpretation is the one obtained in the frame of the conventional models of general relativity theory: It is a relict of a past, very concentrated state of the Universe, as first proposed by Gamow. The consequence of this interpretation is the rejection of models which have not been singular, i.e., those in the lower right portion of the (q_0, λ_0) plane (Fig. 2).

THE AGE OF THE UNIVERSE

The best lower limit for the age of the Universe t_0 is now provided by the age t_g of the globular clusters of our Galaxy. From U, B, and V observations on four globular clusters, Sandage and Tammann constructed their color–magnitude diagrams. They have been interpreted by evolutionary tracks of stars calculated by Iben and Rood, and by Demarque and collaborators.

These calculations depend on the relative abundances Y of helium and Z of heavy elements. Z is provided by spectrographic observations. Fitting of the tracks to the diagrams then gives Y and the age t_g:

$$Y = 29 \pm 3 \text{ percent}, \quad t_g = 12 \pm 3 \text{ billion years}$$

The Y value is quite remarkable, as it is practically equal to the one found in the cosmological interpretation of the $3°$ radiation (nuclear reactions during the first $\frac{1}{4}$ hr).

Various arguments provided by Eggen, Lynden-Bell, and Sandage show that our Galaxy should have formed as a result of contraction of a protogalactic cloud taking place in a time as short as 100–200 million years. Partridge and Peebles conceived of a model for the appearance of such protoclouds in a time on the order of 150 million years after the start of the expansion of the Universe.

Thus, if one thinks of our Galaxy as having been formed in such a way, one gets $t_0 \simeq t_g$, i.e.,

$$t_0 = 12 \pm 3 \text{ billion years} \tag{19}$$

Later, however, I shall show some evidence for the recent production of galaxies; so that t_0 might be larger than t_g.

WHERE DO WE STAND?

We may now combine the collected information to see what can be said about the expansion of the Universe. Briefly, we have: (i) $H_0 = 80 \pm 20$ km sec^{-1} Mpc^{-1}, (ii) $\sigma_0 \simeq 0.01$, (iii) blackbody radiation $\rightarrow$ Universe had a singular state, and (iv) $t_0 = (12 \pm 3) \times 10^9$ yr, hence $H_0 t_0 = 0.7$–1.8.

A good way to represent these results in the (q_0, λ_0) plane is to use semi-logarithmic coordinates in the (σ_0, λ_0) plane (Fig. 5). On the right of curve ∞, space did not have a singular state; so this portion is excluded. The curves $H_0 t_0 = $ const have been calculated recently by Campusano, Heidmann, and Nieto; together with $\sigma_0 \simeq 0.01$, they show where our Universe should be located. This zone is cut into two parts by the $\lambda_0 = 0$ line which separates oscillating from monotonic universes. And it reaches the dashed curve corresponding to Euclidean space, separating the hyperbolic ones from the spherical ones.

From these data it follows that:

1. λ_0 is in the range -1.6 to 1.0; in particular, the value $\lambda_0 = 0$ is possible.
2. q_0 is in the range -1.0 to 1.6; for $\lambda_0 = 0$ it is practically zero.
3. Space is most probably hyperbolic—more exactly, locally hyperbolic. In case it is simply connected, then it is hyperbolic, and infinite. But who knows whether space is simply connected or not? The forms of locally hyperbolic three-dimensional spaces have not yet been mathematically worked out; only examples are known, of which some are closed, as, for instance, the one whose fundamental polyedron is a dodecadron. Then we cannot say whether space is open or closed (Fig. 6).
4. We cannot decide whether the Universe will contract again (oscillating case) or whether expansion will last forever (monotonic case).

If the density of the Universe is increased by eventual hidden masses (ionized hydrogen, black holes, etc.), the results are not essentially different, even for a tenfold increase of σ_0, except that the chance for space to be spherical increases.

One sees here the interest of the globular-cluster age determination and of the Hubble constant value, through the product $H_0 t_0$ (for $t_g \simeq t_0$). Future progress may allow us to distinguish definitely between spherical and hyperbolic cases, lowering of H_0 being in favor of hyperbolic space. But to decide between infinite expansion and recontraction appears more difficult.

THE EARLY PHASES OF THE EXPANSION

Though the future of the expansion of the Universe is not yet clear, some progress has been made recently in the knowledge of its past, and particularly of its very first instants. We owe that to the discovery of the 3° radiation.

In zero-pressure models the density of matter ρ_m varies as $R(t)^{-3}$, and for t small we have $R(t) \propto t^{2/3}$.

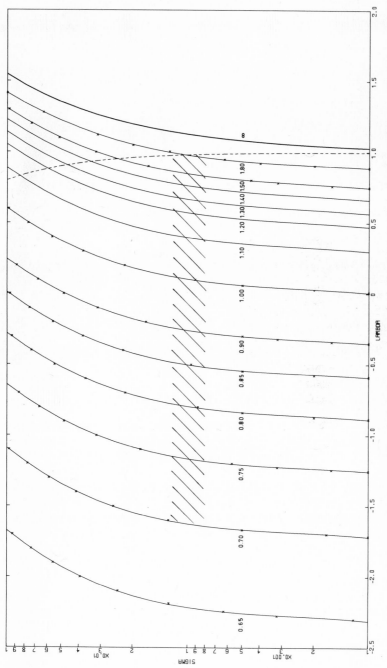

Fig. 5. The (q_0, λ_0) diagram on a semilogarithmic scale, with the density parameter σ_0 as ordinate. The dashed line corresponds to Euclidean space; on its right, space is spherical, on its left, it is hyperbolic. The line $\lambda_0 = 0$ separates oscillating universes (left) from monotonic ones (right), and the ∞ curve separates the foregoing from universes with a nonzero minimum in their variation. Curves for values of $H_0 t_0$ from 0.65 to 1.80 are plotted. The hatched area corresponds to possible actual values for the representative point of our Universe.

Fig. 6. Is space open or closed? René Magritte's painting, "Les Mémoires d'un Saint," 1960 (de Ménil collection, Houston).

For a universe filled with radiation, instead of using the equation of state $p = 0$, we have to use the one for radiation, viz. $p = \rho c^{2/3}$. The density of radiation ρ_r varies therefore as $R^{-4}(t)$, its temperature T as $R^{-1}(t)$, and for t small, $R(t) \propto t^{1/2}$.

These relations, with the actual values for ρ_m, ρ_r, and T, allow the calculation of the physical state of the Universe in the past (Fig. 7).

Physics as we know it today enables us to go back in time up to the first hundredth of a second after the start of expansion.

At that time the Universe was at a temperature of 100 billion degrees; the density of matter, in the form of protons and neutrons, was 10 g/cm^3. The density of radiation was much larger: one million tons/cm^3.

Then, much later, a thousand times later, at $t = 10$ sec, the temperature dropped down to a few billion degrees, this relatively low temperature allowed protons and neutrons to start nuclear reactions, and principally to produce deuterium.

Calculations made by Wagoner, Fowler, and Hoyle, based on nuclear physics data, show that, $\frac{1}{4}$ hr later, more complex nuclear reactions took place (Fig. 8), leading to the formation of 25% helium. This value is the one found in stars and in the interstellar medium and is obtained for calculations made with an actual density of matter of 3×10^{-31} g/cm^3; all this is quite coherent.

After these busy 15 min, the Universe pursued its expansion; it was too cold and too dilute for further nuclear reaction. Then nothing happened for millennia. After 10,000 yr, the temperature dropped below 10,000 deg and the gas, up to then ionized, became neutral.

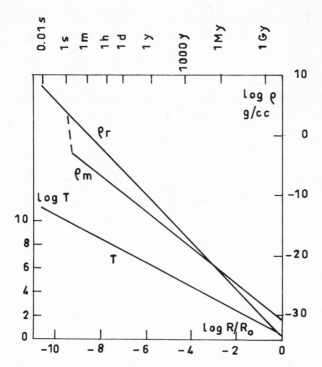

Fig. 7. Variation of density of matter ρ_m, of radiation ρ_r, and of temperature T versus cosmic time and versus dimensions of the Universe $R(t)/R(t_0)$.

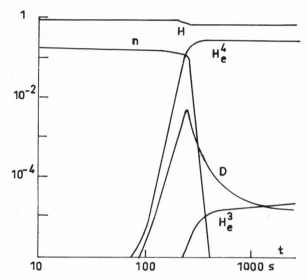

Fig. 8. Variation of the relative amount, by mass, of protons, neutrons, ^{4}He, D, and ^{3}He during the first 1000 sec of the expansion.

After 300,000 yr, radiations lost their preponderance over matter, and it is only after 100 million years of cold blackness that the first condensations appeared in the gas, condensations which gave rise to the first galaxies or quasars.

AND BEFORE THE FIRST HUNDREDTH OF A SECOND?

As we go back in time, the temperature T of the Universe rises more and more. At $t = 10$ sec it is already sufficient for its photons to be energetic enough (1 MeV) to create pairs of positrons and electrons, with the corresponding neutrinos. At $t = 2 \times 10^{-4}$ sec, T is sufficient for μ-meson production; thus, earlier, the radiation is in equilibrium with leptons responsible for weak interactions.

In the same way, before $t \simeq 10^{-4}$ sec, π-mesons enter the scene; we get into the hadronic era, with K-mesons, nucleons and antinucleons, hyperons and antihyperons, etc.

Further back in time we cannot go, because our knowledge of the physics of elementary particles is not complete enough. According to Hagedorn, the mass spectrum of created hadrons increases exponentially with mass, and T would tend to a limit, a "hadronic boiling point," of about 10^{12} deg. Therefore, before $t \simeq 10^4$ sec, T should have been constant.

However, for t small, the cosmological horizon is at a distance $\sim ct$. If the Compton wavelength $\lambda = \hbar/mc$ of a particle of mass m—with λ giving an idea of its diameter—is larger than ct, the particle cannot exist. Thus, according to Harrison, particles with $m < \hbar/tc^2$ should be omitted and hadronic states start to become underpopulated for $t_p \simeq \hbar/m_\pi c^2 \simeq 10^{-23}$ sec. Below this value, T could rise again.

For t smaller still, it appears that quantum effects might appear. From the two constants of general relativity, c and G, and the one of quantum mechanics, $\hbar$, it is possible to derive a universal length, the Planck length:

$$L_p = (\hbar G/c^3)^{1/2} \simeq 10^{-33} \text{ cm} \tag{20}$$

The cosmological horizon is equal to L_p for $t_p \simeq 10^{-43}$ sec. According to Wheeler, before t_p, the Universe may have been subjected to uncertainty relations giving a fuzziness to its geometry. Its state could not be defined precisely but only in terms of probabilities from a universal quantum theory yet to be worked out.

In short, the history of expansion might be divided into successive eras. To give fairer values to their duration, let us express t on a log scale; quite arbitrarily, it should be stated, $t = \log_{10} t$, t being in seconds (Fig. 9). Here 17.5 stands for the present epoch (Mankind started at 17.49990).

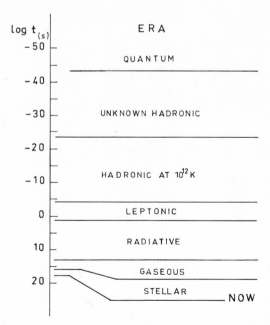

Fig. 9. The successive eras of the expansion of the Universe.

WARNINGS!

Let me end this presentation on the conventional aspect of the expansion of the Universe with some warnings. All of this picture of the expansion is exciting, pleasant, coherent, well in order. But what if the redshifts are not to be interpreted by the Doppler–Fizeau law in the classical mechanical view, or general relativistically, by the fact that the ratio of the wavelength of a photon (as measured by a co-moving observer) to the space radius of curvature is independent of cosmic time?

Not speaking of quasars, the first indications for non-Doppler redshifts *for a galaxy* have been provided recently by Arp and by Balkowski, Bottinelli, Chamaraux, Gouguenheim, and Heidmann, with two independent methods applied to Stephan's Quintet. The level of confidence of the last method is close to 99%. However, this is not yet a proof. In another domain which may be relevant to cosmology: What if not all galaxies were formed at the dawn of the Big Bang; what if some are being formed now? Then, at least, the age of the Universe can be anything larger than the age of our own Galaxy, and not simply 12 billion years.

I would like to report briefly on indications about the recent production of galaxies which I obtained these last two years in collaboration with Arp, Bottinelli, Casini, Duflot, Gouguenheim, Kalloghlian, Khachikian, Lelièvre, and Spinrad. Let me do it in a comic strip style.

Heard that? Ambartsumyan had strange ideas back in 1958: It is possible that galaxies are being formed now by a special activity of their nuclei!

Yeah, and know what? Markarian started to search for galaxies with special nuclei; and he found hundreds of them. . . .

Better start right now to measure them in the 21-cm line, so we can get their dynamical parameters.

Say, for a random distribution we should get three pairs of them on the sky, and I found 18.

Interesting. Must be physical pairs.

I worked out their dynamics: Two have positive energy.

So their components are flying apart. When did they start running away?

Quite simple. A mere billion years ago.

Gee! Quite recent. Is there any parent body somewhere nearby?

Maybe. Look at this unusual double blue object near this Markarian pair (Fig. 10).

Well, this double has the same radial velocity, so it sure is related to the pair. And a better photograph shows a still smaller double inside the double object. Hmm, very exceptional system. . . .

If Markarian galaxies in pairs are recent, this should show up in their morphology. Let us take some large telescope photographs.

Oh, look at this strange filament; never saw that before (Fig. 11).

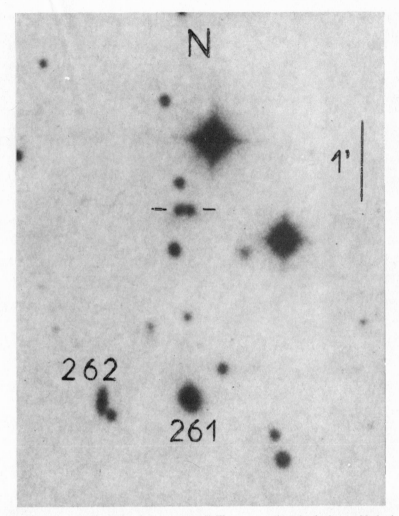

Fig. 10. Three steps in the scale of strangeness: (i) The two numbered galaxies are Markarian galaxies, peculiar because of their strong ultraviolet radiation; (ii) against all probability, these two galaxies form a close pair; (iii) and close to them is a double, very blue, compact object of a quite unusual nature (Palomar Sky Survey photograph).

What about a scheme, just as a working hypothesis: (unknown) parent body → 2 compact objects → 2 Markarian galaxies → 2 normal galaxies?

If so, there should be pairs made up of one Markarian and one normal, one Markarian and one compact. . . . Let us search for them.

Yep! Here are 60 Markarian-normal and 17 Markarian-compact. And Bertola told me he found compact–compact pairs, and Karachentsev normal–normal pairs.

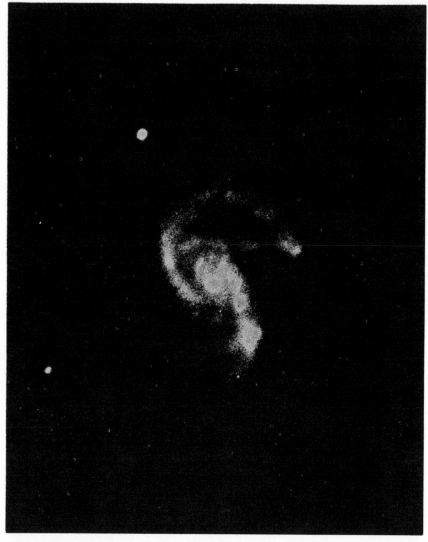

Fig. 11. A long, thin filament connects two bright condensations of the northern spiral arm of the galaxy *Markarian 12*. This most peculiar feature, quite unstable, might be the after-effect of a recent origin of this galaxy (*Asiago Observatory* photograph).

Looks like a nice lead.

Easy, easy! Say we happened upon a bunch of interesting suspects; and we are going to watch them. Maybe they will tell us something new. . . . I dream of white holes, connected to black holes of another universe. I wish our theoreticians could tell us more about the possibilities there.

(To be continued.)

REFERENCES

1. J. Heidmann, *Introduction à la Cosmologie* (Presses Universitaires de France, Paris, 1973).
2. J. Heidmann, in *Stellar Ages*, *IAU Colloquium* (**17**, XXXVII-1), ed. by G. Cayrel de Strobel and A. M. Delplace (1972).
3. A. Sandage and G. A. Tammann, *Ap. J.* **190**, 525 (1974).
4. *Cosmology*, *IAU Symposium* **63**, Cracow, 1973.
5. Second European Regional Meeting of the IAU, Trieste, 1974.

Chapter VI

Models, Laws, and the Universe

William H. McCrea

University of Sussex

This paper is in the context of a colloquium on cosmology, history, and theology. I must begin by stating briefly how, as an astronomer, I venture to view the relationship between these fields. At the end, I shall append some further reflections in the light of the intervening discussion.

The fundamental theme that brings the topics together is that of the creation of the universe. If anyone is concerned scientifically about this, it at least seems obvious that this should be an astronomer or cosmologist, simply because it is his calling to study the universe in the large. There is also a more particular reason. For the astronomer is expected to interest himself in the problem of life on other worlds. If there are beings rather like ourselves in remote parts of the universe, this is generally held to have profound significance for what we believe about ourselves and our Creator. Although this view may be instinctive, it is not clear how the knowledge that there is someone like me a billion miles away should affect my conduct more than the knowledge that there is someone a thousand miles away.

On the other hand, the astronomer deals only with the simplest and most elementary aspects of the universe. Of course, nothing is simple in a naive sense. But in a rather sophisticated sense, most of the universe is simple! In fact, most of it is empty space, and most of what matter there is exists in the form of plasma. Only a tiny part of the matter forms even simple molecules, and only a tiny part of molecular matter forms solid and liquid bodies; a tiny part only of these can be living matter. So far as we know, only an infinitesimal part of living matter feels any concern at all about the rest of the universe—its operation, its significance, its purpose. If cosmology as we know it makes any sense, then it seems that the universe has to be like this. In other words, it requires all the hugeness of the universe to produce man.

It seems, therefore, that we have a spectrum ranging from *cosmology*, dealing with a lot that is simple, to *life science*, dealing with a little that is complex. We might think that if it is at the one end of this spectrum that we get near to creation, it is at the other end that we should get nearer in thought to the Creator. So far as this notion is profitable, it suggests that the astronomer is qualified at best to discuss only the simplest beginning of things, but that he may be the least qualified to discuss the deep questions of significance.

I think this is to a large extent valid. However, it seems not to lay enough stress upon the circumstance that *we*—the observers and interpreters—are involved at every stage throughout. We are dealing, not with a universe in which we are an insignificant accident, but with our experience of that which we choose to call the physical world. So perhaps all of us have about equal qualifications—or disqualifications—to discuss the subject.

MODEL

Nowadays everyone talks about models in every sort of study, but maybe they do not think quite strictly in such terms. In any branch of physical science, all reasoning does concern only some model of some aspect of the physical world. We compare the predictions of the model with experience of the actual world; this may suggest a better model and so on. To some extent we may think of the model as a computer programmed in some particular way. We ask it questions and we put the answers to the test of experience.

A physical theory as such does not make assertions about the physical world. The theory constructs a world of its own, and this we call the *model*. All the mathematics involved is concerned entirely with the model. When we say that we compare the predictions with experience, we mean that we seek to set up a correspondence between the results of the mathematics and some set of physical experiences. To make this more explicit, we may speak of a "mathematical" model or a "theoretical" model.

(The notion is quite different from most people's idea of nineteenth-century "models" of electromagnetic phenomena, mechanical models of the aether, and so on. Some "model" in that sense might serve to *suggest* a mathematical model—Clerk Maxwell somewhere says as much; if it does, it is the mathematical model that then becomes of interest.)

In physical science, in addition to innumerable models for more particular purposes, there have been some models of several general aspects of the physical world; the Ptolemaic system, the Copernican system, Kepler's model of the solar system, Newton's mechanics and gravitation, Maxwell's electromagnetism, Einstein's special relativity, Einstein's general relativity, Rutherford's atom, Bohr's quantum model of the atom, quantum mechanics and developments therefrom. When I refer to general aspects of the physical world, these themselves receive a definite meaning only in terms of a relevant model.

The procedure works. Ancient models would tell where any known planet would be seen at any moment; Kepler's model would tell, fairly well, how any planet known or unknown would move; Newton's model, as used, for instance, by Adams and Leverrier, revealed the existence of a planet nearly 20 times as massive as the Earth that no one had known about before (Neptune); the same model affords split-second predictions about Apollo missions to the Moon, and so forth; among other things, special relativity predicted the equivalence of mass and energy; general relativity predicted the bending of light past the Sun; as discovered by Penrose, Hawking, and others, it predicts also under very general circumstances the existence of singularities in its model universes, i.e., any such model must exhibit something like a big bang followed by an expansion of the universe; using his quantum-mechanical model of elementary particle behavior, Dirac predicted the existence of the positive electron; Rutherford predicted the neutron; Pauli predicted the neutrino; and so on. Thus the prediction of new experience is a regular and successful activity. Also, besides such specific prediction, *every* experiment involves some element of prediction. For the experiment can be conceived only in terms of some model of the physical world, and it is planned because some predictive thought suggests that the outcome should be of interest. We may think, for instance, of Faraday's researches in electricity and magnetism, where the model that he was gradually building up itself guided him to the next step at each stage. Of course, most things that we know about the physical world have been "discovered" by experiment, and it would be a gross exaggeration to claim that they had been predicted in any precise manner. At the same time, I cannot think of any genuine positive prediction that has not been validated!

This successfulness of prediction is expressed by a popular aphorism: What is not forbidden is imperative. However, stated thus it is either a truism or it is false. As examples, special relativity does not forbid the existence of tachyons—particles that move faster than light; general relativity does not forbid the existence of particles of negative mass; electromagnetic theory does not forbid the existence of magnetic monopoles. However, to date there is no incontrovertible evidence that any of these has been observed. On the other hand, these are examples of entities that, while not forbidden, have not been predicted by any theory in the way that the positron, the neutron, and the neutrino were predicted.

We must also distinguish between prediction and *speculation*. While speculation plays a recognized tentative role, it comes at quite a different stage in developing a theory. Indeed, it is part of the process of casting around to determine the nature of the theory to be developed. Prediction, on the other hand, is the exploitation of a well-formulated theory.

Here we have then to add a comment about the term *theory*. I take a *theory* to mean any well-formulated deductive body of theorems in natural philosophy. Thus the theories of special relativity, of general relativity, of classical hydrodynamics, and of magnetohydrodynamics provide examples. A model may be the whole construct of a theory and its applications and interpretations; or it may be a construct of more than one theory as, for instance, general relativity and quantum theory; or it may be a construct obtained as a particular application, such as the Schwarzschild internal and external solutions giving a model of a massive body. I think that *model* and *theory* are here used in commonly accepted senses. (It should only be remarked that *theory* is not used in the detective story sense of police "working on the theory that X killed Y"; this would be a mere *hypothesis*, and nothing like a theory in the sense of natural philosophy.)

PROGRESS AND ABSTRACTION

On any reasonable view, the achievements of the various models mentioned above surely denote *progress*. It can be claimed that we get better models as time goes on. I think *there is no useful sense in which a model is right or wrong; we* can say only that it is better or worse than some other model in some well-defined technical sense. Even when we have a model B that is better than a model A in some accepted sense, we do not in general discard A. Instead, we learn the limitations of the uses of A. For example, in the light of special relativity we know there to be bounds to the usefulness of the idea of the Newtonian rigid body. But it would be absurd not to use Newtonian dynamics of a rigid body in discussing, say, the spin of a space vehicle and its control. Again, we learn what are meant by relativistic effects and gravitational effects. In certain problems, if the first are clearly significant and the second are clearly insignificant, then we naturally use the special relativity model and not the general relativity model.

Every model must be limited in some such way. This is what we mean when we say that a model is an *abstraction* of, or from, reality. We have no idea—I would say—as to why this process of abstraction succeeds. However, I think it is a necessary element in what we call understanding or explaining our experience of the physical world. Indeed, I think it is arguable that, if we seek to construct better and better models, we should ultimately produce the universe itself, which would explain nothing!

I say all this because I believe that science is not an endless pursuit of contradictions. While I cannot see any prospect of finality, I do not think that anyone would devote himself to science unless he believes that we are making progress in understanding the physical world, which means making ever-improved models. In other words, I see no validity in the notion of *falsifiability* of physical theories. For that notion is based upon the postulate that there is some correspondence between the working of the actual universe and our present system of human logic. This is a postulate that cannot be validated or falsified.

As I say, nobody understands why the model-making procedure succeeds and, in particular, why we can abstract certain sets of phenomena and so construct "partial models." For the distinction between, say, inertial phenomena and gravitational phenomena, or between gravitational phenomena and electromagnetic phenomena, is made by the models themselves. We cannot even use these terms intelligibly without invoking a model. Yet we can decide to make a model that will take account of inertial effects but ignore gravitational effects, or one that takes account of gravitational effects but ignores electromagnetic effects, and in this way get models that are useful within their limitations. That there is not likely to be a simple answer as to why all this actually works is indicated by the fact that basic physical models do not form a steady progression from one to the next. Each denotes a fresh start with, usually, quite different mathematical techniques. A striking illustration was provided by the almost simultaneous invention about 1925-6 of matrix mechanics, wave mechanics, and Dirac's quantum mechanics. As mathematical models they looked entirely different from one another. Nevertheless, it was generally believed that they made the same observable predictions—although this is still a somewhat debatable point—where they could be said to deal with the same experiments. And they were all surprisingly different from anything that had been done before in atomic physics.

OBSERVABILITY AND UNCERTAINTY

There are certain obvious rules that have to be followed in constructing models—obvious after some genius has called attention to the need, no one having apparently thought about them until relativity and quantum-mechanical models came to be formulated. In particular, *a satisfactory model must require for its predictions only information that can be got from observation, and it should predict only results that can be checked by observation, where observation means that which is made possible by procedures provided by the model itself.* For example, Bohr's original model of the hydrogen atom was such that, if one attempted to observe the electron in its predicted orbit, either one observed nothing or one knocked the electron out of the orbit by the process of observation. This actually made nonsense of the picture of an electron going around in an orbit. But basically it was the resolve to avoid this imperfection that led to the invention of quantum mechanics, i.e., to a "better" model.

Bound up with this feature of observability is also the feature of *uncertainty*. As is well known, there are irreducible uncertainties in the quantum model for the simple reason that Planck's constant h is not zero. The act of observing a system disturbs it in an irreducible and unpredictable fashion. I conjecture that *better models would be subject to uncertainties associated with other physical constants as well*, although I cannot prove this generally. Such further uncertainties might not have to do with disturbing a system by observing it, but rather with irreducible limitations upon the amount of information that can be obtained by observing a system in order to predict its behavior.

LAWS

People are apt to think of the natural scientist as asserting: The entire universe is governed by law; the same laws hold good at all places and all times; the scientist's sole dedication is to discover those laws one by one.

This attitude is to a great extent traceable to the remarkable success of the Newtonian model. Also, we have laws of chemical combination, laws of quantum transitions, laws of Mendelian inheritance, and so on, that do seem to be precise and permanent items of knowledge.

In a broad sense, we have to adopt some such attitude in order to do any science at all. Were we to have to say that thus and thus is the way things seem to be today but who knows what they will be like tomorrow, then there would be no point in working at them! But the attitude has unsatisfactory aspects; among these are: (a) It tends to put us in the situation of believing in the existence of precise laws that we can never discover. Experience has shown us that every model is sooner or later superseded by what appears to be a better model. Then if we try to express this in terms of a system of laws, we reach the highly unsatisfactory concept that, all around us, nature is faithfully obeying laws of its own, but in the light of the experience stated, we expect never to ascertain exactly what they are. (b) It is based on the contradictory concept that most possibilities are, by law, impossible! If we contemplate, for instance, a comet moving in the depths of space, we contemplate first an infinity of "possible" motions, and then we invoke the law of gravitation that forbids all except one. Surely it would be more satisfactory if we could have a *model*, i.e., some manner of viewing the situation, *that would admit only the one possibility; then, of course, no law is needed to say so.*

Within its scope, this situation has been largely achieved by the theory of *general relativity*. There we may take any Riemannian space-time (with appropriate "signature") and simply read off the whole history of the self-gravitating matter which it represents according to the rules of interpretation that form part of the theory. No laws of motion or of gravitation are invoked or formulated if the model is employed in accordance with its own conventions. Einstein himself scarcely seemed fully to appreciate this situation; he claimed to have a new law of gravitation, but this is merely the convention for a region of the space-time to be said to contain no matter, and he liked to speak of field equations, but these are merely the conventions for reading off the density and stress of the matter present. Einstein's greatest achievement, beyond all his predecessors, was almost undoubtedly to produce a *model of the physical world that makes physical predictions without* (or almost without) *requiring the concept of particular laws of physics.*

LAW AND THE UNIVERSE

In spite of what has just been said, we may usefully employ the term "laws of physics" as a general description of our organized knowledge of the physical world constituted by our set of models. (Various authors use the term in various senses, and I think there is no standard usage.)

One aspect is then specially relevant if we are to try to talk about creation. For there can surely be no meaning in laws of physics without a physical universe to "obey" them. This implies the rather obvious notion of *the universe and the laws being interdependent.* But if the universe was created and is evolving, the universe is changing. So we might well infer that the laws must be changing, too. Either we know how they are changing or we do not. If we do, then the law of their change is itself an unchanging law, and so we arrive at a contradiction. If we do not know, then we have almost no laws at all.

One might, of course, claim that, for example, some observations verify that hydrogen atoms in the Andromeda galaxy are the same as the hydrogen atoms in our laboratory, implying that the relevant laws of physics were the same some two million light-years away, some two million years ago. The aggregate of all such observations might then be quoted as rather convincing evidence that the laws of physics are indeed the same at all places and at all times. Taken literally in conjunction with anything like the big-bang account of creation, this would imply that at the very instant of creation the laws of physics were all there ready to be obeyed by every particle for evermore. While this inference cannot be tested by observation, it does seem to be a *reductio ad absurdum* of the usual view of laws of physics when this view is pushed to the limits of the universe. It might appear that the difficulty could not arise if there are no limits to the universe, i.e., if the universe is in a steady state. Indeed, something of the sort has been used as an argument in support of steady-state cosmology. However, I do not find the argument to be convincing, since obviously the universe could not be in more than a statistically steady state, while the common (perhaps naive) interpretation of the laws of physics is that they are always exactly the same.

Actually the interpretation of the observations of the Andromeda galaxy, in the example, is not correct as stated. *The observation is made by means of a telescope, and what we have checked is that we have a scheme* (concerning hydrogen atoms, in this example) *that correctly predicted the outcome of something we did, not something in the Andromeda galaxy.* This may seem to be a quibble in a case like the example, but the idea seems to be fundamental when we proceed to enquire into the nature of cosmology.

To return to the dilemma concerning changing laws, there appear to be two ways by which we might escape: (a) We might say that it serves further to underline the objections to anything we can call laws, even in the present general sense, and to urge us to concentrate upon the *model* aspect. Thus we ought to point to a model and say simply: That is what the universe is like—its past, present, and future—and that is all there is to it. In fact, this brings out the fundamental feature that anything like a relativistic cosmological model just *is:* it has no aspect of becoming, no element of changing. Hermann Weyl expressed this in 1949:

> "The objective world simply *is;* it does not *happen.*
>
> Only to the gaze of my consciousness, crawling upward along the life line (world line) of my body, does a section of this world come to life as a fleeting image in space which continuously changes in time."

This is an accurate description of the general relativity model, yet it expresses what appears to me to be the least satisfactory feature of that model. For, if we could construct such a model, it would comprise the entire history of the universe, backward and forward; so, in constructing it we should already know everything that it is designed to predict.

(b) We might say that all we want are the laws of physics *here and now;* we cannot say what they were at any other time or place, but that is no concern of ours.

I suggest that we have to adopt some such standpoint, but perhaps we can express it better. We can assert: The laws (models) deal, not with the universe apart from ourselves, but with our experience of the universe. Thus if, for instance, we say that one second after creation the contents of the universe were in such and such a state, we ought not to mean that that is what we should have seen, had we been there, but that the model with those properties best accounts for what we *now* observe. I shall come back to this issue, but here I must emphasize that at present our models are not formulated according to this view.

COSMOLOGICAL MODEL

Before proceeding further with the very wide considerations we have been discussing, we need to have before us a fairly definite cosmological model that is widely accepted as regards its general features. For our immediate purpose, which is to discuss the status of such a model, we certainly do not require to commit ourselves to the model. It is effectively the same model as that described by Penzias.[1] Described and interpreted in somewhat naive terms it is:

About 20 billion years ago the universe exploded from a singular state; this was the so-called "big-bang." At the singularity, the density was infinite; so was the temperature, for we are describing the "hot" big-bang model.

The universe in the large is always homogeneous and isotropic.

It proceeded to expand from the big bang, the density and temperature decreasing.

In the earliest stages, all processes went so fast that the state was determined wholly by the instantaneous density and temperature. In particular, therefore, we do not have to postulate any initial chemical composition.

In the first millisecond or so, the contents were mostly protons and antiprotons, with about one in 100 million more protons than antiprotons. Very soon the protons and antiprotons annihilated each other in pairs, leaving only the slight imbalance to form the matter of the universe. The rest of the mass reappeared in the form of photons, which therefore at that stage far exceeded (10^8-fold) the mass in the form of matter. So there was a radiation-dominated universe.

Most of the photons still survive, but when we observe them they have been enormously redshifted. Instead of being gamma-ray photons, they are now microwave radiation. In fact, they form the so-called $3°K$ microwave background radiation with properties that seem now to be well established.

The matter underwent nuclear reactions at a decreasing rate, until the rate became negligible and a particular chemical composition was "frozen" into the matter. For the model we have in mind, or any other likely to be of interest, the composition by mass was then 75-70% hydrogen, 25-30% helium (^{4}He), with traces of deuterium (^{2}H), ^{3}He, and ^{7}Li. This was the raw material for forming galaxies.

As time went on and the universe continued to expand, the energy in the radiation dropped more rapidly than the energy in the form of matter, the latter being mostly locked up as proper mass. So there came a stage when the energy in the radiation fell below that of the matter, and thereafter the universe has been matter-dominated. At that stage the matter and radiation had temperature about 10^4 deg.

Then, or a little later, the cooling matter became largely transparent to the radiation. After that, the matter cooled more rapidly than the radiation, reaching probably about $1°K$.

This was about the earliest stage when conditions in the matter were favorable for forming gravitational condensations. Suppose such condensations having "galactic" masses did form. Under the conditions, they would have held together gravitationally and would tend to contract, not to disperse. Stars would form. It is now thought that at first very massive, short-lived stars would form and would produce the heavier elements—about 2% by mass—that are included in the second-generation stars, which are what we mostly observe.

The radiation from the galaxies would have heated up any remaining diffuse material, and the resulting higher temperature, together with the reduction in density resulting from the continuing expansion, would yield conditions unfavorable for forming any further condensations, although some of this matter might flow into the galaxies already formed. Those galaxies would all have been formed during one relatively short interval of cosmic time, and (apart from the age–distance effect) all the galaxies we see would have closely the same age.

Thus we may claim that the following are predicted by the model: the big bang and the expansion of the universe, the background radiation, the chemical composition of the universe in its broad features, conditions in which the galaxies could have formed, the age of the galaxies. Qualitatively these predictions depend upon very few postulates. Quantitatively, they depend upon observed present values of the mean matter density and radiation density and of the Hubble "constant." We can claim further that the model makes no prediction that clearly conflicts with observation.

Scientists have grown accustomed to possessing such a model (actually a range of models) and they are inclined to regard it as being only a rather limited achievement in relation to all the effort that has been lavished upon it. Judged, however, in comparison with anything that was imagined until quite recent times, it is a very great advance.

Another impressive feature of the situation is that what is mostly *laboratory* physics should serve so well when applied to the universe in the large. When we pass from the consideration of ordinary gross matter on the scale of grams and centimeters down to the scale of atomic phenomena and then down to the scale of elementary particles, we find it necessary to introduce all kinds of novel concepts. Yet when we pass from the scale of grams and centimeters up to the scale of galaxies and clusters of galaxies—where far larger factors are involved—we appear to need no new physical concepts. If this is so, it is partly just a discovery of the way the universe is made—and a tribute to laboratory physics and physicists. However, it must also be in some way a reflection of what we said earlier about the laws of physics being the laws *here and now*. I cannot follow this through in detail, but it does seem more rational to say that laboratory physics proves adequate to cope with at any rate many of the observations of the universe in the large, because the observations themselves are made here and now on a laboratory scale, rather than to endow laboratory physics with a universality that it seems impossible to test.

This is very far from claiming that we already know all the physics needed to study the universe in the large or in the laboratory. Also, some few astronomers believe that "new" physics is already required to deal with phenomena concerning the nuclei of galaxies, but the arguments are rather vague.

We said that predictions or inferences derived from a model ought to be only such as may be checked by observation. But how does this apply to the past history of the universe? If we look at a part of the universe one billion light-years away, this means that we see it as it was one billion years ago. That is, when we look into the distance, we look into the past. Inferences about the past can thus be checked by observation, i.e., our requirement concerning observability is fulfilled in a general way and our procedure is self-consistent.

Any particular model of the sort described does, however, make very specific assertions about the universe at all stages right back to the big bang, which we may call the instant of creation of the universe. These are inferences that we could never check in detail. There are two sorts of reasons:

(a) Observation becomes more and more uncertain as we go back. For one thing, the redshift tends to infinity at the singularity, which means that no radiation could reach us from the singularity itself, and very little from the very early stages—for this cause alone. But in any case, nearly all the photons we can detect from the early universe come from the end of what is called the "hadron era"; so we do not really look farther back than that.

(b) The situation is complicated by evolutionary effects, irregular distribution of radiation sources, and so on. This is simply to say that it is complicated because the universe is what it is. In other words, we cannot abstract the universe from the universe in order to study the universe.

All we can claim seems therefore to be that a model along the lines described appears to be sensible in a general way. There is bound to be much unavoidable uncertainty, not all of which is built into the existing treatment. As regards creation, in particular, the model indicates that there is nothing in the universe older than about 20×10^9 yr. I incline to think that this is about all we may say about creation, but some colleagues may be prepared to be more specific.

PURPOSE OF COSMOLOGICAL MODEL

In broadest terms, we want a cosmological model to account for what we do observe and to predict what we can observe of the universe in the large.

What we observe are:

(a) General features, like helium abundance, background radiation, etc. Cosmologists are now tending to be able to account for these, as they should wish, not by discovering more about the past state of the universe that could not be checked by observation, but by finding that they can predict a lot without knowing much in detail about the early universe.

I think there is a principle of science that *nature will not satisfy idle curiosity*. If there is something about the past that we do not need to know in order to predict something about our future experience, then I think that is something we can never know and that it is meaningless to ask about it.

(b) Statistical features like apparent magnitude–redshift relations for galaxies, quasars, etc., radio source counts, and so on. This is where we have seen that the situation is complex and almost certainly defies any tidy resolution.

(c) Matter is parcelled into galaxies. This is what we should most wish a cosmological model to explain. For any galaxy that we observe seems to be of the order of between 10 and 20 billion years old. Before the material formed a galaxy it was not in any form that we can observe in detail by looking into the past. Almost as soon as the material became a galaxy, it began its kaleidoscopic career of forming stars and stellar clusters, spiral arms, supernovae, and so on in one generation after another, in fact displaying all the phenomena of astronomy and astrophysics. We have remarked that the history of the universe appears to provide a phase uniquely favorable to galaxy formation. But the model supplies no clue as to what produces the condensations. This is the classical hen-and-egg problem; if the universe starts as perfectly homogeneous, then there can be nothing in the universe to disturb that state and to lead to condensations; if the universe always had some nonhomogeneity, then it is meaningless to ask what causes the nonhomogeneity. Thus the problem of condensations leading to galaxy formation has not yet been even adequately formulated. The best we can hope for at present is to be able to show that *any* sort of nonhomogeneity (patchiness) will necessarily result in the sort of condensation required to form galaxies. This is another example of a type of problem that is different for the universe as a whole from what it is for some part of the universe; thus the problem of galaxy formation is quite different in kind from that of, say, star formation.

What we expect by way of prediction includes:

1. Information about intergalactic matter. Ever since astronomers embarked upon the study of the universe in the large they have faced the question as to whether the galaxies (and quasars) that they observe comprise the predominant material content of the universe or whether the vast tracts of space between the galaxies contain in some diffuse form more matter than there is in the galaxies themselves. Obviously this is an observational problem. But it is also a case where clearly the observations cannot be interpreted without a model.

2. Possibly predictions about large-scale activity in the nuclei of galaxies. Whereas the career of a galaxy mentioned above proceeds without active participation by the rest of the universe, and so is by convention regarded as being largely outside the scope of cosmology, some astronomers think that there is evidence of strong, violent activity in the nuclear regions of galaxies resulting in the ejection of such considerable amounts of energy and matter that this must be treated on a cosmological basis.

3. A study of large-scale inhomogeneities in the universe. On the one hand, continuing study of the microwave-background radiation in the universe tends

to confirm the current interpretation and in particular to confirm the exceedingly high degree of isotropy of the *radiation*. On the other hand, observations of the distribution of galaxies and quasars certainly indicate inhomogeneity on a considerable scale, and some authors think this to be so great that we ought to employ a hierarchical model rather than a homogeneous model of the distribution of *matter*. All this raises deep problems requiring urgent investigation.

4. We have always to remind ourselves that we never really know what the universe of discourse includes! For experience, particularly that of the past decade or so, shows that every fresh technique of observation produces some extension of this universe—we may think of energetic and massive cosmic rays, x-ray sources and background, the microwave background, the outer extensions of galaxies, quasars, etc., as examples relevant to cosmology.

5. The interpretation of the asymmetry of physical phenomena with respect to time. Much has been written about this in recent years and nearly all tends to show that the asymmetry is to be understood only in cosmological terms. It is a big subject, but clearly one of the most important in the whole of natural philosophy.

6. The origin of elementary particles, of the unit of electric charge, of the universal constants of physics (Planck's constant, and so forth). Progress in these most fundamental of all aspects of physics is probably to be expected from physics itself rather than from astronomy and cosmology. But the problems are fundamentally cosmological.

There is another reason for mentioning these topics, and that concerns *precision*. Everything that physics does about elementary particles and the like is enormously precise; all the physics that cosmologists use in computing abundances of elements in cosmological models is precise; all the mathematics of the models is precise, and so on; we have to do, in fact, with "exact science." Nevertheless, when we talk about the universe as a whole, things do seem to become blurred and not at all precise, inevitably so.

The real problem of cosmology is therefore: Can we understand how to proceed from vague beginnings to produce the superlatively precise physics of the laboratory? Or, to express it in still more extreme form: Given that there exists a universe, can we prove that it must be the universe we know?

ALTERNATIVE MODEL

It is well known that there has been what seems to be an extreme rival to big-bang cosmology, namely *steady-state* cosmology with continual creation of matter. This has had much influence upon cosmological thinking for about a quarter of a century. At the present time, however, I think cosmologists would agree that a simple big-bang model proves to be more instructive than a simple steady-state model. But all cosmologists must surely concur, too, that everything we have at present is bound to be superseded; at any rate some of us would expect that an improved model is likely to incorporate some combination of features of both big-bang and steady-state concepts. Also it should be said that the more general aspects of cosmology that we have been trying to discuss are not tied to any particular model.

SCHOLIUM

In this paper I have tried to describe the nature of physical science as applied to cosmology, the production and achievements of a cosmological model, and the more far-reaching aims of such endeavors. No matter how successful those endeavors may prove, at the very least we still have to postulate the *universe*. As a matter of history modern cosmologists have been unable to think about the universe without thinking about creation. Postulating a universe seems to me to be almost the same as postulating a *Creator*.

Scientists someday may be able to infer, given only that there is a universe, that it must produce such and such matter, this must form galaxies, these must contain planets, these must produce life, life must result in consciousness, consciousness must produce us, and we must conceive all these systems of thought. So we should have what ultimately begins to look like a closed system. The feature that we could be most sure about at that stage would be *ourselves*. The postulation of the universe and of ourselves may be the same thing.

As we have seen repeatedly, we cannot formulate any science without reference to the *observer* and, again as a matter of history, progress in fundamental science has been made by increasingly recognizing the role of the observer. It seems to me therefore that we cannot think about the universe without the concept of *personality*. Cosmology requires, I venture to assert, the concepts of Creator and of personality, and together these mean God.

Earlier, I suggested that nature does not satisfy idle curiosity. Here I should take this to mean that we may expect to know about the Creator only what we need to know for the business of living. But we also recalled that we never cease to learn by experience. There can be no cosmology without physical experience, and there can be no religion without religious experience. Men exert themselves to the utmost to gain new experience of the physical world; likewise religious experience has to be sought.

Any such experience that any of us may claim speaks to us surely of *purpose*. Whether it be the glory of a morning in springtime, or the beauty of a human face, or the vastness of the universe producing the specks that are ourselves, can we possibly believe there is no purpose in it?

Some may consent thus far and be therewith content. But to others of us purpose is inseparable from person, and the Person of the Creator is revealed in the person of Christ.

ACKNOWLEDGMENT

The present contribution was written while I held a Leverhulme Emeritus Fellowship, for which I wish to record my gratitude to the Leverhulme Trust.

BIBLIOGRAPHICAL NOTES

Further information about the current state of cosmology is given in several other contributions to this volume, in particular those of Heidmann,[2] Misner,[3] and Penzias.[1] Reference may be made also to John[4] and Davies.[5]

The author has written elsewhere on topics mentioned here.[6,7]

REFERENCES

1. A. A. Penzias, this volume, Chapter VIII.
2. J. Heidmann, this volume, Chapter V.
3. C. W. Misner, this volume, Chapter VII.
4. Laurie John (ed.), *Cosmology Now* (British Broadcasting Corporation, London, 1973; reprinted 1974).
5. P. C. W. Davies, *The Physics of Time Asymmetry* (Surrey Univ. Press, London, 1974).
6. W. H. McCrea, Natural Philosophy, *Math. Gazette* 58, 161–177 (1974).
7. W. H. McCrea, The Interpretation of Cosmology, *La Nuova Critica* (III) 11, 11–20 (1960).

Chapter VII

Cosmology and Theology

Charles W. Misner

University of Maryland

In this paper I will discuss three different aspects of the relationship between cosmology and theology. The first part of the paper takes up the first of these. There I will adopt my most familiar role as a professional theoretical physicist and will try to describe models of the very early universe near the Big Bang which seem plausible and central within the dominant current theory, Einstein's theory of relativity. Theology plays a role in this Big-Bang description in a most peripheral way; it enters only in the choice of the subject matter. Cosmology has always appeared as a portion of the conceptualized physical world that is particularly fascinating to broad audiences and to theologians. This may be because cosmology cannot dismiss broad and universal questions as lying outside its special domain, or it may be because of the exhilaration of surveying such a broad and dramatic landscape. It is in any case appropriate that theology not be developed in complete disregard for the physical world revealed to us by common experience and further by scientific observation and experiment. Therefore, I will present some modern conceptualizations of the physical Universe in the hope that they may provide useful raw material for philosophical and theological discussion.

In the second part of this paper I adopt a completely different role and write no longer as the professional astrophysicist but rather as the amateur theologian. Here I try to consider the question, "Did God create the Universe?" and describe my reactions to it based on an unsystematic familiarity with the Roman Catholic tradition in addition to my background as a theoretical physicist. The final part considers still another aspect of the relationship between cosmology and theology, one of the relationship between their methodologies. Here I ask whether aspects of the scientist's treatment of mystery and truth could be useful to the theologian.

EXTRAPOLATING GEOMETRICALLY TOWARD THE BEGINNINGS
OF THE UNIVERSE

This section is an exposition of some current and central themes in astrophysical cosmology. As several other excellent expositions in this same line also appear among conference papers, I feel that some defense is appropriate for including further elaboration, particularly since my title is not simply "cosmology" but "cosmology and theology." In the discussion following Peat's paper,[11] Ehlers referred to the need for reducing all discussions ultimately to the common language of ordinary discourse derived from our shared experiences. I would contend that the scientific conceptualizations by which we understand the physical Universe need to become part of our shared common language. This is particularly true when one is approaching such a subtle and refractory a subject as theology. Here, if advances are to be made, they require for their expression rich and subtle pictures. The scientist's deepening insight into nature, and the wealth of experience that this implies, may be able to provide these richer pictures to carry, by analogy, theological content that would not be expressible in cultures poorer than our own.

Peacocke[10] mentioned another way in which familiarity with the scientific understanding of nature may be an aid to theology. It is an approach that fits comfortably into my own theological background, which is extremely elementary but rather traditional. This is the idea of accepting nature as part of revelation. If we see the same God as the God of nature and the God of theology, it is reasonable to expect that something of theological significance may be gained by our improving understanding of nature.

Spacetime Geometry

Let me begin my exposition of some standard views of nature of our Universe on a large scale by echoing the viewpoint of Ehlers in saying that spacetime geometry is the central concept. Figure 1 introduces or reviews some of the ideas of flat spacetime geometry that were brought into physics through Einstein's special relativity theory. It shows how measurements of distance by the use of a radar set are reduced to measurements of time intervals. This method using time measurement is in fact regarded as a fundamental one in present-day physics. The radar operates by sending out a pulse of radio waves and measuring the time required for an echo to return. Every distance measurement must in some way involve time, since two distant objects must communicate with each other in the process of establishing their spatial relationship, and this communication proceeds not faster than the velocity of light. In the caption of Fig. 1 some of the basic notions of spacetime geometry are introduced, such as the ideas of world line, light ray, event, and simultaneity.

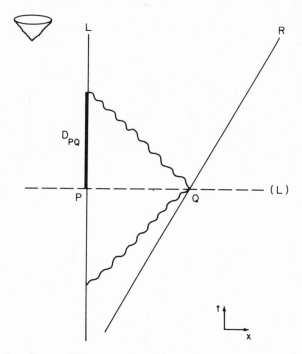

Fig. 1. Spacetime diagram. Time *t* runs from the bottom toward the top, while some space dimension *x* is measured from left to right. The vertical line *L* represents the history of one object whose space location does not change as time proceeds. This line is called the object's *world line*. A moving object has a world line such as *R*, along which the space (horizontal) location changes as one follows the line forward (upward) in time. The scale is chosen so that a photon (quantum particle of light) has a world line (represented for distinctiveness as a wavy line) which runs at a 45° angle in the diagram. A map scale of, say, 1 cm/yr vertically (time direction) and 1 cm/light-year horizontally (space direction) could achieve this. A spacetime map such as this does not carry the compass rose found on nautical charts to give its orientation, but the future light cone (upper left) instead indicates which are the time-light or lightlike directions.

With these interpretations, the diagram can be seen to represent one object *L*, a radar set, emitting a pulse of radiation which is reflected from the second object *R* at *Q* and returns as a radar echo to *L*. A point such as *Q* in this diagram identifies an *event* in spacetime (radar pulse hits object *R*) which has both a specific location *x* and time *t* for its occurrence. Half the proper time interval between emission and return of the radar pulse as registered on a standard clock built into the radar set *L* (i.e., the time segment D_{PQ} shown) is called the *distance* from *L* to *R* at a particular instant of *L*-time identified by the reflection event *Q* or the midpoint *P* of the radar pulse emission–return sequence. These events, *P* and *Q*, are by definition *simultaneous* for *L*. This definition is motivated by the idea that light should take just as long to get to the object at *Q* as it does to return, i.e., that the velocity of light is the same in all directions.

Charles W. Misner

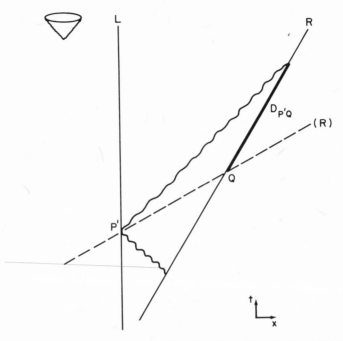

Fig. 2. Another distance measurement. Here the active radar set is the right-hand world line
R. The pulse of radiation it emits is reflected in the event P' from the object L. Because no
(absolute) motion can be ascribed to the radar set, one again assumes the waves took just ás
long to travel out to their target L as to return, and says P' is simultaneous (for R) to the
midpoint Q of the time required for the round trip. If this half-return-time interval $D_{P'Q}$ is
5 sec, one calls the distance 5 light-sec.

Now, in Fig. 2 all considerations are identical except that it is the world line
on the right on the diagram that represents the radar set. The distance measurement is carried out according to the same prescription, and using the same conventions, as in Fig. 1, but we see that it leads to a different concept of simultaneity. From the point of view of the radar set marked R, the events Q and P'
are said to be simultaneous. Also, the distance measured will not be the same as
in the previous case, although it is determined by the same convention, namely
half the time required for the radar pulse to go to the distant object and return.
No allowance is made in the procedures for determining distance, or in the convention for what one means by simultaneity, for the possibility that the observer making these measurements may be in motion. The principle of relativity
forbids that any uniform motion have more right to be called "absolutely at
rest" than any other. In Fig. 3 both measurement processes, and both lines of
simultaneity, are shown. The distances assigned by the two measurements are
different and the relationship is that $D_{P'Q}$ is less than D_{PQ}. The inequality looks
just the opposite in the diagram, but this is a consequence of the paper's being a

substance satisfying Euclidean geometry while we are attempting to describe spacetime, which is something that obeys a different set of geometrical laws, those of Minkowskian geometry.

There is a rule for reading these spacetime maps which allows this difference in geometry to be correctly taken into account. Information is coded into maps in a variety of different ways. A topographical map represents horizontal distances in a straightforward way which can be read off the map with the use of a ruler. But vertical distances are shown through a system of contour lines, and this altitude information is decoded from the map using a different rule. Instead of measuring with a ruler, one counts the number of contours from one point to another to determine the altitude difference. The time intervals measured by standard clocks on world lines in a spacetime diagram have to be decoded from the diagram by a slightly more complicated formula. The vertical distance Δt

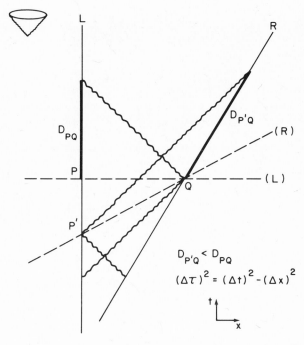

Fig. 3. Both distance measurements are here superimposed for comparison. The dashed line (L) connects events such as P and Q judged simultaneous by the observer represented by world line L; (R) is similarly the simultaneity line for R. Clock times cannot be read directly from the diagram except for clocks with vertical world lines like L. For any other world line (such as R) the clock interval $\Delta \tau$ is related to map intervals Δt (vertical) and Δx (horizontal) by the Minkowski-geometry analog of the Pythagorean rule for Euclidean triangles, namely $(\Delta \tau)^2 = (\Delta t)^2 - (\Delta x)^2$. Thus $D_{P'Q}$ is a smaller $\Delta \tau$ interval than D_{PQ} even though Euclidean rulers laid on this sheet of paper would show it drawn as a longer line.

and the horizontal distance Δx can be measured directly on the paper and converted into length or time using the standard scale factor of the map. Then the clock interval along the hypotenuse of the triangle is given by the Minkowski formula $\Delta\tau^2 = \Delta t^2 - \Delta x^2$, which has a peculiar minus sign that distinguishes it from the usual Pythagorean theorem for Euclidean right triangles.

Using this map-reading rule in Fig. 3, one finds that the distance $D_{P'Q}$ measured by one observer is smaller than the distance D_{PQ} measured by the other observer, and this difference is at the heart of the Lorentz contraction phenomenon of special relativity. In these diagrams, then, one can see some of the fundamental viewpoints about the nature of spacetime which characterize special relativity.

Figure 4 is a map of the expanding Universe. This map incorporates some of the simpler features of the expanding Universe as described at this Colloquium by Heidmann and others. The description is in the framework of the geometry which Ehlers presented. This map again has the light cone of spacetime replacing the cartographer's compass rose. It is drawn so that light rays are always at a 45° angle in the page. The heavy black line marked "us" is the world line of our own Galaxy evolving through time, and is regarded as a fixed location. In this, and in most elementary models of the Universe, there is a natural idea of simultaneity, indicated in Fig. 4 by the line marked "now." This present epoch contains many other galaxies besides our own. The life history of each is shown by a (vertical) world line. To interpret this map, however, we have to put in an additional feature not uncommon from maps of the Earth: The length scale is not the same in

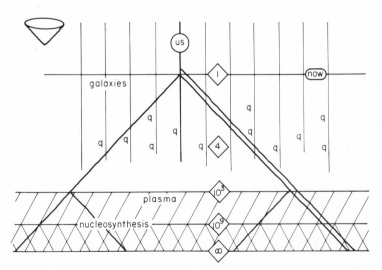

Fig. 4. A spacetime map of the standard Big-Bang cosmological model. See text for discussion.

all parts of the diagram. (One is familiar with Greenland's appearing excessively large on most projections of the Earth.) The scale factors appropriate to different times are shown in the diamond-shaped boxes in Fig. 4. The scale is shown as unity at the present time ("now"), and as, for instance, four at an earlier epoch. This indicates the relative change in scale from perhaps 1 mm on the map per megaparsec (Mpc) in the Universe now, to 4 mm/Mpc earlier. Thus two galaxies which are 10 Mpc apart now are only $2\frac{1}{2}$ Mpc apart at the earlier epoch because of the changes in map scale. (The world lines as drawn maintain their 1 cm separation on the page.) The distances between galaxies are described as increasing as time goes on, so this map incorporates the expansion of the Universe in the variable scale factors.

The map also shows that there was (consistent with current probable interpretations) an "era of quasars"—most quasars existed at a characteristic stratum in the Universe. We now have astronomical evidence of these fossils, but they are relatively rare now, whereas a larger population existed at an earlier epoch, thus implying a truly evolving, and not a steady-state Universe. Farther back in time there was the ionized hydrogen plasma, which last scattered the blackbody or microwave background radiation; the map scale there is 1000 times the scale for the present epoch. Further back there is an era of helium formation, when the primordial helium (as found in the oldest stars) was, in this picture, to have been produced; the map scale factor there is 10^9 times that used for the present epoch. Even though the bulk of the matter in the Universe remains in the form of hydrogen after this nucleosynthesis epoch, the helium and deuterium produced at this time are of great interest because of the possibilities they provide for comparing this picture of the Universe with observations. Still further back there is shown a map scale factor of infinity. This is the singular beginning of the Universe, which we will consider in more detail a little later.

This map has been drawn so that light rays move at $45°$ angles. On the diagram one may follow a light ray arriving at our galaxy at the present epoch from the past, where it may originate in some distant galaxy. Toward the right-hand side of the diagram one may follow two such parallel light rays, which illustrate the cosmological redshift effect. These rays may be regarded as marking the beginning and the end of a single wavelength of light. The separation between them is constant on the map, but the varying scale factors imply that this separation translates into a larger separation now, when we observe the light, than it corresponded to in the past, when the light was emitted. The ratio of wavelengths is just the ratio of the scale factors on the map, and allows us to interpret the scale factors printed on this map as the quantity $1 + z$, where z is the traditional redshift parameter. The fact that the map can describe the cosmological redshift phenomenon in this relatively simple way is, of course, part of our basis for adopting this map as a reasonable description of the Universe. By following light rays drawn on this map, we can also describe a rather puzzling feature of current standard models of the Universe. This is the phenomenon of horizons which prevent us from seeing all the matter in the Universe at the present time. If one follows light rays back from the event corresponding to us now, these light rays will intercept the world lines of more and more matter as they reach further back in the past. However, as can be seen on the map, these light rays reach the singularity at the beginning of the Universe before they have intercepted all the world lines of matter that could be drawn vertically on this map. This phenomenon implies that at some later time we will be able to see—that is, be subject to causal influences emanating from—matter in the Universe which at present has no causal relationship to us. This phenomenon of a limited horizon is, of course, not unique to the present epoch. We can also consider the horizon limitations as of some past epoch, such as that when the scale factor $(1 + z)$ is shown as 1000.

This is the epoch at which the bulk of the matter in the Universe, in the form of ionized hydrogen, recombined into neutral hydrogen gas that was then transparent to electromagnetic radiation. The microwave background radiation which can be detected at centimeter wavelengths at the present time was last scattered by the hot, ionized plasma at this epoch. By following light rays back from us now one can see in the diagram two different regions of the plasma (on the left- and right-hand sides of the diagram) from which we can observe such radiation coming to us from different directions in the sky. Consider the past horizons for these two different emitting regions. Tracing light rays back at $45°$ from each of these two regions, one sees that, within the limitations that physical influences propagate at velocities less than the velocity of light, they are influenced only by entirely disjoint parts of the Universe. Nothing influencing the state of one of these emitting regions could have had anything to do with the state of the other. The fact that the radiation observed from two such causally disjoint regions should have identical properties is therefore a mystery. (Radio telescopes can "see" this primordial fireball and those causally disjoint regions which have been observed have never shown measurable temperature difference beyond the observational uncertainties of about 1/10%.) Thus in the best current models of the Universe the homogeneity is a puzzle or a paradox. The fact that the properties of the very distant regions of the Universe are the same is incorporated within the model, but it is denied any physical foundations. This homogeneity paradox is probably a symptom of our inadequate intellectual command of the cosmological singularity. If so, it would be playing a role in the present century comparable to that which the Olbers paradox was available to play in the late nineteenth century with regard to the homogeneity in space and time then implicit in most conceptions of the Universe.

When variable scale factors are used as in Fig. 4 to present our conceptions of the Universe on a small map, it is done for the same reasons that cause us to use variable scale factors in drawing maps of the Earth. The need for variable scale factors on flat maps of the Earth is an indication that the surface of the Earth is genuinely curved. Similarly, the fact that it is impossible to sketch our present ideas of the Universe on a map without using variable scale factors is a restatement of the proposition that spacetime is genuinely curved. The three-dimensional geometry of space alone, at a fixed time, could be flat; this is consistent with the data and observations that are available at present. The four-dimensional geometry cannot be presented without the use of variable scale factors and is inherently curved. The curvature of four-dimensional spacetime is an essential feature of the Einstein view of the Universe and of gravitation. It is not so much the mathematics of the Einstein theory as its fundamental conceptions that bring in the idea of curvature. Even if one retains the equations of Newtonian gravitation, one finds that spacetime becomes curved whenever the Einstein viewpoint that identifies free fall with inertial motion is used in interpreting the Newtonian equations. (Curved Newtonian spacetime does not have the Riemannian geometry of general relativity, but is an example of a more general class of geometry. See, for example, Misner *et al.*,[9] Chapter 12.)

One must avoid the temptation to imagine that variable scale factors are introduced merely by the circumstance that we are attempting to represent the Universe on a two-dimensional sketch. A four-dimensional small-scale model of the Universe would require variable scale factors just as much as our two-dimensional sketches. The familiar analogy to the point we are making here is that variable scale factors on the maps of the Earth do not arise because a three-dimensional globe is being presented on a two-dimensional paper; they arise because the two-dimensional surface of the globe is curved and the paper is flat. It is relatively easy to imagine three-dimensional maps of the Universe to replace the simpler two-dimensional one presented in Fig. 4. One possible form for a three-dimensional map would be a block of clear plastic in which the world lines of representative galaxies and light rays have been imbedded. Another way to present the same information would be by using a movie. Then time in the movie could represent ages of the Universe provided we used an appropriate scale factor, and distances horizontal and vertical on the screen could represent two of the spatial dimensions. To achieve such a presentation of the simple Friedmann cosmological models that we have in mind at present, it would still be necessary to use variable scale factors. In such a motion picture presentation a time scale of one frame per million years (40 sec of movie to 10^9 yr in the universe) might be appropriate near the current epoch, but at early stages, closer to the singularity, larger scales would have to be used. Most models of the spatial maps appearing on the screen would also require variable scale factors from center to edge of the screen. The models with flat space sections could of course be presented using a constant scale factor at each time from one edge of the screen to the other. It is possible in this manner to conceive also of making four-dimensional maps of the Universe. One simply needs to produce a movie in which the individual frames are projected as three-dimensional images. This should in principle be possible through the use of holography, although I am not aware that any such technology has yet been developed. It is useful to imagine such a holographic movie, however, to make the point that spacetime is essentially different on the small scale from what it is on the grander cosmological scale. No simple, fixed scale factor allows a large part of the Universe to be modeled consistently within a small spacetime region such as the 1 m^3 for 10 min that one might construct in a holographic movie. Only the device of using variable scale factors allows one to take the curved spacetime on a cosmological scale and flatten it out onto a small local piece of spacetime in the lecture hall.

Limits on Extrapolation toward the Beginnings

The presentation above was limited to the most standard picture of the Universe within general relativity theory, the same model that figured centrally in the papers by Penzias[12] and Heidmann.[6] But this is not the only possible model even within general relativity theory. It is just one theoretical extrapolation of present-day observations. We now want to consider this extrapolation process so that we can reasonably estimate its reliability. There are two basic questions here. One is the appropriateness of using general relativity theory for the extrapolation. The second question is the stability of the extrapolation within relativity theory.

How sure are we that general relativity is the proper instrument for extrapolating our conceptions of the Universe to, and beyond, interpretations of current observational data? One must say that general relativity is not as firmly founded as special relativity, quantum mechanics, or most of the conventional theories of physics. In fact, it is not as well founded as any other of the theories that are as routinely used as is general relativity. But its foundations should not be considered from a too narrow experimental viewpoint, because physical theories cannot be adequately judged in isolation. A large component of the empirical foundations for a theory may arise through correspondence principles rather than through experiments designed to confront directly the theory in question.

By correspondence principles, here, I mean limiting relationships between one theory and another—the fact that general relativity grew out of Newtonian mechanics and special relativity and bears formal mathematical relations to them by which it can reproduce those theories in suitable circumstances and limits. A correspondence principle was first formulated under that name for the relationship that quantum mechanics has to classical mechanics, but analogous relationships can be found throughout physics.

It is very characteristic of improper theories to be deficient in correspondence principles. Relativists see many "crackpot" theories; people write letters to relativists proposing why special relativity is wrong because they can rethink the Michelson–Morley experiment, or the Lorentz contraction, from some other viewpoint. The reason one regards most such proposals as "crackpot" is that they are not born within the milieu of evolving physical theory; they do not have roots and branches reaching out, securing them into the other, more firmly established, theories of physics. One knows that if he begins working on such a theory he will have to reconstruct his whole world view, rather than just revise a current line of development. Every experiment or observation from centuries past becomes a possible crucial test of the "crackpot" theory, because current standard theories, and the previous theories they improve upon, are not incorporated into the newly proposed theory by correspondence principles. Thus one makes a demand that any theory be at most "conventionally revolutionary," rather than "crackpot," and this demand is based on a requirement for thorough confrontation with experiment. The conventionally revolutionary theory (such as special relativity at its inception) may discard previously fundamental concepts and change the basic laws, but it does so in a way that leaves its testing against all previously satisfactory experiments under the control of the previous theory whose domain of authority or validity it clarifies.

Thus most of the discussion of a conventionally revolutionary theory properly focuses on the small group of experiments where it differs from the previous theory. The "crackpot" theory usually directs its attention to a similar small group of critical experiments, but this limitation is now unjustified since the crackpot theory is—in my use of the word here—by definition deficient in correspondence principles and must, for an adequate testing, also discuss a host of experiments that were nonproblematic in standard theories with which the new theory has lost touch.

General relativity is just the opposite of a crackpot theory; it is a theory which is unusually rich in correspondence principles. Even were one to ignore the famous "tests of general relativity," and a number of similar important modern experiments (see Chapters 38 and 40 of Misner *et al.*[9]), there is still a wide basis of experimental support for general relativity from the experience incorporated in general relativity through correspondence principles. Thus Maxwell–Lorentz electrodynamics with special relativity fails to describe the motions of planets in the solar system, and Newtonian gravitational theory fails to describe, for instance, the propagation of radio waves and the energy losses of relativistic charged particles moving in strong magnetic fields. Innumerable experiments, then, show that each of these theories fails where general relativity succeeds. This major success is simply to admit that electrically charged and rapidly moving objects may also have gravitational interactions. (In contrast, Newtonian theory is known to be erroneous in dealing with rapidly moving particles, while special relativity does not include gravitational fields without pointing to general relativity as a possible method for doing so.) The rich interconnection of general relativity with other theories is a fundamental basis for the widespread acceptance of it to play the primary role in directing current investigations on questions of cosmology and gravitation. But the area is still rather small in which general relativity successfully describes phenomena outside the ranges it delineates for special relativity or for Newtonian gravitation, so that one cannot be confident that it is a permanently important synthesis of these theories. Its powers of extrapolation are, then, relatively untested, and its predictions should be regarded as the most plausible presently available, but not firmly founded.

In addition to the basis just described for adopting general relativity as the theoretical framework within which we shall attempt to extrapolate present observations into a plausible description of the most distant past, there remains the question of how accurately that theory claims to be able to construct a picture of the past. Figure 5 shows the problem that I am now worrying about—the instability of the extrapolation process. One's main theoretical reconstruction, as to how things occurred in the past, is to be suggested by the central line

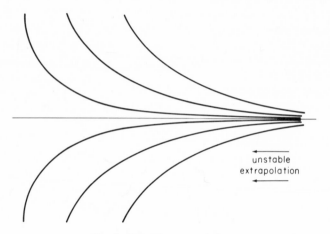

Fig. 5. Unstable extrapolation.

in that figure. But if present observational data are a little insecure, it may be that the correct fit to the present data could be better represented by one of the other curves which lie nearby at the right-hand ("present") side of the diagram. Then calculations done within some theoretical framework may begin to diverge in the past, as suggested by the varying extension of these curves toward the left side of the diagram (toward the past). If errors are larger, the divergence is sooner and more severe. Thus calculations done within an adopted theory may show that the predicted, or retrodicted, picture of past events allows a wide variation as a consequence of small changes in the possible present fitting of theory to observation. One then says that the theory gives an unstable extrapolation into past mysteries, and one would not find the extrapolations useful. Figure 6 shows the opposite case, in which the variations in present observations produce limited amounts of variation in the computed previous picture. Then

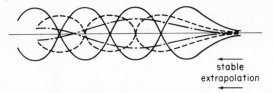

Fig. 6. Stable extrapolation.

one has, in some sense, a stable extrapolation. For a question as to whether some parameter plotted here were positive or negative in the past, this behavior would not be stable. But for the more common question, whether the quantity remains bounded, or lies near a central or standard estimate, this behavior represents a stable application of the theory.

Our concern is: Is the application of general relativity to present observation (as regards certain questions about the past) a stable, that is, reliable, process? This is a technical mathematical question, and it has been addressed by Hawking and Ellis[4] following lines pioneered by Penrose. Figure 7 shows the outcome of that study. Present estimates of observational facts may be represented by the initial slopes of the curves in this diagram. The rest of the curve represents mathematical extrapolation from these data using an adopted physical theory.

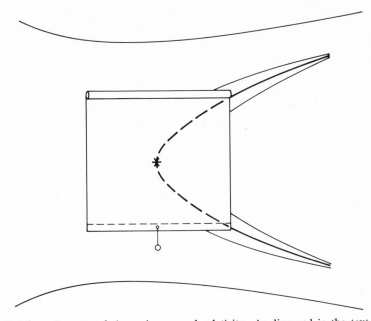

Fig. 7. Schematic extrapolation using general relativity. As discussed in the text, a full knowledge of the importance of Einstein's equation for the past behavior of the Universe is not available, but "singularity theorems" limit the possibilities significantly.

For some estimates of present conditions, the Universe will have done things very different from the standard model of Fig. 4. It may, for instance, never have had a highly condensed phase—a possibility suggested by the upper and lower curves in Fig. 7. For the simplest mathematical models, the Friedmann-Robertson-Walker models that have been described most completely here, one gets a very specific calculation that tells exactly what happened all the way back to time zero. In Fig. 7 this calculation is represented by the heavy lines leading back to an asterisk that represents the initial Big-Bang singularity in these models.

But in addition to these standard, central extrapolations, one wants to consider extrapolations of less simple, but plausible, descriptions of present conditions. The discussion by Hawking and Ellis treats not only the extrapolation process, but also (especially in their 1968 paper[3]) the range of uncertainty in present observational estimates. For extrapolations using plausible, but not incontrovertible, interpretations of the observational data, the variation is within a realm represented schematically by two lighter curves on Fig. 7, which indicate that one can extrapolate toward the past, not deviating too widely from the central, standard models until very early times, over which the shade is drawn. The reason the shade is drawn over the Big Bang is that the mathematical extrapolation process as yet refuses to reveal all the details that we would like about those earlier times. But it does contain the feature that the shade is only so big in time, and that—while these variant lines enter the undescribable region—they do *not* emerge. The statement is mathematically precise that these lines cannot be extrapolated into *any* description of the Universe at times sufficiently far in the past. We do not know what happened in the past, but we know, according to these predictions, that it is very wild. There is some kind of hell, but we are not given a detailed picture of it.

I would next like to present some of the most plausible pictures that can be constructed, with a certain amount of conjecture and idealization, within Einstein's theory, about what went on behind the shade. The most plausible picture seems to be that the Universe does in fact collapse to infinite density in the past, but that is not a rigorous mathematical conclusion. If one allows for some uncertainties in the present description—the fact that there are certain irregularities, that the expansion of the Universe might not be precisely as fast in one direction as in another (that is, Hubble's constant might vary as a function of direction)— then one can produce models of an early Universe in which the curvature of the Universe appears as the dominant feature. The energies and tensions in the curvature of spacetime, governed by Einstein's equations, overwhelm the uncertainties that are introduced by the increasingly unknown equation of state of matter near the initial singularity.

These models with curvature-dominated singularities arise in the following way. The behavior of matter in an expanding universe will depend, as does the motion of matter everywhere, on the forces acting on it. The gravitational force (which can be described by spacetime curvature) is normally important only on the large scale. It is the only significant force affecting the motion of planets or of galaxies. But as we project the expanding Universe farther into the past, matter is more compact, and other forces can be involved, just as the pressure of gas and radiation counterbalance gravity inside of stars. Thus we anticipate past eras where mechanical pressures, within the then more concentrated matter, influence the expansion. Present knowledge of the properties of matter is quite good until very extreme conditions are met—until temperatures over 100 million electron-volts (10^{12} °K) give such a density of strongly interacting mesons and other particles that our limited knowledge of the nuclear (strong) forces becomes a factor. At this stage of the extrapolation no single "best guess" is plausible, but the extrapolation process appears sufficiently stable as regards qualitative properties such as the nature and existence of singularities that the full expected range of possible matter pressures can be accommodated. For almost all general relativistic model universes whose qualitative behavior is known back to the singularity, the initial era is curvature-dominated. The properties of matter, whatever they may be within the presently plausible wide range of uncertainty, do not influence the evolution of spacetime significantly. Instead, this early evolution consists of a dynamic interplay of the curvature and motions of the basic expanding spacetime fabric, with the matter and radiation carried along for the ride on these waves of curvature.

An infinite oscillation is the most interesting feature of the curvature-dominated singularity that appears to be the most typical starting point for a Big-Bang universe in Einstein's theory. Various parameters measuring aspects of the spacetime geometry oscillate increasingly rapidly in the past in these models. Figure 8 shows an elementary (mathematical) example of the kind of oscillation that the Einstein equations typically produce near the initial singularity. The known solutions of Einstein's equations are much more elaborate, with a parameter set not just oscillating regularly, but ergodically exploring its domain in a quasiperiodic oscillation of infinitely many cycles. The oscillation in Fig. 8 is perfectly regular; in each decade of t the function $\sin \log t^{6\pi}$ oscillates through three cycles. The three cycles between $t = 0.1$ and $t = 1.0$ are precisely identical

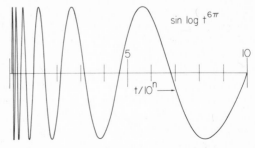

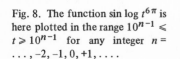

Fig. 8. The function $\sin \log t^{6\pi}$ is here plotted in the range $10^{n-1} \leqslant t \geqslant 10^{n-1}$ for any integer $n = \ldots, -2, -1, 0, +1, \ldots$.

to the three between $t = 1.0$ and $t = 10$, except for the difference in the scale for t. In fact, the abscissa of Fig. 8 need not be read as t, but it could be $10^6 \ t$ just as well, so that the oscillations that are resolved and visible in the diagram would then lie in the range $t = 0.0000001$ to $t = 0.00001$. In the same way, the graph displays any other two decades of t values, however small or large. The function has a singularity at $t = 0$, but it shows structure in every small interval near the singularity.

It is the idea of an infinite number of distinct phenomena (here simple oscillations) occurring within a finite interval that Fig. 8 is intended to illustrate. A much more beautiful and elaborate illustration was used in place of Fig. 8 at the oral presentation of this paper in Denver. There, through the generous assistance of Nelson Max of Carnegie–Mellon University, I was able to show the last five minutes of his computer-animated color motion picture[7] "Space Filling Curves," which gives a vivid impression of the infinite degree of tortuosity nested within tortuosity that is achieved as a curve snakes its way about sufficiently to cover every point on a square. Max's illustration is at once more complex and more accessible than Fig. 8. Both illustrate the infinite complexity that can be fitted into a finite domain. An illustration of this same feature in the Big-Bang beginnings of cosmological models in Einstein's theory of gravitation can be found presented in Box 30.1 of Misner *et al.*[9] The argument that such behavior is typical of a large class of Einstein's models is developed by Belinsky *et al.*[1]

The words "Big Bang" have, in my opinion, induced a widespread oversimplification of the beginnings of the Universe that may be expected on the basis of classical general relativity. As a stowaway, hidden in those words, there may be an image of someone lighting a firecracker. In addition, the words usually designate the simple Friedmann solutions of the Einstein equations rather than the models of the initial singularity that have a better claim to be thought to be generic. As counterpropaganda, let me propose the phrase "steady-state Big Bang" for the asymptotic state of the Universe in early times, which I find a plausible expectation from classical general relativity theory—a view of the initial singularity I will try to develop, beginning from Fig. 8.

The steady-state Big-Bang idea keeps "Big Bang" in its name because it does not differ from the standard Friedmann–Gamow ideas in their current development except in the first fraction of a second in the beginning of the Universe. The standard models for the primaeval fireball at temperatures below 10^{12} °K apply, which includes everything that has yet made any contact with astrophysical observations. The difference in viewpoint and emphasis arises only when the current standard theory, classical general relativity, is asked to extrapolate, from knowledge and from plausible, partially testable conjecture, on into pure speculation. In this realm of pure speculation the current best theory is not a certain guide but it is likely to be closer to the truth than undisciplined, random speculation. That is, after all, a considerable part of the respect that science commands—its ability to predict behaviors not yet arranged for observation. It is another ability of good scientific theories, however, that I rely upon more in the present discussion of the ultimate beginnings of the Universe. This is the ability to pose questions in ways that are fruitful for further development. For I do not seriously think that Einstein's classical general relativity theory will prove competent, by itself, to give us a firm scientific comprehension of the earliest stages of cosmology. But I do fear that if we misread the script for the early Universe that Einstein's theory, accurately developed, writes for the very early Universe, then we may not ask the right questions or search in the right area to find a competent theory.

Let us return to the firecracker that a Big-Bang picture may have brought us. The firecracker leads us to consider its fuse, and the pyrotechnic engineer who lights it. Some physicists may doubt the Big Bang because of the implicit indignity (or magic) of deriving a beautiful Universe from an engineer's punk. Many physicists focus their attention on the fuse. These see the Einstein–Friedmann–Gamow Big-Bang model as a firecracker without a fuse, and begin to search for a theory to supplant Einstein's general relativity that would provide a description of the fuse, that is, provide a nonsingular early Universe that would hopefully lead to a physical explanation, in pre-Big-Bang times, of how the fuse came to be burning. I find it not unreasonable to be alert to attractive speculations with this motivation, for plausible answers to the question "How does the fuse work?," but I think it shortsighted to believe that this is the unique and best question posed by the necessity for a cosmological singularity within classical general relativity.

The more complex and probably more typical behavior of most known relativistic model universes near their initial singularity gives rise to the steady-state Big-Bang concept. Here, below every level of our exploration into the past Universe, there lies another level that continues to show activity. In the simplification of Fig. 8, before every time t, the function $\sin \log t^{6\pi}$ shows three new cycles of oscillation before one pushes the inspection back to $t/10$. In Max's cinematographic exploration of a space-filling curve, the examining microscope zooms in to ever-increasing magnifications, and reveals sinuosity on increasingly fine scales as one explores better and better approximations to the limiting curve.

I think we are justified in saying that the curve in Fig. 8 keeps oscillating forever as we search for its beginning. The suggestion then is that as we understand better and better the very early Universe, we may learn that complexities are to be found at every smaller and smaller level in the remote past. This need not mean that each underlying or previous level is a scale model of the adjacent one (as in the simple models that can be presented explicitly)—it may ultimately prove novel and interesting in various ways—but the infinite regress should be a part of our picture of the cosmological singularity. Much remains to be learned of what Einstein's general relativity, adequately investigated, actually suggests for the beginning of the Universe. But at the present stage of that investigation,

I find that the steady-steady Big Bang appears favored over the simple Fried-mann firecracker.

In the steady-state Big-Bang models from relativistic cosmology, I find that the usual statement, that the Universe is ten or twenty billion years old, is not accurate for philosophical or theological discussions. There is a precise technical sense in which it is correct; but we now propose using the word "time" in much broader contexts, and I fear that some nontechnical aura surrounding the word "time" enters inaccurately into most interpretations of the standard statement of the age of the Universe. There is no necessity for using only the physicist's technical *proper time* to talk about the past, especially when one is discussing the whole Universe; the whole Universe is not a small, local region where one can guarantee that special relativity, with its concepts of time, is applicable. I think that more accurately, toward the past, we should use some of the time variables that appear naturally in the process of solving Einstein's equations for the in-finitely complex models to which I have referred. In terms of these time vari-ables, the past history of the Universe must be said to be infinite. This could have physical consequences. The infinite number of past phenomena or actual occurrences ascribed to early eras in the steady-state Big-Bang models may have an influence on the future that could imply equilibrium, or some other definite statistical result, because of the infinite complexity of the past.

In the final paragraphs of Box 30.1 in Misner *et al.*[9] I have compared the problem of defining the age of the Universe to Zeno's paradox (where the cru-cial swing of Achilles' leg as he strides past his goal is dissected conceptually and declared impossible). Let me quote a portion:

" 'The cosmological singularity occurred ten thousand million years ago.' In this state-ment, take time to mean the proper time along the world line of the solar system, ephemeris time. Then the statement would have a most direct physical significance if it meant that the Earth had completed 10^{10} orbits about the sun since the beginning of the universe. But proper time is not that closely tied to actual physical phenomena. The statement merely implies that those 5×10^9 orbits which the earth may have actually accomplished give a standard of time which is to be extrapolated in prescribed ways, thus giving theoretical meaning to the other 5×10^9 years which are asserted to have preceded the formation of the solar system.

"A hardier standard clock changes the details of the argument, but not its qualitative conclusion. To interpret 10^{10} years in terms of SI (Système Internationale) seconds as-signs a past history containing some 3×10^{27} oscillations of a hyperfine transition in neutral Cesium. But again the critical early ticks of the clock (needed to locate the singularity in time by actual physical events) are missing. The time needed for stellar nucleosynthesis to produce the first Cesium disqualifies this clock on historical grounds, and the still earlier high temperatures nearer the singularity would have ionized all Cesium even if this element had predated stars.

"Thus proper time near the singularity is not a direct counting of simple and actual physical phenomena, but an elaborate mathematical extrapolation. Each actual clock has its 'ticks' discounted by a suitable factor—3×10^7 seconds per orbit from the Earth–sun system, 1.1×10^{-10} seconds per oscillation for the Cesium transition, etc. Since no single clock (because of its finite size and strength) is conceivable all the way back to the singularity, a statement about the proper time since the singularity involves the concept of an infinite sequence of successively smaller and sturdier clocks with their ticks then discounted and added. 'Finite proper time,' then, need not imply that any finite sequence of events was possible. It may describe a necessarily infinite number of events ('ticks') in any physically conceivable history, converted by mathematics into a finite sum by the action of a non-local convergence factor, the 'discount' applied to convert 'ticks' into 'proper time.'

"Here one has the conceptual inverse of Zeno's paradox. One rejects Zeno's suggestion that a single swing of a pendulum is infinitely complicated—being composed of a half period, plus a quarter period, plus 2^{-n} *ad infinitum*—because the terms in his infinite series are mathematical abstractions, not physically achieved discrete acts in a drama that must be played out. By a comparable standard, one should ignore as a mathematical abstraction the finite sum of the proper-time series for the age of the universe, if it can be proved that there must be an infinite number of discrete acts played out during its past history. In both cases, finiteness would be judged by counting the number of discrete ticks on realizable clocks, not by assessing the weight of unrealizable mathematical abstractions."

DID GOD CREATE THE UNIVERSE?

This section addresses the question, "Did God create the Universe?" from the point of view of a physicist familiar with the main thrust of theoretical astrophysics in questions of cosmology, and with extrapolations of that theory within general relativity theory. The first thing I should do in this discussion is give a disclaimer, to point out that I am not a theologian, and that I do not have any professional training that would allow me to enter the discussion on a level as challenging as the very carefully prepared and erudite statements that others have contributed. In addition to claiming amateur status, I should point out that my background here is influenced by the Roman Catholic tradition, which I accept as an evolving interpretation of a valid ancient tradition. I have been influenced by various scattered contacts with philosophy, but nothing terribly systematic, including some Aristotle and St Thomas. As a graduate student in physics I encountered the ideas of Pierre Duhem from the end of the last century, which I find attractive and useful as a kind of Declaration of Independence for the scientist, but not ultimately satisfactory in view of the barriers he erects to keep the scientist's curiosity safely away from a satisfying grasp of reality. I think Duhem worried not only about the compatibility of his physics with his religion, but also about the incompatibility of proposals and concepts being used in physics at that time with his faith in the nonexistence of the atom. He wanted to explain how the atom could be used as a sensible scientific concept in spite of the fact that, in his view, it did not exist. Scientists need the freedom to talk about things for which they are not prepared firmly to assert existence. They are somewhat like actors who put on a cloak and are prepared to play one or another role. It may happen that if one of their plays gets a very long run, the role they have been playing leaves an imprint that they later find impossible to shed.

Thus I see scientific exploration as initially unfettered by *a priori* constraints on what style of thought or speculation is admissible, but as ultimately settling upon some incomplete but certain and accurate comprehension of portions of external reality.

Let us return to creation and ask "Does the Universe need creating?" I am happy to approach this in the framework of the "creator as potter" analogy. I find creation in this simple view still compatible with the models of the Universe I described earlier. The model is a mathematical structure for which a map like Fig. 4 serves as a simplified outline. The scientist feels that his mathematical structure has seized some essential features of the real Universe, or that it will have, should he ultimately prove to have succeeded. Many details need to be filled in, and rough edges need to be smoothed, but the best outline model of the Universe is not far removed from the map of Fig. 4. Whatever elaborations, improvements, and revised physical foundations I try to imagine, my model of the Universe as intelligible through theoretical physics still seems to float in the realm of the possible—conceived but not born. Any extrapolated, hoped, or imagined understandings that I ascribe to physics in future eons are still not coincident with the Universe they are describing.

I suggested, in the preceding section, how four-dimensional models of the Universe could be produced by holographic cinematography. But these rely on given raw materials, the spacetime and matter of the existing Universe. The creator-as-potter analogy now has God puttering around in his studio to produce merchandise analogous to the holographic movie. But when He runs the projector, the product is the real thing, the spacetime we use as raw material for our models. The need for creation is a reflection of a distinction I perceive between a conceptual, mathematical visualization of a possible universe, and the alternative state where there actually exists a spacetime to be modeled, and to model other conceptual universes with.

Note that God as Creator on the model-maker analogy is noticeably different from God as pyrotechnic engineer lighting the fuse. God as model-maker is not limited to producing Big-Bang universes. He might well think that unfinished edges on his handiwork were poor craftmanship, and prefer to produce universes without edges. The Big-Bang Universe in my view is neither more nor less in need of creation than a steady-state Universe, but it is more difficult to achieve an astrophysically comprehensive conception and explanation of a Big-Bang model than of a steady-state model. The steady-state Big Bang discussed in the previous section serves to emphasize that any conception of creation on a "light the fuse" analogy will have a long time to wait before the scientific preliminaries are settled. The steady-state vs. Big-Bang controversy among cosmologists of previous decades appears settled, but that Big Bang referred to the evolution of the Universe after its first second of expansion. Whether the earlier expansion had been going on forever, or was preceded by something else, or simply started from nothing, is, and will probably for long remain, a scientific mystery.

But I do not see that this scientific mystery bears on the question of creation. The difference between actuality and potentiality, between design and construction, between conceptualizability and existence, is the heart of the creation mystery for me. I can extrapolate progress in theoretical physics in my imagination to a degree that allows for a future human culture approaching, or perhaps achieving, the ability to outline the construction of an accurate and comprehensive model of the early Universe. Nothing I know, however, points to a human ability to bring into existence other universes based on variants of that model.

In all these thoughts on creation, the object of the most impressive creative act is spacetime itself. Energy implicit in the curvature of spacetime (once there is a spacetime) can convert into other forms such as photons, proton–antiproton pairs, etc. All manifestations of physics and existence are interrelated, so it suffices to focus on the one component that appears in every simplification, the spacetime itself. But spacetime is not the void that worried people in the past; it is much more an ether. It is a substance, with properties. It can be curved. It bears the electromagnetic field as some characteristic of the vacuum that we know has even a commercial importance. We know that there is a billion dollar industry—the TV industry—which does nothing except produce in empty space potentialities for electrons, were they to be inserted there, to perform some motion. A vacuum so rich in marketable potentialities cannot properly be called a void; it is really an ether. The entire spacetime fabric is something that is as subject to God's creation, or the denial of it, as trees, people, or rocks. But it is not space that is to be created, and then allowed to persist automatically, as one might think for a rock, it is the full spacetime fabric from beginning to end that needs to be saved from oblivion in a library of unused designs, and enacted into existence.

In order to recognize that our Universe, at least to the extent that it is currently intelligible, has a design that is distinct from its existence, it may be useful to consider other designs for which we cannot assert a comparable existence. One such design caused us difficulties in the exposition of some fundamentals of special relativity in Fig. 3. There it was desired to simplify spacetime geometry by cutting it back from four to two dimensions. Yet the model, the printed page containing Fig. 3, was awkward to interpret—obviously longer lines were to represent shorter intervals. The mathematician has no difficulty designing a two-dimensional Minkowski spacetime, but the printer finds it impossible to make or buy one to bind into this volume in place of the page on which Fig. 3 appears. Because of our inability to bestow existence on this simple conception, Fig. 3 had to be drawn as a two-dimensional Euclidean space and supplemented with a "pony" explaining how it was to be translated conceptually into a two-dimensional Minkowski spacetime. Were existence not distinct from conceptualization, this awkwardness could have been avoided. A parallel, but perhaps simpler, insoluble problem is the request to produce (not necessarily create from nothing and nowhere) a geometrical region with a hyper-Minkowski–Pythagorean relationship $(\Delta\tau)^2 = (\Delta t)^2 + (\Delta z)^2 - (\Delta x)^2 - (\Delta y)^2$ of signature ++−− to replace the fundamental geometrical formula $(\Delta\tau)^2 = (\Delta t)^2 - (\Delta z)^2 - (\Delta x)^2 - (\Delta y)^2$ of Minkowski signature +−−− which governs the spacetime of normal laboratories and bookshelves. To say that one of these formulas is a correct physical law and the other incorrect is to focus attention on a great mystery, but in a way that is disrespectful of the depths of that mystery. It sets the difficulty on the same level of that of a child who failed to memorize the sevenses multiplication table in time for the quiz. To say instead that God blessed one formula in some creative act, and dismissed the other, describes our own scientific mastery of nature in a more accurate and more humbling light.

In another place[8] I have given a more elaborate example, geometrodynamics, in the same vein. Geometrodynamics is a theory of pure empty space satisfying certain laws. It is a self-consistent theory whose laws allow the existence of uncertain universes bearing interesting but emaciated resemblances to our Universe. When judged for *a priori* quality as a physical theory by internal criteria (excluding, therefore, observational comparison with our Universe), it does well. It appears to satisfy most, perhaps all, the metaphysical criteria that physicists seem to use in deciding whether a theory is competently formulated. Yet we are utterly powerless to create a universe satisfying those laws.

From considerations such as these, I find reinforcement for the traditional teaching that God created the Universe. Physics does not even appear to be approaching an understanding of the Universe that would make its existence necessary (except as a prerequisite to our discussing it). Saying that God created the Universe does not explain either God or the Universe, but it keeps our consciousness alive to mysteries of awesome majesty that we might otherwise ignore, and that deserve our respect.

It may be that physics is more Christian than biology in the lessons its practitioners are in a position to learn concerning a Creator. Hefner at this Colloquium mentioned religions whose creation myth involves a primordial time inhabited by a pantheon of divinities. The biologist might find this description congenial to the experience of creation that falls within the purview of his profession. The biological creation myth, as perceived by an outsider, is that the god Chemistry and the goddess Time copulate in a ritual called evolution and bring forth Life. I am even happy to accept this as an important component of a fuller understanding of Creation, but I would insist that Chemistry and Time are merely angelic messengers of the true God, as indeed are also Relativity and Spacetime or the theories that will later refine and supplant them. The difference here between the biologist and the cosmologist is not the manner in which they organize their domain, but the basis on which it rests. The biologist uses chemical ideas and knows that astronomy (through cosmochemistry or nucleosynthesis) can supply all the material required. Most sciences, including biology, chemistry, and atomic physics, handle their domain without need for miracles.

It would not surprise me if these scientists then presume that this is possible in all fields of science. Physics, however, includes fundamental laws whose justification cannot at present be given by calling upon work in a still more fundamental field. Physics uses spacetime and relativity, electromagnetism and quantum electrodynamics, and other more tentative postulates, receiving them as magic gifts of nature, not as fruits of workers in other fields proceeding in nonmysterious (but to us only partially familiar) ways from nonmysterious foundations. The organic chemist, in answer to the question, "Why are there ninety-two elements, and when were they produced?," may say, "The man in the next office knows that." But the physicist, being asked, "Why is the Universe built to follow certain physical laws, and not others?," may well reply, "God knows."

METHODOLOGY IN PHYSICS AND THEOLOGY

In considering the relationships of physics and theology as regards methodology, it seems to me there are two features that could be usefully focused on. Those are the concepts of mystery and of truth. I believe these ideas occur in both physics and theology, although theology here is rather a corner of religion, and one that may not even be present in all religions. Someone quoted the proposal that theology asks people on occasion to believe things that they know to be impossible. That certainly was not the Thomistic tradition which came out of the Greek and Christian background; both believe that theology is rational. But I am nevertheless prepared to believe the impossible on occasion. I think in physics we have discovered that this is a very rational thing to do—believe in things that are impossible—because although physics began by rejecting mystery and tying itself to the observational data, it has been led to preach many previously impossible concepts or ideas. Examples include saying that a wave can be a particle or *vice versa*, or saying that the velocity of light is not a relative velocity, or saying that spacetime can be curved. There are numerous ideas which, if anyone had the imagination to propose them sufficiently far in advance of the culturally fertile time, would have been clearly rejected as total impossibilities. And it is only by exposing ourselves to much wider areas of empirical information, and through painful reconsideration of the possible modes of thought in the face of these, that we have found it quite reasonable to believe in the previously impossible or to conceive the previously inconceivable. I would imagine that a theologian would like, at least in some cases, to try to use these examples of the elasticity of human thought in trying to unravel what prospects and possibilities and ultimate intelligibility there might be in theological ideas of mystery. A central feature is a nested and interwoven intelligibility that I think is the nexus of mystery in both physics and what I know of theology. This is the idea that only through the complete and total solution of one mystery will you even gain the language to discover the statement or the location or focus of some further, deeper mystery. The essence of mystery is this, that although every facet is ultimately intelligible, one is not able, nevertheless, to get completely to the bottom of it. One step reveals everything that one wanted to know or could question at one level, and yet one has not answered all questions, but just discovered some further ones.

Let us now turn to the question of truth. I hope later to expand the following remark into a paper with the title, "Eternal and infallible truths are to be found in the mythical theories of physics." Some theologies and religions make an important point of immutable, unchangeable, absolute truth. I think physics has a model for immutable truth. A truth that we do not expect is ever going to change. A truth that represents a permanent and final grasp of some limited aspects of nature. Most people would say that is incompatible with the expectation that our theories will be falsified. I adhere to the expectation that our theories will be falsified, and look for the immutable truth only in those theories that have already been falsified, that is, in what I would call mythical theories. Newtonian mechanics, special relativity, Maxwell's equations, the Schrödinger-equation theory of atomic structure, and the like. Every one of these theories remains in constant everyday use in spite of the fact that it is known to be false in certain applications. In some cases, its concepts have been utterly discarded or eliminated in more refined theories which we believe to be more correct. Nevertheless, I think these theories adequately grasp certain aspects of nature. For instance, I expect geometrical optics will continue to be used by people who use hand lenses, align binoculars, and so forth. It is much easier to think of a light ray as moving in a straight line for certain specific applications than to try to operate in that context with the wave theory of light.

For these applications the quantum electrodynamical, field operator theory of light would simply not clarify one's grasp of the relevant physics but only becloud it with so many complications that one would be unable to focus on those aspects of reality that could be helpful guides to action. Consider an astronomer sitting at a telescope trying to photograph a distant galaxy. It requires an exposure of several hours. During this period the telescope must be constantly aimed at a fixed point on the heavens in order not to blur the photographic image. I suspect strongly that an astronomer working on telescopes he himself guides, feels that what he is doing is following a star that moves across the sky. I doubt that his analysis of the situation is that motors upon command are rotating the base of the telescope below him to follow the rotation of the Earth so that the telescope tube stays fixed. That picture, although we know it to be true, is unnecessary, distracting, and an inappropriate insight into the reality of the situation. Ptolemaic astronomy summarizes the relevant relationships more directly than Copernician astronomy. It is easier to move a telescope to follow a star drifting across the sky than it is to counteract the rotation of the earth by twisting the telescope base to keep the mirror fixed. Sometimes a myth in the reflex is better than a theory in the cortex.

I believe that once a theory has been falsified and its limits are known, it is a better theory. It is a theory whose truth is more clearly defined and in that sense more true. The analogy here might be someone driving down a superhighway in midwinter who sees a sign "bridge out ahead." Such a person will continue to move through the driving snow at a very slow pace. If the sign says "bridge out 73.2 miles ahead," he will make much better progress for the first part of that distance. Knowing the limits of what you know is an important part of your grasp of reality.

In some well-known theories of physics, such as Schrödinger-equation atomic structure, we have an important insight that will continue in use even after better theories (quantum electrodynamics) are in their turn falsified. Physicists expect to continue teaching the old, immutable, infallible myths of Newtonian mechanics, special relativity, and Maxwell's equations while searching for failings in the theories that have supplanted these. I want no suggestion here that the myth of Newtonian mechanics is false or inadequate. It is an example of the most certain and permanent truth man has ever achieved. Its only failing is its scope; it does not cover everything. But within its now well-recognized domain it conceptualizes its portion of external physical reality better than its successors, and that is why it continues in use.

Perhaps theologians have developed a similar idea of permanent truth fixed in myths or theories where the limits or failings of the statement are an important part of its claims to credibility. In any case I would think that myths in theology and in science (as defined above) bear comparison, especially since there is such a wide area of agreement about the credibility of scientific theories. The creation myths certainly have long since been falsified in various fundamental interpretations, and nevertheless continue to be expounded as conveying important information, beautifully expressed. Their value and validity might be easier to clarify in some cultures if it were assessed in a comparison with what I have here called the mythical theories of physics. I would expect fo find important differences as well as similarities, however, for the antiquity of religious myths suggest that they have hidden values that would disappear if the traditional statement were dissected and reassembled in a new literary style. Only the components that survive rewriting in texts and manuals of systematic theology would likely be fully digested, analyzed, and catalogued truths comparable to the mythical theories of physics that still survive banally in physics texts but spring to life in a stream of new applications in fundamental research, technology, and commerce.

REFERENCES

1. V. A. Belinsky, I. M. Khalatnikov, and E. M. Lifshitz, *Usp. Fiz. Nauk* **102**, 463–500 (1970) [English transl. *Adv. Phys.* **19**, 525–573].

2. J. Ehlers, this Colloquium.
3. S. W. Hawking and G. F. R. Ellis, *Astrophys. J.* **152**, 25–36 (1968).
4. S. W. Hawking and G. F. R. Ellis, *The Large Scale Structure of Space-Time* (Cambridge Univ. Press, Cambridge, England, 1973).
5. P. J. Hefner, this Colloquium.
6. J. Heidmann, this volume, Chapter V.
7. N. Max, *Space Filling Curves* (16 mm sound color motion picture, 30 min), distributed by Educational Development Center, 39 Chapel Street, Newton, Mass. (1974).
8. C. W. Misner in *PSA 1972, Proceedings of the 1972 Biennial Meeting of the Philosophy of Science Association*, ed. by K. F. Schaffner and R. S. Cohen (D. Reidel, Dordrecht, 1974), pp. 7–29.
9. C. W. Misner, K. S. Thorne, and J. A. Wheeler, *Gravitation* (W. H. Freeman and Co., San Francisco, 1973).
10. A. R. Peacocke, this volume, Chapter XXIII.
11. D. Peat, this volume, Chapter XXIV.
12. A. A. Penzias, this volume, Chapter VIII.

Chapter VIII

An Observational View of the Cosmos

A. A. Penzias

Bell Laboratories

It is a commonplace that our Sun is one of some hundred billion stars which make up our Galaxy, one of billions of galaxies spread over a space so vast that distance is measured in the thousands of millions of light-years. This knowledge, however, was gained only within the adult careers of people still alive today. The geniuses of earlier centuries could question, reason, and speculate; but the vastness of intergalactic space was beyond the power of their instruments. Countless milestones in humbler callings had to be passed before a technology equal to the task could be reached. This myriad technology has nurtured a variety of observational disciplines whose results fill an ever-growing literature. Although much work remains to be done, the broad and apparently durable outlines of a picture of the Universe has emerged. A description of the main features of this picture is the subject of this paper.

This description of the Universe is based upon three observationally derived features: (1) The constituent components of the Universe, the clusters of galaxies, are receding from one another with relative velocities proportional to their separation, implying expansion from a common origin, (2) there is a universal radiation whose spectrum is that of a black body at about 3° above absolute zero, which is a direct consequence of the hot initial stage of the expansion, and (3) the present mass density of the Universe is too low to exert a gravitational attraction great enough to overcome the expanding motion.

THE EXPANDING UNIVERSE

In 1929 Edwin Hubble published a short paper entitled, "A Relation Between Distance and Radial Velocity Among Extra Galactic Nebulae," which provided clear evidence of a fundamental property of the Universe: The component parts are observed to be moving away from each other with velocities proportional to their separation. The constant of proportionality between distance and velocity is then, assuming the speed to have remained constant, just the time interval between the time at which they were next to each other and the time of observation. Since the proportionality is observed to be more or less the same for all objects, we infer that they were all "next to each other" at the same time in the past, at which time they started moving apart. Because of the enormous energy imparted to them in this initial condensed state, this event can be described as a cosmic explosion.[1]

In determining the velocity–distance relation, the observer should separate out his purely local motion, e.g., the Earth's motion around the Sun, the rotation of our Galaxy, and its motion within the local group of galaxies. Since the sum of these motions is smaller than the recession velocity of even a nearby cluster of galaxies, the problem is not serious. Once local motions have been taken into account, however, the same proportionality between velocity and distance will be determined by observers on any galaxy. Since only relative velocities enter into this determination, the question of a preferred location in the Universe is not important. The only restriction is the assumption that the region we sample is in fact representative of the Universe as a whole. Although the validity of this assumption cannot be proven, it is nonetheless comforting that as the improvements of our instruments permit the measurements of an ever-larger sample of the Universe, the basic result remains the same.

This apparently universal proportionality between velocity and separation fixes the time at which all the component parts began their motion from a common origin. The time interval between this event and the present is called the Hubble age of the Universe and is about 17 billion years. From completely separate data it has been determined that the oldest known stars have about the same age, lending support to the idea that the Hubble age is in face *the* age of the Universe; this is the central feature of the evolutionary cosmology described in this paper.

MICROWAVE BACKGROUND RADIATION

Although the existence of a cosmic background radiation in the microwave portion of the electromagnetic spectrum had been predicted some 20 years earlier, this radiation was first discovered in 1964 and then only by accident. Robert W. Wilson and I were carrying out a study of the microwave spectrum of

[1] A proponent of an alternative interpretation, the steady-state theory of the Universe, used the term "Big Bang" to describe the evolutionary cosmology described above. While the controversy has been settled in favor of the latter view, the name has proven to be a durable one.

galactic radiation. Superimposed on the faint galactic radiation we found a stronger component which we determined to be of extragalactic origin.

The measurement consisted of a comparison between the radiation collected by an antenna and that radiated by a reference standard cooled nearly to absolute zero by immersion in liquid helium. Since all absorptive elements also radiate noise, the effects of such elements in the measurement path must be carefully taken into account. Usually the most difficult of these problems come from the atmosphere, especially at shorter wavelengths. In our case, a difficult problem was a whitish substance deposited on the inside of our horn antenna by a pair of pigeons which had to be removed before final measurement could be made.[2]

The probability that the radiation originated in our solar system could, we felt, be very strongly limited by the fact that our measurements had detected no day-to-night or seasonal variation. This observed isotropy, when combined with the restriction that any proposed source of radiaton could not produce observable radiation at meter wavelengths, made that probability very small indeed. Radiation of galactic origin was perhaps the most attractive candidate, especially in view of the fact that the observed interstellar dust was thought to have a temperature appropriate to this radiation. Consequently, we carefully searched for variation with galactic latitude and investigated the structure of galactic radiation near the plane. We found no evidence indicating that any but an extremely small part of our radiation could have a galactic origin. A more direct argument against such an origin, however, was that any such effect should have been observable in the microwave spectrum of nearby galaxies with similar populations, especially in the Great Galaxy in Andromeda, M31, where an effect one order of magnitude smaller would have been clearly visible in published maps.

The results of a number of microwave background-radiation measurements are plotted in Fig. 1. Two prominent characteristics are immediately evident, its strong intensity and the blackbody character of its spectrum. The intensity of this radiation overwhelms that due to our Galaxy and that due to other galaxies, represented by the dashed line at the left of the figure. Indeed, it accounts for the major portion of intergalactic radiant energy, being some 100 times more powerful than the integrated starlight from the galaxies. That this radiation escaped the notice of terrestrial observers is due in large measure to the masking effect of radiation from our warm ($\sim300°K$) environment. Thus experimenters are led to the use of cryogenic techniques and, especially at the shorter wavelengths, to measurements above the atmosphere.

The interfering effects due to ambient temperature radiators do not diminish in intensity for $\lambda \leqslant 1mm$ as the $\sim3°K$ blackbody does; thus these effects become proportionally more difficult to deal with at short wavelengths. Precise

[2]Because of the wide variety of measurements that have by now been made, detailed account of the experimental techniques required to remove the possibility of an instrumental or atmospheric origin for this effect is not included in this treatment. The interested reader can find an account of this work in our original paper.[13]

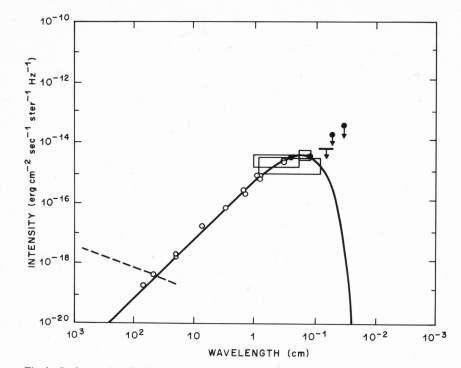

Fig. 1. Background-radiation measurements superimposed on a 2.8°K blackbody spectrum. Open circles and boxes are direct measurements. Filled circles are indirect measurements from molecular excitation. The data are taken from the review by Thaddeus[18] and modified to include the work reported by Muehlner and Weiss[11] and Hegyi et al.[6]

data from this portion of the spectrum are vitally important, however, to determine the blackbody character of the radiation. Our present knowledge of the spectral character of this radiation was obtained in large measure through the painstaking efforts of workers carrying out these difficult direct measurements. In addition to this direct technique, there exists also an alternative method which measures the temperature indirectly by observing the rotational excitation of interstellar molecules, notably CN. Taken together these data provide convincing evidence of the blackbody character of this radiation, a characteristic which is a prediction made uniquely by the theory which we now describe.

Theory

The existence of a universal radiation is a central feature of the evolutionary cosmology in which the Universe expands from a hot, condensed state. It was George Gamow[3] who derived a relation between temperature and density in an

---

[3] A fine, short review of this work together with a complete list of references is contained in the memorial article by his students Alpher and Herman.[1]

early state of such a universe from which the present "three degrees" can be predicted. Since this important result is treated extensively in the literature (e.g., Harrison[8]), we will only sketch its broadest outlines in the treatment below. This treatment has three main elements: (1) The Universe expanded from a very hot initial state in which radiation rather than matter dominated, and as the expansion proceeded, the radiation cooled. (2) From present nuclear abundance data it is possible to infer the density of matter at the time when the radiation temperature was 10^9 deg. (3) From that time on, the temperature of the radiation varied as the cube root of the matter density, thus providing a relation between the present density of matter and the temperature of the radiation.

At the earliest stages the temperatures are so high that the energies of the thermal photons are many times the rest mass energy of particles. Here, matter and radiation exist in thermal equilibrium through pair production and particle-antiparticle annihilation processes. As the expansion proceeds, the temperature decreases until even the production of electron-position pairs from thermal photons is no longer energetically possible (about 10^9 deg). Matter is uncoupled from the radiation and there is no longer a mechanism for maintaining the relative population of neutrons at its thermal equilibrium value. Now, since neutrons are unstable and decay to protons in a few hundred seconds outside of a nucleus, only those neutrons that collide with protons during this short period to form deuterons survive. Since most of these surviving deuterons combine in pairs to form helium nuclei, one may calculate the fraction of neutrons that encountered protons from the resulting helium-to-hydrogen ratio. The probability that a neutron will collide with a proton can be crudely estimated from

$$\frac{\text{neutrons forming deuterons}}{\text{total neutrons}} = n_p \sigma_{np} v_n \tau_n \tag{1}$$

where n_p is the number density of protons, σ_{np} the neutron-proton cross section for forming deuterium, v_n the thermal velocity of the neutrons, and τ_n their mean lifetime. Since the last three quantities are known, an estimate of the left-hand quantity yields the proton density and hence the mass density. When proper account is taken of the fact that some of the resulting deuterium is photodisintegrated before it can form helium, the above relation yields a proton density corresponding to a mass density of matter of about 10^{-5} g/cm^3. For comparison it should be noted that the mass density of the thermal radiation is much larger,

$$\rho(\text{radiation}) = \sigma T^4 / c^2 \approx 1 \text{ g/cm}^3, \quad T = 10^9 \text{ deg}$$

From this time on, the ambient thermal radiation expands essentially adiabatically with the Universe. In the early stages the energy density of the radiation is much greater than that of the matter. Later, when the cooling due to expansion has lowered the energy of the radiation relative to that of the matter, the matter is no longer ionized and is hence transparent to radiation. Thus in neither case can any appreciable portion of the radiation energy be transferred to the matter and the effect of expansion upon the radiation can be treated independently of the matter in the calculation that follows.

Taking advantage of the isotropy of the Universe, we can develop a convenient description of the effect of the expansion on the radiation by dividing the Universe into volume elements whose imaginary boundaries form a co-moving coordinate system expanding with it. Then, from the assumed symmetry and uniformity of the Universe we are free to concentrate on a single such element without troubling ourselves with problems like the dimensions of the Universe as a whole. From the symmetry it is easy to see that as much matter leaves this element as enters and the matter density is proportional to the inverse of the volume. However, the escaping photons are replaced by lower frequency photons which have been Doppler-shifted by the recession of their sources in the expansion. To evaluate this effect, it is convenient to replace the boundaries of our volume element with mirrors, which we are entitled to do by the assumed isotropy. This step simplifies the problem because we need then only consider the photons within our element and their collisions with the mirrored walls.

If we consider a parallel pair of mirrors separated by a distance l and moving away from a common center with velocity v, we have

$$l = l_1 + 2v(t - t_1)$$

Now, at each reflection the frequency of the incident photon is reduced by the Doppler effect. (When a source of radiation is viewed through a receding mirror, it appears to be moving away at twice the velocity of the mirror.) Thus we have

$$\nu_2 = \nu_1 \exp(-2n v/c)$$

where n, the number of reflections, is given by[4]

$$n = \int_{t_1}^{t_2} \frac{c}{l} \, dt$$

But,

$$dt = dl/2v$$

Thus, the integral yields

$$n = (c/2v) \ln (l_2/l_1)$$

and by substitution,

$$\nu_2 = \nu_1 (l_1/l_2) \tag{2}$$

Since the number of photons in the volume element remains unchanged, energy density, the product of the number density and the energy per photon,

[4]In the interests of simplicity, this relation supposes that the photons travel on a path normal to the mirrors. An identical final result can be obtained by considering an arbitrary angle and the sum of the effects of all three pairs of receding mirror boundaries.

is proportional to l^{-4} (i.e., the number density is proportional to l^{-3} and, as we have just seen, the frequency, and hence the energy, of each photon is proportional to l^{-1}). Furthermore, the frequency of each photon is changed the same fractional amount and the radiation retains its blackbody spectral shape; only its temperature changes. Now, since the energy density of the radiation is proportional to the fourth power of its temperature, we combined this T^4 dependence with the l^{-4} dependence we have just obtained to derive our result that *the temperature of the expanding radiation goes as the reciprocal of l.*

We can use the inverse dependence of radiation temperature on the scale size of the volume element to obtain a relation between this temperature and the density of matter. Once pair production has ceased, ρ, the matter density, varies simply as l^{-3}.

Thus,

$$\frac{T_1}{T_0} = \frac{l_0}{l_1} = \left(\frac{\rho_1}{\rho_0}\right)^{1/3} \tag{3}$$

If we take T_1 and ρ_1 to be the radiation temperature and matter density at the time of deuterium formation (10^9 deg and 10^{-5} g), we have the relation first used by Gamow to predict the present temperature of the microwave background from the density of matter. Using 10^{-30} g/cm^3 for the present density of the Universe (see the next section), we have $T_0 \approx 5°K$, the value first obtained by Gamow and his co-workers. The agreement between the predicted and measured values of T_0 should be regarded as excellent, given the uncertainties in the values of the numerical quantities used. This elegantly simple prediction of a blackbody radiation is surely one of the great triumphs of modern physical reasoning. Because the temperature of the microwave background can be accurately measured, expression (3) can be profitable employed as a predictor of ρ_0, a determination which will be taken up in the next section.

Although the evidence for the cosmological origin of the microwave background is compelling, it is instructive to consider at least briefly the alternative explanation, namely that the observed radiation comes from the superposition of a number of discrete sources. In principle, one is free to postulate an array of otherwise unobservable discrete radio sources which produce the observed background. This situation can be illustrated in the simple static case in which all sources are identical and uniformly distributed.[5] The number of such objects which would be detectable in a discrete source survey is just $4\pi\, nR_{max}^3/3$, where n is the density of sources and R_{max} is the maximum distance at which the source is detectable, i.e.,

$$S_{min} = \frac{L_s}{R_{max}^2}$$

---

[5] For a complete treatment of the problem see the book by Peebles,[12] which provides an excellent introductory review of the entire subject for the interested reader.

where $S_{\min}$ is the sensitivity limit of the survey and L_s is the intrinsic luminosity of the source. Thus the number of detectable sources N is

$$N \propto n \left(\frac{L_s}{S_{\min}}\right)^{3/2}$$

Therefore, since the observed background radiation is proportional to the product of n and L_s, one can make N arbitrarily small by making L_s small and still have the background radiation by increasing n proportionally. Thus a discrete-source explanation of the microwave background requires the postulation of a very large number of very faint sources.

A strong constraint on such models is imposed by the observed small-scale uniformity of the background. Sensitive observations with high-resolution antennas have failed to reveal significant point-to-point variations in the microwave background. If this background were made up of discrete sources, one could expect a fractional fluctuation in intensity of the order of the reciprocal of the square root of the number of discrete sources within the antenna beam at each direction. From the observed lack of such variations one can place a general lower limit on the number of such sources which is much greater than the total number of galaxies. Thus the discrete-source model requires a completely new class of objects which dominate the Universe but are observable only in the property for which they were invented. The likelihood of such a contrived circumstance is obviously quite low.

Our original measurements showed the radiation to be isotropic to at least a few percent, the limit of our sensitivity. This radiation field defines a rest frame "tied to" an average over the present locations of the points at which its component photons were last scattered. Motion with respect to this rest frame will Doppler-shift the observed photons, yielding an increase in intensity from the direction toward which the observer is moving. A number of isotropy measurements have been made to search for such motion, with the result that the motion of our local group of galaxies is limited to no more than a few hundred kilometers per second with respect to the larger frame. This lack of motion (by cosmic standards) is consistent with a general homogeneity on a scale larger than the clusters of galaxies.[6] The immediate significance of this result is its support for the notion that the expansion we observe is characteristic of the entire Universe, rather than some "local" phenomenon in a region only a little larger than the present range of our optical telescopes.

----

[6] From the equipartition of energy the component parts of a gravitationally bound inhomogeneity will have velocities of the order of, $(GM/r)^{1/2}$, where M and r are the mass and radius of this large object. Since its average density is known from observation, the observed upper limit on the velocity corresponds to an upper limit on the radius, equal to about the size of a cluster of galaxies.

"MISSING MASS" OR ABSENT MASS?

Given that the universe is expanding according to the Hubble relation, one can determine whether the expansion will go on indefinitely or be reversed by the mutual gravitational attraction of its constituent elements. Such a determination may be made in the following way. We consider a small mass m to be located on the surface of a large spherical volume element of radius r_0.[7] and compute its gravitational and kinetic energies with respect to the center of the sphere. The kinetic energy is given by the velocity due to expansion, i.e.,

$$\text{kinetic energy} = mv^2/2 = mr_0^2 H_0^2/2$$

and

$$\text{potential energy} = mMg/r_0$$

where M is the mass of the sphere. If we have taken care to select a volume element large enough to average out local density fluctuations such as clusters of galaxies, M is just the product of the volume and its mass density ρ_0. Thus

$$\text{potential energy} = 4\pi m \, \rho_0 r_0^3 G/3r_0 = 4\pi m \rho_0 r_0^2 G/3$$

yielding

$$\frac{\text{potential energy}}{\text{kinetic energy}} = \frac{\rho_0}{\rho_c}$$

where

$$\rho_c = 3H_0^2/8\pi \, G \tag{4}$$

and is referred to as the critical density, i.e., the minimum density required to overcome the expansion of the Universe. For a Hubble constant of 60 km sec Mpc^{-1},[8] we have

$$\rho_c = 8 \times 10^{-30} \text{ g}$$

about 100 times the density due to observed galaxies. Thus, if the additional mass required to provide a critical density exists, it must be located in intergalactic space. This problem has been the subject of a number of investigations. An extensive review of this work by Field[2] and a recent paper by Gott et al.[4] are recommended to interested readers and form the basis for the qualitative outline of the problem presented below.

Because the list of the forms that intergalactic mass[9] might take is limited only by the ingenuity of the theorist, it is never possible for the observer to

[7]The subscript denotes the present epoch.

[8]1 Mpc $\approx 3 \times 10^{24}$ cm $\approx$ three million light-years.

[9]Contrary to the situation in the early Universe, the present mass density of the radiation is much smaller than that of matter ($3°$K $= 8 \times 10^{-35}$ g/cm^3).

prove the nonexistence of the "missing" mass. All that one can do is to examine the most plausible cases to the limits of the available data. Since hydrogen is by far the most abundant element, it seems most profitable to search for this element because it must be the principal component of any large amount of intergalactic matter. Furthermore, we can neglect particulate matter (i.e., dust grains, rocks, etc.) because such objects can only be made of chemical compounds largely made up of other much less abundant elements. An agglomeration of hydrogen molecules alone would not be stable in the presence of starlight from the galaxies even if it could be formed, leaving us with hydrogen gas whose various forms may be searched for in a variety of ways.

Neutral atomic hydrogen has been searched for through its 21-cm line emission,[14] 21-cm line absorption,[15] and Lyman-α absorption.[5] This last determination depends on a cosmological origin of quasar redshifts, as did a similar search for molecular hydrogen.[3] In all cases the results were upper limits well below the critical density. This is perhaps not surprising since the diffuse intergalactic gas has very likely been ionized by the light from the galaxies. Because cooling processes are collisional, cooling depends upon the square of density, while heating, i.e., the amount of energy absorbed, is directly proportional to the density. On the other hand, if an intergalactic hydrogen density $\geqslant \rho_c$ existed but was so completely ionized as to have no detectable neutral component, it would most likely be detectable through its x-ray emission.[4]

Additionally, Gott et al.[4] persuasively argue against the existence of intergalactic hydrogen on theoretical grounds related to the formation of the clusters of galaxies from the intergalactic medium.

Finally, one may make another independent estimate of the present density of the Universe based upon considerations introduced in the previous section. If we take the cosmological origin of the microwave background radiation to be proven, we may then use its present temperature to derive the present matter density by inferring a relation between temperature and matter density in the early Universe from the amounts of helium and deuterium produced. It turns out[19] that deuterium is by far the more sensitive indicator, as may be seen from Fig. 2. Furthermore, it has the advantage of an entirely primordial origin, i.e., no additional deuterium is thought to be produced by stellar processes.[16]

Despite the considerable effort generated by the strong theoretical interest in the problem, detection of deuterium outside the solar system was not accomplished until recently. The first such detection was of deuterated hydrocyanic acid in a large interstellar molecular cloud[10] and more recently in a number of other such clouds. Deuterium has now also been detected in its atomic and molecular form in the interstellar absorption spectra of some nearby stars obtained with the Copernicus satellite.[17] The interpretation of all these results requires consideration of various processes, such as chemical fractionation (see Reeves[16] for a review) in order to infer the primordial abundance. When such an analysis is carried out, one sees that the observed interstellar deuterium im-

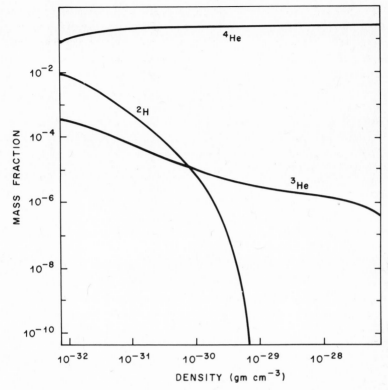

Fig. 2. Relative abundance of helium and deuterium as a function of the present density of the Universe.[19] Note the strong density dependence of deuterium abundance. Clearly the primordial deuterium abundance of $\sim 10^{-5}$ indicated from observations is inconsistent with the density of $\sim 10^{-29}$ g/cm^3 required to overcome the expansion.

plies a present density of the Universe in agreement with the values indicated by the methods described above; i.e., about one-tenth the critical density required to overcome the expansion.

CONCLUSIONS

Taken together, the observed features we have discussed provide a simple, consistent picture of the Universe. The fragments of the cosmic explosion which took place some 17 billion years ago are moving ever further apart; a Universe with a beginning but no discernible end. The observational support for this view seems compelling. In particular, there are phenomena, notably the cosmic black-body radiation and interstellar deuterium, which are natural consequences of the basic processes in this scheme and which other cosmological theories can only treat as unrelated phenomena for which no satisfactory explanations yet exist.

The work remaining for cosmologists includes awesome questions, such as, "How did the galaxies form?" and "What is the role of antimatter?" The answers to such questions are being sought largely within the framework whose outlines I have described. It thus seems appropriate to commend it for study by workers in other disciplines as well.

REFERENCES

1. R. A. Alpher and R. Herman, *Cosmology, Fusion and Other Matters*, ed. by F. Reines (Colorado Assoc. Univ. Press, Boulder, Colorado, 1972).
2. G. B. Field, *Ann. Rev. Astron. Astrophys.* **10**, 227 (1972).
3. G. B. Field, P. M. Solomon, and E. J. Wampler, *Ap. J.* **145**, 351 (1966).
4. J. R. Gott, J. E. Gunn, D. N. Schramm, and B. M. Tinsley, *Ap. J.* **194**, 543 (1974).
5. J. E. Gunn, and B. A. Peterson, *Ap. J.* **142**, 1633 (1965).
6. D. J. Hegyi, W. A. Traub, and N. P. Carleton, *Ap. J.* **190**, 543 (1974).
7. E. Hubble, *Proc. Nat. Acad. Sci.* **15**, 168 (1929).
8. E. R. Harrison, *Ann. Rev. Astron. Astrophys.* **11**, 155 (1973).
9. E. R. Harrison, *Physics Today* **21**(6), 31 (1968).
10. K. B. Jefferts, A. A. Penzias, and R. W. Wilson, *Ap. J.* **179**, L57 (1973).
11. D. J. Muehlner and R. Weiss, *Phys. Rev. Lett.* **30**, 757 (1973).
12. P. J. E. Peebles, *Physical Cosmology* (Princeton Univ. Press, Princeton, New Jersey, 1971).
13. A. A. Penzias, and R. W. Wilson, *Ap. J.* **142**, 419 (1965).
14. A. A. Penzias, and R. W. Wilson, *Ap. J.* **156**, 799 (1969).
15. A. A. Penzias, and E. H. Scott, *Ap. J.* **153**, L7 (1968).
16. H. Reeves, *Ann. Rev. Astron. Astrophys.* **12**, 437 (1974).
17. J. B. Rogerson, and D. G. York, *Ap. J.* **186**, L95 (1973).
18. P. Thaddeus, *Ann. Rev. Astron. Astrophys.* **10**, 305 (1972).
19. R. V. Wagoner, *Ap. J.* **179**, 343 (1973).

Chapter IX

The Generation of Matter and the Conservation of Energy

Karl Philberth

Puchheim, West Germany

INTRODUCTION

Cosmology is a strange science. The farther away the phenomena, the more intense is our interest in them. The decisive questions concern the maximum spatial and temporal distances, the limits of the Universe, and its beginning and its end.

Philosophers boldly direct their abstract reasoning to the very ends of the world. But physicists have the problem that their science is based on experience, and we have direct experience only of the nearest cosmic vicinity. The comprehension of more and more distant objects requires more and more hypothetical extrapolations.

Hubble's law, for example, which relates the recession velocities of the galaxies to their distances, is based on the assumption that the measured redshifts are Doppler effects: Intragalactic experiences are extrapolated to cosmic distances. For distances of less than one billion (10^9) light-years, the Doppler effect is simply a linear function of the redshift; for more than ten billion light-years, however, the result depends on what type of Doppler effect is admitted, a classical or quasiclassical one, the special relativistic one or a general relativistic one.

The basic cosmological question is whether and how our judgments about physical experiences can be extrapolated to the whole cosmos; or, in other words, whether and which physical laws have general cosmological validity. With regard to this problem it is useful to subdivide the basic physical laws in two groups: laws that contain physical "constants" and laws that do not.

Well-known universal "constants" are the gravitational factor of Newton's law, the electrostatic factor of Coulomb's law, Planck's quantum of action, the elementary charge, the velocity of light, and others. It is a decisive cosmological question whether one or more of these "constants" are not truly constant, but functions of the cosmic age.

There are laws without such "constants." For example, Newton's law, according to which the inertial force is equal to the temporal change of the momentum; or the laws of conservation of energy, of electric charge, of number of baryons, and so on. With regard to the problem of cosmological universality, these laws are more fundamental. They cannot vary as a continuous function of cosmic age. Either they have unlimited cosmic validity or they do not. Within this group, the law of conservation of energy is most important.

Is the law of the conservation of energy—the first principle of thermodynamics—a universal cosmological law? Does it hold for all parts and ages of the cosmos, even back to the very beginning? Is energy or matter eternal? Or is there a finite beginning which involved a violation of the law of energy conservation? Did energy come into existence "somewhat" before the beginning? Or could energy and matter come into existence within a finite temporal period without violation of the energy conservation law? A lot of questions! But one has to reflect on them in order to avoid basic misconceptions.

At first glance, the idea of eternal matter seems to offer an easy solution to the problem. However, it is no solution at all. What seems to be an answer is in reality the systematic refusal of the answer. The question about the origin is just pushed away to infinity. Moreoever, the astronomic facts are against this idea.

Another attempt is to attribute by definition the beginning, the cosmic age zero, to a moment when energy or even matter already existed. Under such a premise, energy originates from "somewhat" (a period of infinitesimal or finite length) before the beginning. An example is the famous "original sun" or any other high-density conglomeration of the total energy, which is assumed to explode—by definition—at the beginning. But this statement, too, is a pseudo-answer; it refuses a comment on the true origin, on the generation of energy.

There are only two honest alternatives: Either the law of energy conservation is a general cosmological law, valid without any exception for every period and every space; or it has been violated at least once, regardless of whether this violation is assumed to be within the "normal" cosmic age, to be within a "prebeginning" stage, or to be infinitely long ago.

In view of this alternative one is inclined to say: of course, there must have been a situation in which the law of energy conservation was violated; for no energy existed before the beginning of the cosmos, while gigantic mass energies exist now—and this means an increase, not a conservation of energy. This conclusion, however, proves to be wrong.

It is the purpose of this paper to demonstrate that the development of the cosmos in all its stages, including the very beginning and the very end, can be explained without any violation of the law of energy conservation. This leads to a cosmological theory which is plausible and in accordance with current astronomical facts. Some basic features of this theory will be exposed at the end of this paper; for more details, reference is made to the work[1] in which this theory was originally developed.

Before we enter the discussion of the cosmological aspects of energy conservation, some philosophical reflections may be allowed.

PHILOSOPHICAL REFLECTIONS—CREATION AND PHYSICAL LAWS

"Where do things come from?" is one of the oldest and ever-new questions of mankind. It is the question of the child who learns about his immediate surroundings; it is the question of the philosopher who studies the varieties of existence; it is the question of the astronomer who realizes the gigantic distances and masses of the cosmos. The child asks for comprehensive information; the philosopher asks for final reasons; the astronomer asks for physical explanations. Thus there are quite diverse viewpoints with their specific realms of questions and answers to be distinguished, but which should not be treated without a certain regard for each other. How bloodless are discussions of final reasons without relationship to direct experience; how senseless is the storage of physical knowledge without regard to final meaning.

Moreover, exaggerated one-aspect thinking is menaced by a special hazard: that subconsciously, in an uncontrolled way, mere feelings get interconnected, where comprehensions should be correlated in precise clarity. In fact, there is inevitably a certain interconnection between philosophy and cosmology.

On the one hand, a judgement has to be made about which physical laws are assumed to be absolutely universal, which ones vary, and which ones have local character only. This judgement is more or less influenced by our feeling for what is "reasonable," by our practical or philosophical attitude. On the other hand, the physical models of the Universe imply philosophical relevance. Classical materialism, for example, is based on the idea that matter is absolute and eternal. It is not compatible with cosmological models stating a finite age of all matter and even spacetime itself. The theory of the "oscillating cosmos" stands under the patronage of materialism; it is a desparate and unconvincing attempt to maintain the thesis of eternal matter in the face of modern astronomical results.

Creation means the work of a creator. It would be senseless to use another definition of this word. I am convinced that the Universe is creation in the proper sense of the word: produced by the free will of a personal creator, of God, who Himself is not subject to space, time, or matter. But this conviction is to be distinguished from purely physical statements. Therefore, the expression "generation" instead of "creation" will be used, where no explicit reference to the philosophical or theological concept is intended.

In the last centuries the confrontation between half-true results of an immature science and the narrow-minded interpretation of religious doctrines has caused serious conflicts. Narrow-mindedness in the interpretation of the idea of creation would be unwise and even dangerous. Some relaxed considerations may help to get rid of some prejudices.

The biblical idea of creation may be summarized by "in the beginning was the creation out of nothing." This is a clear statement as long as we understand it in the realm of metaphysical knowledge. Creation is the step from not-being into being; nothing was before or is besides the Creator and His creation. Now, superficial reasoning could derive the rash conclusion that the physical cosmos started with a sudden generation of all matter, that there was no energy "before," but there is a positive amount of energy "after."

But these conclusions are not valid because they confuse theological and physical concepts. In the theological sense, "beginning" means *causa prima*, the primary cause; in the physical sense, "beginning" means the point zero on the time axis. "Creation out of nothing" means that existence took place, where no existence had been, but it does not state whether the total amount of cosmic energy is zero or different from zero.

How long does the "beginning " last? With regard to a book, for example, it is quite common to say, it begins with the first chapter. It would be logically rigorous but unrealistic to conclude that the first chapter consists of one letter only because any following letter is not the beginning. "Beginning" generally means a first section or a first phase. And there is no reason at all to confine the interpretation of the beginning of the cosmos to an interval on the order of seconds, hours, or years.

The beginning of the cosmos may have taken place at an extremely short interval, comparable with the elementary time unit, or it may have been an extremely long interval, comparable with the cosmic age. Periods far less than a second, periods on the order of years, periods far more than one billion (10^9) years—each could be understood to be the cosmic beginning, and there is no theological or purely logical reason to prefer the one to the other interpretation. Only scientific argumentation can provide the answer on the physical definition of the beginning phase and on its length.

Speaking of creation in the context of cosmology, some more or less unreflected physical interpretations are often presupposed. In this case, too, we must be careful to avoid prejudices. Are there objective criteria for the judgement about which physical processes or stages are "creation" and which ones are not? Obviously there are no objective criteria. We have to look for "reasonable" criteria. There are two extreme viewpoints which may be acceptable from the philosophical point of view, but which yield no reasonable or productive physical application.

The one extreme viewpoint is the idea that only the very first step from nothing deserves to be called creation; all further steps develop this created being but are not creation in the proper sense. A physical application is to choose the first or one first particle and to call it the "creation-particle," and to describe all other particles and processes without reference to the problem of creation. This is a possible but sterile viewpoint.

The other extreme viewpoint is to take all processes at any time as creation or part of the creation. A physical application is to call the transformation of boiling water to steam an act of creation. Of course, in a certain sense steam is created; but it is not a creation out of nothing. If every change is called creation, there are no physical consequences at all. This, also, is a sterile viewpoint.

A clear discrimination between real creative processes—in the sense of creation out of nothing—and mere transformations seems to be simple as far as the physical plane is concerned. The famous law of conservation of energy offers itself as establishing the difference. The application of this law to our problem seems to be easy. All physical processes subject to the law of conservation of energy are mere energy transformations; hence, true processes of creation or generation out of nothing are those which violate this law. Or, more specifically: Generation out of nothing is the increase of energy in an isolated system; disappearance into nothing is the decrease of energy in an isolated system.

Now, such reasoning is attractive on two counts: On the physical plane, it is directly based on a law which is solidly supported and corresponds directly to our natural feelings. On the philosophical plane, it seems to prove directly the existence of a creator or at least of a nonphysical principle. The attractiveness of both reasons is seductive. Let us turn first to the philosophical point. The physical one will be treated later in a more detailed way.

Philosophically and theologically, we must not restrict the Creator's work to those parts or stages of the Universe where laws are violated. This would be a primitive metaphysical or theological understanding of what laws are and what the Creator is. For, what conclusions can be drawn, after all, if any stage of the cosmic development shows processes which are contradictionary to our physical laws? Does it prove the existence of a transcendent power? Not at all. It only proves that our physical laws are not general enough to cover all phenomena. In former cosmic stages a law of today may have had a different form or may not have been valid at all. This fact itself can in turn be formulated as a physical law: the law of time dependence. One could conceivably argue that the systematic description, finally, of whatever happens could be called a law—by definition.

This, too, is an extreme and sterile viewpoint. It is more realistic and helpful to establish laws on the overwhelming majority of observations and then to admit that transcendent or other reasons may bring about exceptions. Examples are the well-attested levitations and bilocations of some saints. The simplest and most reasonable explanation is that, in these special cases, God has invalidated His physical laws in order to acknowledge and recommend the ways of life of the individuals involved.

With regard to the initial generation of matter, one could apply an analogous argument: that the Creator wanted to demonstrate His superiority to laws. But this argument is not convincing. Should we really conceive of God that He created the Universe by breaking His own fundamental laws, in order to evidence His existence?

The following sections are intended to demonstrate that the whole creation of the cosmos, from its very beginning to its very end, can be understood without any violation of the law of conservation of energy. It is an awesome idea that God's Word has created some universal laws, and that within these laws the whole variety of the Universe is comprised.

THE CONSERVATION OF THE TOTAL COSMIC ENERGY

The principle of conservation of energy means that in an isolated system the total amount of energy is constant. All the following reflections and calculations are based on the premiss that this principle is valid from the very beginning to the very end of the cosmos. This premiss will prove to be quite in favor of a recently developed cosmological theory—which will be outlined below—while most of the well-known theories cannot be reconciled with it.

The principle can be subdivided in two laws: the conservation of total cosmic energy and the conservation of local energies. Let us begin with the first. The cosmos as a whole is an isolated system. One could even say it is the best isolated system or, even more, the only absolutely isolated system. Therefore, our premiss necessitates that the total energy of the cosmos has a constant value.

Is this constant value positive, negative, or zero? Zero is the only proper possibility: No energy exists "before" the beginning and "after" the end of the cosmos; and no energy exists during all cosmic stages. It is only by unrealistic formalism that a finite cosmic energy value can be adjusted to the nonexistence of energy "outside" the cosmic existence. Examples of such pseudoanswers have been discussed in the Introduction: the period of energy generation is put infinitely far back or is put to a "prebeginning" time or is reduced to an infinitely small period at the beginning. We shall treat the only reasonable case—that the total cosmic energy is always zero.

Exact and generally accepted physical definitions exist for "energy" and for "mass." "Energy" and "mass" are not identical; but the Einstein equation $E = mc^2$ connects them so directly to each other that they may often be used in the same sense. The expression "matter," however, is not a technical physical term and should not be used as a synonym for mass. It is reasonable to apply "matter" or "materiality" as a general term for particles with nonzero rest mass. Under this definition, the positive energies of photons and the negative gravitational energies are not "matter." More than 99% of the materiality of the cosmos consists of nucleons (protons, neutrons). The total energy and the total mass of the cosmos are zero, but its materiality is a gigantic amount.

Total energy zero—how is that realized? It is improbable that the cosmos contains considerable quantities of the so-called antimatter; but, however that may be, no explanation of the energy-zero budget is to be found in this direction, because both matter and antimatter have positive Einstein energies mc^2 which could not compensate each other.

The cosmic energy budget contains the positive mass energies mc^2 and the negative gravitational energies. Together they compensate each other to zero. The negative sign of the gravitational energy is not a mere mathematical formality; it corresponds to a physical fact. All potential energies of attractive forces are negative; all potential energies of repelling forces are positive. Two equal electric charges, for example, repel each other; positive energy is necessary to get them closer to each other, and this invested positive energy is their potential energy. However, a positive and a negative charge, a proton and a neutron, the sun and the earth—such pairs attract each other by electric, nuclear, and gravitational forces, respectively. In these cases, the coming together of the two components of the pair produces energy, and once this energy is removed, the pair is less energetic: its potential energy is negative.

The quantitative expression for the negative potential energy is the well known mass defect. The nuclear mass defects are about 1%; the gravitational mass defect of the sun is one per million; the gravitational mass defect of degenerate stars (e.g., neutron stars) is up to 50%. And the cosmos as a whole has a mass defect of exactly 100%—because its total energy is zero.

A generally accepted assumption is the so-called cosmological principle. It states the cosmos to be homogeneous and isotropic for sufficiently large regions (order of ten million light-years radius). From each and every galaxy the cosmos has the same structure and the same radius. This principle admitted, the cosmic energy-zero budget implies that, not only for the cosmos as a whole, but for every galaxy, the sum of its mass energy and of the energy of its gravitational attraction is zero.

QUASICLASSICAL MASS DEFECT AND MUTUALITY OF ATTRACTION

The total energy and thus the total mass of the cosmos being zero, could one expect its gravitational attraction and its potential to be zero, too?

Yes and No. Yes, in a symbolic "outside aspect"—insofar as one tries to imagine—which is unrealistic—the viewpoint of a fictive object "outside" the cosmos; because such an object would not be subject to a gravitational force or to a potential of the cosmos. With regard to such an extracosmic object, the whole cosmos seems not to exist. No, in the real "inside aspect"—insofar as real objects, objects of the cosmos, are concerned. These intracosmic objects have a quasiclassical gravitational interconnection with the whole cosmos. We will now demonstrate this.

For the discussion of gravitational problems it is useful to compare the cosmos with the surface of a sphere which is homogeneously covered with mass particles. All these mass particles are equivalent to one another with respect to geometrical position and gravitational potential, just as the galaxies of the cosmos are according to the cosmological principle. Let us study the mass defect of such a spherical surface.

Many equal particles, each with the mass dM and all together with the mass M, are brought radially, one after the other, from very far to a spherical surface with radius s. This process produces gravitational energy which is assumed to be removed. The resulting mass of the surface is M'. The absolute mass defect is $M - M'$, the produced and removed gravitational energy is $(M - M')c^2$. Every particle produces energy dE equal $(dM - dM')c^2$, and this energy is equal to the product of dM and the difference of the gravitational potential very far away and at the surface of radius s. This difference of the potential is given by general relativity; it is $1 - \sqrt{|g_{00}|}$, where g_{00} is the first coefficient in Schwarzschild's line element:

$$dE = (dM - dM')c^2 = dM (1 - \sqrt{|g_{00}|})c^2 = dM (1 - \sqrt{1 - 2M'G/sc^2})c^2 \quad (1)$$

M' being the "defected" mass, M the nondefected mass, G the gravitational factor, s the radius, and c the velocity of light.

The integration of this differential equation yields

$$M' = M - \tfrac{1}{2} M^2 G/sc^2 \quad (2)$$

Differentiation with regard to M results in

$$dE = (dM - dM')c^2 = (MG/s) \, dM \quad (3)$$

The comparison of the cosmos with a spherical surface is quite attractive, but it yields no full proof. That is why a new calculation is added, one which applies to every gravitational system in which all particles are equivalent with regard to the gravitational potential. This equivalence is realized for the homogeneously covered spherical surface and for the cosmos if the galaxies are taken as its "particles."

The first part of the consideration follows the preceding argumentation; the produced and removed energy dE is $(dM - dM')c^2$. This energy dE results from the fact that the particle being brought to the system is introduced into the gravitational field of every particle which is already part of the system *and* that every particle which is already part of the system is introduced into the gravitational field of the particle being brought to the system. Both groups of effects are energetically equal to each other; each produces the energy $dE/2$. The new particle loses the energy $dE/2$ and all other particles of the system also lose the energy $dE/2$. Having lost the energy $dE/2$, the new particle has the mass defect $(dE/2c^2)/dM$. And having this mass defect, the new particle has become an equivalent member of the system of all the other particles, which, as a whole, has the mass defect $(M - M')/M$. In other words, both mass defects are equal:

$$(dE/2c^2)/dM = (M - M')/M \quad (4)$$

Remembering that $dE = (dM - dM')c^2$, Eq. (4) leads to the differential equation

$$dM'/dM = 2M'/M - 1 \quad (5)$$

with the general solution

$$M' = M - CM^2 \tag{6}$$

where C is a constant.

Comparison of Eqs. (2) and (6) shows C to be $G/(2sc^2)$ for a spherical surface homogeneously covered with mass. For other surfaces, for example, a toroid the path diameter of which is small compared to its rotational diameter, and for the whole cosmos itself, the constant C is equal to G/dc^2, where d is a "diametrical factor" which has the dimension of a length (for the spherical surface it is equal to its diameter) and which depends on the geometrical form but is independent of M and M'. Thus

$$M' = M - M^2 G/dc^2 \tag{7}$$

Differentiation with regard to M results in

$$dE = (dM - dM')c^2 = (2MG/d)\, dM \tag{8}$$

Equations (3) for the spherical case and Eqs. (8) for the general case show the situation to be quasiclassical: "quasi" insofar as classical physics has no Einstein relation $E = mc^2$ and thus no mass defect at all; "classical" insofar as the gravitational energy dE has the classical value, i.e., "nondefected" mass dM multiplied by the nondefected total mass M and the gravitational factor G, divided by the radius s or half the "diametrical factor," respectively. This factor is determined by Newton's law, because it is independent of M and M' and for $M \to 0$ the gravitational potential approaches Newton's potential.

There are several astrophysical and cosmological consequences. The first one is the demonstration of what we have stated at the beginning of this section: That, although the total cosmic mass M' is zero, the galaxies dM have a negative gravitational potential $-dE/dM$ which is determined in a quasiclassical way by the nondefected value of the total cosmic mass M, the gravitational factor G, and a diametrical factor. This factor can be expected to be on the order of the cosmic radius R.

A second consequence is the relation between the total and the infinitesimal mass defect. Equation (5) yields the infinitesimal mass defect $(dM - dM')/dM$ to be twice the total mass defect $(M - M')/M$. For example, in the case of a spherical surface of radius s identical to the Schwarzschild radius, the total mass defect is 50%, the infinitesimal mass defect is 100%; this means that infinitesimal change dM (approach or generation of a particle with the mass dM) corresponds to $dM' = 0$, that is, to no change of the defected mass M' and the total energy $E = M'c^2$. In the case of the cosmos, the defected mass M' is zero, the total mass defect is 100%, the infinitesimal mass defect is 200%; this means that the generation of a particle with the nondefected mass dM produces a change of $dM' = -dM$, the energy increment is $dE = c^2\, dM' = -c^2\, dM$: The total energy has not increased, but decreased.

Let us emphasize: These consequences and Eq. (5) from which they are derived are fundamentally connected with the mutuality of the gravitational attraction between every pair of objects: One object is in the field of the other object and the other object is in the field of the first. Both effects yield the same amount of gravitational energy.

GENERATION OF MASS AND CONSERVATION OF LOCAL ENERGY

The infinitesimal mass defect of any object in the cosmos with mass dM was explained to be 200%. Because of the mutuality of gravitational attraction, this mass defect can be subdivided into two parts of 100% each. One part belongs to the gravitational energy of the object in the "collective" gravitational field, that is, the field of all other masses of the cosmos; the other part belongs to the gravitational energy of all other masses in the "individual" gravitational field, that is, the field of the object. The first part corresponds to a collective cosmic potential $\phi = -c^2$ to which the object is exposed: $dE = \phi\, dM = -c^2\, dM$; the second part corresponds to the infinitesimally small individual potential to which the cosmic masses are exposed. The individual potential can be regarded as a reciprocal aspect of the collective potential. That is why the integration of all mass from the very first moment of its being with the negative energy $-c^2\, dM$, Each object as an equivalent part of the total cosmos has a mass defect of 100%; as an individual vis-à-vis the cosmos it has a mass defect of 200%.

As shown in the last section, the spontaneous generation of a particle with mass dM on a spherical surface at the Schwarzschild radius would not change the total energy $E = M'c^2$. Such a generation would be possible without violation of the conservation of energy. What about the cosmos? Would a spontaneous generation of mass at any intracosmic point be possible without violation of the conservation of energy? At first glance, one is inclined to say, No; on second thought, one wants to say, Yes; but in the final analysis, one has to say, No.

At first glance, the infinitesimal mass defect of 200% seems to be sufficient argumentation against the possibility of spontaneous mass generation. Such a generation seems to disturb the energy budget by 100% and therefore to violate the conservation of energy.

On second thought, however, one may call to mind that the new piece of mass becomes an equivalent part of the total cosmos, thus sharing in its 100% mass defect. But this is a deceptive notion because Eq. (5) leaves no doubt about dM'/dM being -1 for $M' = 0$. A more sophisticated reasoning would be that the spontaneously generated mass is "born" in the collective cosmic potential $\phi = -c^2$ which exists at the place of birth and which invests the generated mass from the very first moment of its being with the negative energy $-c^2\, dM$, thus compensating its mass energy $c^2\, dM$ to zero. The infinitesimally small individual potential, however, needs billions of years until it reaches the majority of cosmic masses, that is, until its energetic effect becomes considerable; meanwhile, other effects, e.g., cosmic expansion, have compensated its effect. Indeed, some cosmological theories (P. Jordan, P. A. M. Dirac, F. Hoyle) are based on such or similar argumentation and state that spontaneous generation of supernovae or hydrogen atoms is happening even now.

But the final analysis of the problem is this: As far as only the conservation of the total cosmic energy is concerned, spontaneous mass generation is or may be possible. The local conservation of energy, however, admits no chance of such an event in the current stage of the cosmos. What does the conservation of local energy mean?

An observer inside the cosmos is himself part of the cosmos and thus subject to the collective gravitational potential $-c^2$. Since other objects are on the same potential level as he is, the collective gravitational potential is not relevant for their relation with reference to such an observer. This potential $-c^2$ is an "existential," not a "relative," potential. From a fictive absolute standpoint outside the cosmos, it compensates the energy of every object to zero; from every relative standpoint inside the cosmos, it is without true reality. That is why gravitation and gravitational mass defect are based in a quasiclassical way on the nondefected masses as stated in the last section.

The law of conservation of local energy means that the energy of every intracosmic isolated system is conserved with reference to every intracosmic observer. And this is nothing else but the good old "normal" law of energy conservation. For with reference to the inside observer, the existential gravitational potential must not be taken into account. There is a remarkable contrast between our two laws of energy conservation. The conservation of the total energy refers to the cosmos as a whole, from a standpoint which does not belong to the cosmos; the conservation of local energy refers to any part of the cosmos, from any standpoint which itself is part of the cosmos. And both laws are valid.

One could try to reduce one of the laws to the other. However, this is impossible. The local conservation of energy does not imply that the total cosmic energy is zero or at least constant. If, for example, the total nondefected mass M, the gravitational factor G, and the rate of expansion were all constant, the total cosmic energy (corresponding to the defected cosmic mass M') would change without the conservation of energy being violated anywhere locally.

On the other hand, the conservation of total cosmic energy does not imply the conservation of local energy. Let us imagine two sealed bottles standing on a table. One is filled with water, the other is empty. Everybody agrees that the water content cannot exchange from one bottle to the other unless there is a breaking of the seals. But such an exchange would not contradict the law of cosmic energy conservation. The sum of mass energies would be the same; the sum of gravitational energies would also be the same; and, in sufficiently long distances, the gravitational field itself would be approximately the same. Now, instead of the two bottle-shaped containers, two concentric containers with spherical surfaces could be imagined. In this case an exchange of water from the outer to the inner container (or vice versa) would not change at all the gravitational field outside the outer spherical surface.

Such an exchange of the water content would be possible as far as the conservation of the total cosmic energy alone is concerned. It is the conservation of local energy which makes it impossible, because each of the containers is an isolated system and, with reference to an observer near the containers, they have no potential energy, but only mass energy.

The law of local energy conservation does not allow the exchange of matter between two containers. Spontaneous generation of matter would produce even more of an energetic disturbance. The above-mentioned "individual" gravitational field of a spontaneously generated piece of matter would spread with the velocity of light; and this again cannot be reconciled with the conservation of local energy.

Obviously the law of local energy conservation is very "strict." Still more restrictive is this law together with the law of total cosmic energy conservation. Under the validity of both laws, is there any possibility of spontaneous generation of matter, that is, of generation out of nothing? The next section will deal with this question.

THE HOMOGENEOUS PHASE OF GENERATION

In our current cosmological situation no matter or energy can be created spontaneously. If, for example, a hydrogen atom, a quantity of water, a supernova, or an electromagnetic field were generated out of nothing at any point in the cosmos, one could assume a fictive surface around this object and could further assume any appropriate instrument near this surface. The instrument means an "observer"; the surface comprises a "system" and, for appropriate choice of the surface, even an isolated system. Under these circumstances, spontaneous generation would mean increase of the mass and energy in the isolated system with reference to an intracosmic observer. In other words, the law of local energy conservation would be violated.

Relativity and quantum physics have taught us to be suspicious about fictive experiments which cannot be realized. But in our case the experiment is based on a "system" and on an "observer" which are real or can be realized (at least in principle). The two bottles or containers of water cited in the last section represent systems; the hydrogen atom may be generated in a box or inside compact material, representing a system; the empty space around a new supernova can be understood as the limitation comprising a system. All these systems are isolated or at least sufficiently isolated for the validity of the argumentation.

The "observer" is not the cognizing human mind or an imagined perceiving being. the "observer" with regard to a process is any physical item which "observes," that is, which responds to this process. In our case, the "observer" may be: a balance—for the bottles of water; a device measuring the rotational inertia—for the spherical containers; an absorbed photon—for the hydrogen atom; a gravimeter—for the supernova.

The first phase of cosmic development, however, must not be assumed to have had conditions similar to the current ones. In that phase there was not yet any formation; the whole cosmos was but one homogeneous substratum, the "original gas." Let us reflect on such a situation, under the reasonable assumption that this gas had an ideal form of homogeneity, the absolute nonindividuality of the particles. Such a state is well known under the name "Bose gas." Is the energy conserved in a cosmos with Bose-gas conditions? The absolute homogeneity and nonindividuality have no special consequences as far as the conservation of the total energy of the cosmos is concerned; this energy has the constant value zero. But the law of conservation of local energy will be shown to be neither violated nor maintained; it is inapplicable to such a special state.

The very idea of an isolated system inside the cosmos cannot be reconciled with absolute nonindividuality. The fictive subdivision of the cosmos in two equal halves, for example, separates the particles of the original gas into two groups and makes a clear distinction between particles of one group and particles of the other group. Belonging to this group or belonging to that group is an individual feature of every particle, and this individual feature contradicts the assumed absolute nonindividuality within the whole cosmos.

There is an analogous problem about the observer. The very idea of an observer, of an intracosmic item responding to processes of interest, is not compatible with nonindividuality. The original gas as a whole may "observe itself as a whole"—bringing about statements on the cosmos as a whole. But a part of the cosmos cannot observe unless as observer it is clearly distinct from the other part of the whole. This distinction again is an individuating feature of the observing items which contradicts the assumed absolute nonindividuality.

Let us forget for a moment the preceding argumentation on the system and on the observer and let us admit that a part of the original gas could be observed by a local observer. What is the nature of this local observer? It would be absurd to think of a real material observer like an instrument or a small body. There is no such thing in the first phase of the cosmos, and its imagined introduction would falsify the conditions. The other alternative is to think of an "abstract" observer in the sense of a theoretical "standpoint."

A real local observer is itself subject to the gravitational potential and observes—as shown in the last section—only the quasiclassical mass energies of the observed objects. But a "standpoint" as such is in principle not subject to a potential; it cannot be on the same existential potential $-c^2$ as are the observed objects of the system. Therefore, with reference to the "standpoint as such," the observed system has both its mass energy and its existential potential energy, which compensate each other to zero. In reality the "standpoint as such" represents not an intracosmic but the extracosmic observer; and the part of the original gas observed from this standpoint represents not a true intracosmic system, but stands for the whole cosmos. The conclusion is, then, that the attempt to apply the law of local energy conservation to the original gas leads either to an absurdity or to the law of the conservation of total cosmic energy.

As shown above, it is the law of local energy conservation which blocks the spontaneous generation of matter. This law, however, is not applicable to a cosmos comprising a homogeneous, nonindividualized gas. For such a cosmos, the principle of energy conservation is reduced to the conservation of the total cosmic energy. This means that in the original gas cosmos, generation or disappearance of matter does not violate the principle of energy conservation as long as the total cosmic energy is zero. A direct cosmological application of this conclusion will be given in the next section.

The de Broglie matter wave offers a way for the intuitive understanding of the original gas cosmos, its nonindividuality, and its not being subject to local energy conservation. The matter waves of the particles are spread homogeneously all over the cosmos. The number of particles corresponds to the intensity and volume of the collective matter wave field. The mass energy of this matter wave field and its negative gravitational energy compensate each other everywhere existentially to zero. These energy forms are but two aspects of one phenomenon, they cannot be seperated.

The nonapplicability of local energy conservation must not be interpreted as "the" generation of matter. Let us study this with the help of an interesting microphysical parallel.

The Heisenberg uncertainty principle means an indeterminacy, a certain "play," in microphysical processes. Such indeterminate processes, amplified by physiological mechanisms, seem (P. Jordan) to be important for the brain functions of mammals and even man. But one must not conclude that the animal's exercise of options and the personal freedom of man are nothing more than the uncertainties of some atoms. Personal freedom is a spiritual fact, and the uncertainty of controling microphysical processes is nothing but the method, the mode of this freedom's realization without violation of physical laws in the body.

The nonapplicability of the law of local energy conservation in the original gas cosmos can be regarded as a cosmic indeterminacy, a "play," concerning the amount of matter. Such a play is neither generation nor disappearance of matter. Matter was generated because the Creator wanted its generation; and the nonapplicability of local energy conservation is nothing but the method, the mode of this generation's realization without violation of physical laws in the cosmos.

THE FIRST COSMIC PHASE—COSMOLOGY

In the first cosmic phase, the phase of the original gas, mass was generated without violation of the law of local energy conservation. What was the functional relation of the nondefected absolute cosmic mass M, the absolute radius A, and the gravitational factor G versus the cosmic age T? There are cosmological theories which admit M and G to be constant and A to be determined by the kinetic and potential energies. Such theories are based on models of exploding gases on a stellar scale. They do not seem to grasp what the cosmos really is.

The cosmos is characterized by the dualism of an interior and an exterior aspect. From the interior aspect there is existence: the energy Mc^2. There is disconnection: No object experiences gravitational adherence to other objects. There is openness: the c-velocity expansion in all three directions. And these three features are but different descriptions of the fact that the substratum of particles expands like an uncurbed energy field with quasiinfinite velocity in a quasi-Euclidean space.

From the exterior aspect there is nothingness: The total energy $M'c^2$ is zero. There is interconnection: Objects subject one another to the collective gravitational potential $-c^2$. There is seclusion: Cosmic spacetime is closed on itself like a spherical surface with Schwarzschild radius. And these three features are but different descriptions of the transcendent fact that self-relatedness is closed on itself and has no absolute existence. The cosmos has absolute existence only in a spiritual sense, as the loved creation of the Creator.

These considerations have quantitative physical consequences. Fundamental are the zero energy and the totally closed spacetime of the cosmos. The mesocosmic gravitational potential, e.g., of the sun, is nothing but the direct issue—in principle proportional to the ratio of mass to radius compared with M/A—of the existential cosmic potential $-c^2$. From another point of view, the mesocosmic gravitational spacetime deformation is a direct issue of the closed cosmic spacetime. And there is still another point of view, understanding gravitation as a correlation of quanta of action. Whatever the point of view is, the gravitational factor G is not a matter-immanent constant, but a descriptive factor based on a "total-cosmos-determined" situation. A is not given by M and G; but A is proportional to cT and G is a function of T, dependent on A and M.

Again, there are some physical constants which govern the microcosm and the macrocosm. Such fundamental constants are c (velocity of light), h (Planck's quantum of action), e (elementary charge), G_e (Coulomb's electrostatic factor), m (elementary mass $\approx$ proton mass), τ ($=h/mc^2$, elementary duration), λ ($=h/mc$, elementary length), V_1 [$=(4\pi/3)\lambda^3$, elementary volume], and h/V_1 (action density unit). Two other useful units are R ($=cT$, classical cosmic radius) and M_k [classical mass, contained in the classical cosmic volume $(4\pi/3)R^3$].

The generativity of the original gas in the first cosmic phase and the c-velocity expansion of the cosmos are not sufficient to pin down M_k as a function of T. There is another decisive principle involved: The cosmic action density [that is, the cosmic action $M_k c^2 T$ divided by the classical cosmic volume $(4\pi/3)R^3$] is constant and is equal to the action density unit h/V_1. The law of constant action density and the law of local energy conservation replace each other in a curious way. In the homogeneous original gas phase, the first one is valid, the second one is inapplicable; in the following phase of nonhomogeneity, the first one is invalid, the second one is applicable. However, the expansion velocity is constant at any time of the cosmic existence.

The law of constant action density means that $M_k T/R^3$ and thus M_k/T^2 are constant, and hence that M_k is proportional to T^2. The equality of the cosmic mass energy $M_k c^2$ with the negative potential energy yields $G \propto 2Rc^2/M_k$. In the original gas phase G was proportional to $1/T$.

At $T = 0$, space, time, and energy did not exist. At $T = \tau$ one neutron existed, its volume V_1 is the cosmic volume; the gravitational factor G and the electrostatic factor G_e (both referred to a pair comprising one proton and one electron) were equal. Expressed in the above-stated basic units, all decisive magnitudes were "one"; differentiations were not yet created. For $T/\tau = 2, 3, 4, \ldots$, the number of neutrons was $4, 9, 16, \ldots$. The "total-cosmos-determined" gravitational factor G decreased, the "elementary-particle-determined" electrostatic factor G_e, however, kept its value. Neutrons, being unstable with a mean life of 932 sec, decayed to protons and electrons. The original gas became a mixture of the continually "new-born" neutrons and the protons and electrons, but the homogeneity and nonindividuality were kept.

The considerations of this section have intended to demonstrate the cosmological relevance of what has been explained on mass generation and energy conservation. It is not the purpose to explain or prove cosmological details. This has been done elsewhere.[1,3],[1] However, some interesting points may be mentioned briefly.

At the cosmic age T of about 10^9 years a wave mechanical process introduced the end of the homogeneous original gas phase. By feedback from gravitational effects, the gas became more and more nonhomogeneous, the law of local energy conservation became applicable, and mass generation ended. The peak value of cosmic materiality was reached: $M_k \approx 10^{80}$ protons. The progressing gravitational collapse of the gas clouds produced the protogalaxies. This process was counteracted by the velocity of the protons, which was about 10^7 cm/sec = $c/3000$, a result of the neutron decay. The conflict of both influences determined the average size of the clouds and hence the number of the later galaxies; this number turns out to be $3000^3 = 27 \times 10^9$, in good accordance with observations.

The general potential equation $G = 2Rc^2/M_e$ holds for all cosmic phases. After the end of the first homogeneous phase, however, its application becomes more difficult, because M_e means an "effective value"; it is calculated by an integration of the whole cosmos, taking into account that distant parts of the cosmos present themselves to us in a younger stage of development. Therefore the gravitational factor G continued to decrease for about another 10^9 years after the end of mass generation. Later on it began to increase. Today it increases, but less than linear with T.

---

[1] For a brief report in English, see the cosmological supplement to Philberth.[2]

The future fate of the cosmos will be determined by the exhaustion of nuclear energies, by the redshift fading of photons and neutrinos, and, more and more, by the increase of the gravitational factor G. This increase forces smaller and smaller masses into gravitational collapse. There will be a stage when the materiality consists of nothing more than gas molecules, atoms, and elementary particles. The individuality of the objects tends toward zero and it reaches zero when nothing but elementary particles are left. And even these particles will lose existence. No individuum, no "system," no "observer," and finally no matter at all is left, and G reaches the value hc/m^2 of the very beginning: An amazing analogy to the beginning of the cosmos. Space and time do not contract, they become "empty," unreal. The Creator has taken back the existence of His creation.

REFERENCES

1. B. Philberth, *Der Dreieine*, 4th ed. (Christiana-Verlag, Stein am Rhein, Switzerland, 1976).
2. K. Philberth, *Elektron, Pion, Proton und Elementarlänge* (Christiana-Verlag, Stein am Rhein, Switzerland, 1974).
3. J. J. Knappik, *Dynamiczna ekspansja kosmosu* (Veritas, London, 1976).

Chapter X

On a Chaotic Early Universe

Kenji Tomita

Hiroshima University

INTRODUCTION

Man's concept of the Universe has changed with the development of observational apparatus. Its content has expanded from a system of the earth and the heavens to the solar system, our Galaxy, the local group of galaxies, and regions including many clusters of galaxies. The global isotropy and homogeneity of the spatial distribution of galaxies and their clusters has become a characteristic of the Universe. Moreover, the discovery of the velocity–distance law by Hubble revealed the nonstatic aspect of the Universe: Evolution has become an essential part of our concept of it. The evolution of the Universe appears not only as that of the whole system, but also as changes in the state of the materials and astronomical objects that comprise it. With the expansion of the Universe, the chemical composition of materials changes and various objects are born, interact with each other, and die.

One of the most self-consistent and beautiful theories for describing the whole system of the Universe is the gravitational theory as created by Einstein. On this basis the isotropic and homogeneous models of the Universe were derived by Friedmann, Lemaître, and Tolman. The cosmological principle, requiring global isotropy and homogeneity of spacetime and the spatial distribution of matter, plays a vital role in the derivation of these models.[1] When we go back to the past in these models, the density of matter increases due to the contraction of spatial scale, and ultimately reaches infinity. The "early Universe" means the Universe at an early stage, when the density was so high that the physical situation was much different from the present one. This term is of course also employed in anisotropic or inhomogeneous cases.[2]

The cosmological principle is a reasonable one in the present state of the Universe, but in the past, far beyond our capacity for observation, it need not hold necessarily. So models without isotropy or homogeneity have also been derived. However, the evolution of the constituents of the Universe was first studied in the simplest case, where the Universe is assumed always to have been isotropic and homogeneous from the beginning of cosmic expansion (Big Bang). At the stage of densities so high that any local irregularities such as galaxies could not exist, admissible deviations from mean values are only statistical or due to quantum fluctuations of the matter density and the corresponding fluctuations of spatial curvature, or their growth owing to gravitational instability. The chemical composition of matter at the earliest stage should be considered to have consisted of the most fundamental particles, contrary to the present compounds, i.e., complicated compositions. The theories for the origin of galaxies and elements were proposed on these assumptions. Let us now review them briefly.

Let us assume that the fluctuations of the matter density were statistical at the epoch t_N when the baryon mass density was equal to 10^{15} g/cm^3 (nuclear density), i.e., $\delta\rho/\rho \sim 10^{-34}$ at t_N, and assume that they have grown owing to gravitational instability. Then, their present values would be so small that they could not have separated from the uniform expansion and become protogalaxies. So there must have been far larger fluctuations ($\delta\rho/\rho \sim 10^{-5}$) at t_N, or the growth must have started with $\rho_b \sim 10^{72}$ g/cm^3 (an extremely high density). Moreover, in order for the galaxies to be well formed in the present epoch, the amplitude of the fluctuations must have had extraordinarily restricted values. Otherwise, most of the fluctuations would have collapsed far in the past, or could not be separated in the present from the uniform expansion. However, if we consider only the most favorable, ideal fluctuations, the further evolution of gravitationally bound clouds can be connected with the explanation of some observational features of galaxies. It has been shown that galactic rotation can be formed discontinuously through shock waves aroused in anisotropically collapsing protoclusters, and galaxies can be formed by the fragmentation of the protogalaxies.[3] There is also another standpoint, in which one assumes that density perturbations have derived from the turbulent motions at the pregalactic early stage, but since these motions change considerably the physical situation in the earlier Universe, this is not included in the present case.

As for the origin of the elements, this depends strongly on when the cosmic radiation appeared. If this radiation is taken to exist uniformly at least after the epoch corresponding to temperature $T \sim 10^{10}$, that is, if we assume a so-called "hot" Universe, the constituent particles reach thermal equilibrium and the number densities of protons and neutrons become comparable at $T \sim 10^{10}$.[4] As matter was cooled due to expansion, ^{2}H ^{3}H, ^{3}He, ^{4}He, and heavier elements were formed irreversibly through the process of nuclear fusion. The final mass ratios X of these elements to H depends on the value of the present baryon mass density $\rho_{bo} = 10^{-29} f$ g/cm^3, and are: $X(^4\text{He}) = 0.270, 0.277$; $X(^2\text{H}) = 10^{-12}, 2.2 \times 10^{-4}$; and $X(^3\text{H}) = 2.7 \times 10^{-6}, 4.3 \times 10^{-5}$; for $f = 1.27$ and 1.27×10^{-2}, respectively.[5] The observed values are 0.20–0.35, $\sim 10^{-4}$, and $\sim 10^{-5}$ for ^{4}He, ^{2}H, and ^{3}H, and so are consistent with the low-density models of $\rho_{bo} \sim 10^{-31}$.

Moreover, this density is comparable with the mean density of galaxies. Such good consistency for low-density models and the fact that the cosmic radiation predicted by Gamow[6] was discovered by Penzias and Wilson[7] as the 2.7° isotropic radiation make plausible the assumption of a hot Universe. However, it is difficult to understand why the time-independent ratio of the photon number to the baryon number should be so large (10^8–10^{10}) if the cosmic radiation has existed from the epoch when $T \sim 10^{10}$. In order to avoid this difficulty, we may assume that the Universe at the initial stage was so cold that this ratio was $\ll 1$, and at some later stage it was heated through exothermic reactions and dissipation processes. In this "cold" Universe, the resultant composition of elements depends on the kinds of fundamental particles, their relative abundance, and their degeneracy, and it seems possible that consistent final abundance ratios of the elements can also be derived in this cold case.[8] So it seems hasty to rule out a cold early Universe.

Here it is significant to indicate two undesirable properties in the isotropic, homogeneous models. First, among the solutions of the Einstein equation, the solution representing these models occupies a special place and displays behavior quite different from the other general solutions. Second, when we consider two arbitrary points in these models, their distance L decreases with the decrease of the cosmic time t, but more slowly than t. By some epoch, accordingly, L becomes larger than ct (the maximum distance where any information can have propagated since the Big Bang), and so these two points are not causally connected. This situation seems to be incompatible with the strong condition of isotropy and homogeneity.

For these models the anisotropic but homogeneous models also were applied to describe the early Universe.[9] This seemed plausible, because the discoveries of helium-deficient stars and the systematic polarization of radio waves from remote sources were interpreted to suggest no helium in the early Universe and the existence of a uniform magnetic field.[10] However, all this can be interpreted in other ways. The magnetic field can be nonuniform.[11] Accordingly anisotropic models were studied rather in order to clarify the behavior of more general solutions near singular points.[12] Moreover, the problem of the isotropization of arbitrarily anisotropic models was investigated in connection with the remarkable behavior of the neutrino at the epoch ($T \sim 10^{10}$) when it decoupled from the other particles.[13,14] However, it has not yet been answered whether the anisotropy damps out in the neutrino processes to the level required by observation.

It is inhomogeneous models that are more general and reflect the above noncausality. When spatial inhomogeneity is realized maximally, the term "chaotic" in the title is justified.

COSMIC TURBULENCE AND THE WEAK COSMOLOGICAL PRINCIPLE

Besides the gravitational instability theory of galaxy formation, there is the turbulence theory, according to which the separation of perturbed regions from the expanding motion arose through compression due to nonlinear (inertial) forces appearing in hypothetical early turbulent motions.[15]

The turbulence assumed here has no special direction and center, and its physical quantities averaged spatially are isotropic and homogeneous everywhere. This is consistent also with the requirement of the cosmological principle.

Now we know that, when we go back into the past in an isotropic and homogeneous model, any rotational perturbations increase and reach the nonlinear region, so that the spacetime structure itself is nonlinearly perturbed at the same time. If turbulent eddy motions at an epoch are the consequences of continuous evolution of past fluid motions, the past spacetime structure must have had a turbulent character.[16] In density and gravitational-wave perturbations, too, a similar situation arises from each one of their two independent components. Therefore it is reasonable to assume for the early Universe an inhomogeneous model whose spacetime curvature is of a turbulent character. This assumption is compatible with the noncausality in the early Universe and can be described by most general models.

Here it seems significant that we impose a weak condition on this turbulence, in order to develop our theory. It is the requirement that the turbulence of the spacetime curvature is isotropic and homogeneous in a similar sense to fluid turbulence. This requirement can be expressed: the average values of the physical quantities such as curvature invariants are invariant for any spatial rotation and translation. Compared with the ordinary cosmological principle, the condition on spacetime curvature is weakened in this requirement. So we shall call it the "weak" cosmological principle in the early Universe.

INHOMOGENEOUS MODELS

Weakly Nonlinear Perturbation Theory

In order to connect a chaotic early model with the isotropic and homogeneous model, it is useful to analyze deviations from the latter by the perturbation method. The case when a strongly nonlinear process plays an important role in their connection will be treated in the last part of this section.

The deviation in the linear approximation is classified into three types, i.e., density, rotational, and gravitational waves, as was shown by Lifshitz.[17] In the second-order approximation, they are coupled with each other and themselves through the nonlinearity of the gravitational interaction. The second-order density perturbations aroused by linearized rotational and gravitational waves have amplitudes of $\sim (L/ct)^2 (v/c)^2$ and $\sim h_\mu^\nu h_\nu^\mu$, respectively, which are inversely proportional to t. Here h_ν^μ is the metric perturbation corresponding to gravitational waves. An epoch when the gravitational nonlinearity becomes essential is given by the condition that these amplitudes are of the order of unity, so that it depends on the velocity, the size, and the amplitude of the metric perturbations.[18]

Anti-Newtonian Approximation

At the stage of full nonlinearity, the condition $ct/L \ll 1$ is satisfied, e.g., for rotational waves we have $ct/L \ll v/c < 1$, so that the dynamics of irregularities at this stage can be treated by means of the approximation of $c \to 0$ (anti-Newtonian). On this approximation the Einstein equation is simplified and solved. A characteristic property of this solution is the continuous change from the Kasner-type vacuum solution to an isotropic and homogeneous solution. By adjusting this approximate solution to the perturbed solutions at the overlapping stage, the parameters in Kasner-type solutions are related to the size, amplitude, etc., of various perturbations at the later stage.[16]

When we go further into the past, the term P_α^β ($\alpha, \beta = 1, 2, 3$) representing three-dimensional curvature, which is neglected in the above approximation, becomes effective because of the sharp anisotropy of the Kasner-type model. Before this epoch, the local motion of irregularities is similar to that of an anisotropic but homogeneous model, in which the Kasner-type models with different directions of anisotropy appear alternatively or in an oscillatory way.[19]

Post-Anti-Newtonian Approximation

At an intermediate stage between the Kasner-type model and the isotropic and homogeneous model, the solution linear (or of first order) with respect to ct/L can be derived by taking into account the term P_α^β evaluated by use of zeroth-order quantities and again solving the Einstein equation. The mathematical treatment of this linear solution will be shown in another paper, and here only its qualitative result is written down. The first-order solution generally has a density perturbation which increases with time. If the isotropization and homogenization of the zeroth-order solution proceed far enough, the density perturbation and its corresponding velocity perturbation of first order are expressed as

$$\delta\epsilon/\epsilon \sim (ct/L)^2(L/r) \propto t^{2(1-n)}$$

$$\delta v/c \sim (ct/L)^3(L/r) \propto t^{3(1-n)}$$

where r is the three-dimensional curvature radius in a relevant irregularity and $r \propto L \propto t^n$ ($n < 1$). Therefore, the first-order solution gives rise to inhomogenization, in contrast to the zeroth-order solution. At an epoch where $L = ct$ (or $t = t_e$), the first-order velocity perturbation reaches a maximum value, while the relevant irregularity collides with neighboring irregularities. Accordingly, if the maximum velocity exceeds the sound velocity c_s at this epoch, the perturbation becomes a shock wave, and, if not, it becomes a sound wave which propagates toward the outside after the epoch t_e. The sound velocity depends on whether the early Universe is hot or cold: $c_s = c/\sqrt{3}$ or $\ll c$, respectively. Shock waves thus aroused dissipate, and sound waves are left and propagate. It seems that, after t_e, the Universe may be described by an isotropic and homogeneous model with sound-wave-like perturbations. Dynamical treatment of this fully nonlinear stage would be very interesting, but has not yet been performed.

SMOOTHING-OUT PROCESS AND ORIGIN OF COSMIC MICROWAVE RADIATION

The sound waves appearing at the epoch t_e superpose with random phases, amplitudes, and directions, and the region of their propagation ($\sim ct$) expands with time. According to the weak cosmological principle, all types of irregularities appear everywhere with the same probability, so that the average amplitude of superposed sound waves is the same everywhere.

The epoch t_e is a function of L, and t_e is smaller for smaller L. Therefore, whatever the size spectrum is, and even if the irregularities of larger size coexist with or include those of smaller size, small ones are reduced to sound waves earlier than large ones in general, and so the largest ones are left behind. When this process stops depends on the upper limit L_m of the sizes, but we cannot derive L_m in our present theory. Phenomenologically we can consider that L_m is the size of the largest cluster of galaxies, because a trace of L_m will be left as the size of the largest astronomical objects. Then, using $ct_{em} = L_{c1}(t_{em})$, $M_{c1} = \frac{4}{3}\pi\rho_b(t_{em})L_{c1}^3(t_{em}) = \frac{4}{3}\pi\rho_b(t_0)L_{c1}^3(t_0)$, and $6\pi G\rho_b t_0^2 \simeq 1$, we can estimate the redshift at the epoch t_{em} when the last shock waves arise:

$$1 + z_{em} = (c^3 t_0/6\pi G M_{c1})^{2/3} = 10^5 (M_{c1}/10^{14} M_\odot)^{-2/3}$$

In this equation t_0 is the present cosmic age ($=10^{10}$ yr). This redshift is nearly equal to the redshift $z_{eq} = 2 \times 10^4$ at the epoch t_{eq} when the baryon mass density and the cosmic radiation mass density are equal. (Here the Einstein-de Sitter model is employed as the isotropic and homogeneous model. If irregularities of both positive and negative spatial curvature appear at the same probability, it is reasonable to consider that the average spatial curvature vanishes, or the smoothed-out model is flat.)

Rees[20] proposed a theory in which the cosmic radiation is formed by the heating due to shock-wave dissipation at the epoch t_{em}. As shock waves are relativistic for $r \sim L$, the released thermal energy ρ_r becomes comparable with the rest mass energy at this epoch. This conclusion is consistent with the above fact that $z_{em} \sim z_{eq}$. Moreover, the spectrum becomes blackbody-like because of effective absorption due to bremsstrahlung. After the epoch t_{em}, the radiation cools adiabatically because of the lack of heating sources. However, no quantitative analysis for the efficiency of this mechanism has been performed yet.

(For the origin of cosmic radiation, there is another theory, of Layzer and Hively[21] in which it is formed through grains at later stages.)

ORIGIN OF GALAXIES

From the chaotic viewpoint regarding the early Universe, we can assume the following situation at the pregalactic stage;

(a) The cosmic radiation existed at least after the epoch t_{eq}.

(b) Sound waves of considerable amplitude existed from the epoch t_{eq} to the epoch t_D, when the radiation decoupled from matter.

(c) Rotational motions appeared at the initial explosion time or at least at the epoch t_e.

At $t_{eq} < t < t_D$, the Universe was filled with plasma consisting of photons, electrons, and protons, and the radiation mass density was comparable with or somewhat smaller than the matter density, so that the sound velocity c_s was near the light velocity c. Now let us express the characteristic size and velocity of irregularities by L and v. If $L < vt$, hydrodynamic nonlinearity plays an essential role, and so fluid motions are turbulent (the Reynolds number is very large.) If $L > vt$, individual motions are frozen into the dominant expanding motion.[22, 23]

At $t > t_D$, c_s is reduced to much smaller values, comparable with the thermal velocity of matter only, so that many irregularities become supersonic. If $L < vt$, they are compressed by strong inertial force, but the compression can be prevented by the turbulent pressure[24] and another pressure due to the magnetic field which may be produced in the turbulent medium.[25] If $L > vt$, the nonlinear force is not effective and so the density perturbation grows owing to gravitational instability. The dynamical evolution of these supersonic turbulent motions to protogalaxies or protoclusters has so far been estimated in various ways, but is not clarified, because of its complexity. In spite of this incompleteness, however, it is remarkable that the turbulence theory of galaxy formation can derive the observational parameters of galaxies and their relations.[24, 26]

ORIGIN OF CHEMICAL ELEMENTS

The process of nucleogenesis of elements in the early Universe is quite different according to the initial thermal conditions. There are two extreme cases: The Universe was so cold that all fermions were degenerate initially, or it was so hot that it was filled with the cosmic radiation corresponding to the present $2.7°$ radiation. Of course, intermediate cases can be considered.

In the simple case when the temperature in each irregularity is adiabatically connected with the present value $(2.7°)$, the abundance of elements formed depends on the values of L and v of each irregularity, and if L and v are small enough, the final abundance is the same as that in an isotropic and homogeneous model. As L and v increase, the abundance of He and heavier elements decreases. This indicates the possibility that the abundance becomes similar to that in an isotropic and homogeneous model if the Universe was filled with irregularities of small L and v at the stage of nucleogenesis.[27] In the case of a cold Universe, the dependence of the abundance on the motion of irregularities has not yet been analyzed.

CONCLUDING REMARKS

Historically the connection between the concept of chaos and the origin of astronomical objects or the Universe itself is quite old. Chaos or a fully disordered state appears often in the oldest traditions about the initial state of the Universe. As a scientific concept it was taken up by Kepler, Descartes, Swedenborg, Kant, and Laplace.[28] A hydrodynamic turbulent state at the pregalactic early stage of an expanding universe was introduced by von Weizsäcker in the 1940s, followed by Gamow. Furthermore, a disordered state of spacetime structure at the early Universe appears to have originated in the work of Ozernoi and Chernin[22] and Misner[14] in the 1960s.

Many problems closely connected with the early Universe have not been touched upon here, such as the matter–antimatter asymmetry, "time's arrow," the avoidance of a singularity at the earliest stage, the modification of gravitational theory, and the quantization of the gravitational field. They have not yet been taken into account in our present theory.

REFERENCES

1. H. Bondi, *Cosmology* (University Press, Cambridge, 1960).
2. E. R. Harrison, *Ann. Rev. Astron. Astrophys.* **11**, 155 (1973); A. D. Chernin, *Astrophys. Lett.* **8**, 31 (1971); **10**, 125 (1972).
3. R. A. Sunyaev and Ya. B. Zel'dovich, *Astron. Astrophys.* **20** 189, (1972); A. G. Doroshkevich, *Soviet Astron.–AJ* **16**, 986 (1973).
4. C. Hayashi, *Prog. Theor. Phys.* **5**, 224 (1950); see also H. Sato, T. Matsuda, and H. Takeda, *Suppl. Prog. Theor. Phys.* No. 49, p. 11 (1971).
5. R. V. Wagonar, *Astrophys. J.* **179**, 343 (1973).
6. G. Gamow, *Phys. Rev.* **74**, 505 (1948); *Rev. Mod. Phys.* **21**, 367 (1949); *Kgl. Danske Videnskab Selskab Mat.-Fys. Medd.* **27**(10) (1953).
7. A. A. Penzias and R. W. Wilson, *Astrophys. J.* **142**, 419 (1965).
8. M. Kaufmann, *Astrophys. J.* **160**, 459 (1970).
9. K. S. Thorne, *Astrophys. J.* **148**, 51 (1967).
10. K. Kawabata, M. Fujimoto, Y. Sofue, and M. Fukui, *Publ. Astron. Soc. Japan* **21**, 293 (1969).
11. A. H. Nelson, *Publ. Astron. Soc. Japan* **25**, 489 (1973).
12. V. A. Belinski, I. M. Khalatnikov, and E. M. Lifshitz, *Adv. Phys.* **19**, 525 (1970).

13. A. G. Doroshkevich, Ya. B. Zel'dovich, and I. D. Novikov, *Soviet Phys.–JETP* **26**, 408 (1968); *Astrofisika* **5**, 264 (1969); R. A. Matzner, *Ann. Phys.* **65**, 438 (1971).
14. C. W. Misner, *Astrophys. J.* **158**, 431 (1968).
15. C. F. von Weizsäcker, *Astrophys. J.* **114**, 165 (1951); G. Gamow, *Phys. Rev.* **86**, 251 (1952); H. Nariai, *Sci. Rep. Tohoku Univ.* **39**, 213 (1956); **40**, 40 (1956).
16. K. Tomita, *Prog. Theor. Phys.* **48**, 1503 (1972).
17. E. M. Lifshitz, *J. Phys. USSR* **10**, 116 (1946); E. M. Lifshitz and I. M. Khalatnikov, *Adv. Phys.* **12**, 185 (1963).
18. K. Tomita, *Prog. Theor. Phys.* **37**, 831 (1967); **45**, 1747 (1971); **47**, 416 (1972).
19. V. A. Belinski, I. M. Khalatnikov, and E. M. Lifshitz, *Soviet Phys.–JETP* **35**, 838 (1972).
20. M. J. Rees, *Phys. Rev. Lett.* **28**, 1669 (1972).
21. D. Layzer and R. Hively, *Astrophys. J.* **179**, 361 (1973).
22. L. M. Ozernoi and A. D. Chernin, *Soviet Astron.–AJ* **11**, 233 (1967); **12**, 901 (1968).
23. H. Sato, T. Matsuda, and H. Takeda, *Prog. Theor. Phys.* **43**, 1115 (1970); K. Tomita, H. Nariai, H. Sato, T. Matsuda, and H. Takeda, *Prog. Theor. Phys.* **43**, 1511 (1970).
24. L. M. Ozernoi and G. V. Chibisov, *Soviet Astron.–AJ* **14**, 615 (1971).
25. E. R. Harrison, *Month. Notices Roy. Astron. Soc.* **147**, 279 (1970); **165**, 185 (1973); T. Matsuda, H. Sato, and H. Takeda, *Publ. Astron. Soc. Japan* **23**, 1 (1971).
26. N. Dallaporta and F. Lucchin, *Astron. Astrophys.* **19**, 123 (1972).
27. K. Tomita, *Prog. Theor. Phys.* **50**, 1285 (1973).
28. S. Arrhenius, *Das Werden der Welten* (Japanese transl., Iwanami Shoten, 1951); J. L. E. Dreyer, *A History of Astronomy from Thales to Kepler* (Dover, New York, 1953).

Chapter XI

Cosmological Implications of Non-Velocity Redshifts—A Tired-Light Mechanism

Jean-Pierre Vigier

Institut Henri Poincaré

VARIATION OF THE HUBBLE "CONSTANT"

Since Arp's original paper[1] "anomalous" redshifts (henceforth denoted ARs) have been showing up experimentally (see Ref. 2 for a recent review), not only in distant objects (such as companion galaxies,[3] NGC 7603,[4,5] or Stephan's Quintet[6,7]), but also in those in our own Galaxy (such as HD 217312[8] and WR binaries[9]) and even in our own Sun,[10,11] where they seem to be associated with a significant increase of the deflection of light exceeding Einstein's prediction.[12]

Evidently, any particular AR example can be contested or explained separately by specific characteristics.[13] However, the evolution of the experimental situation is progressively weighting the scales (in the author's opinion) in favor of AR existence. We thus think that the time has come to evaluate some of the possible cosmological implications of ARs and to briefly discuss various objections[14-16] against a semiheuristic proposal[17] to explain the ARs in terms of photon interactions.

Clearly, the main strong point of the conventional "Big-Bang" model is the compatibility of Hubble's law with observed blackbody microwave-background radiation. Despite some famous contradictors,[18] most people seem to think that the situation is now airtight in favor of this model. Recent experimental data, however, do not support such an attitude.

Our main argument is that, contrary to the observed isotropy of the 2.7°K radiation, the *isotropic and universal proportionality of redshift to distance, predicted for all distant objects by the expanding-universe model, cannot be con-*

sidered as an established fact at the present stage of experimental knowledge. Indeed, the different values obtained for the Hubble constant raise serious doubts as to its isotropy and uniformity in depth.

Let us briefly review the situation. Since Hubble[19] first described the relation of redshift to distance as "roughly linear" and added[20] a small positive quadratic term, suggestions that the redshift is not so regular have been constantly recurring, up to the present day, in the specialized literature. The most important studies of that effect were made by de Vaucouleurs,[21-24] who concluded that significant velocity perturbations exist, related to the inhomogeneous distribution of galaxies in the Ursa Major, Coma, and Virgo regions (see Shapley and Ames,[25] Reiz,[26] and de Vaucouleurs[27]), called the Local Supergalaxy (LSG). De Vaucouleurs attempted to explain them in terms of local Doppler motions, i.e., as a superposition of a general rotation of the LSG (of ~ 500 km/sec) around a center located in the Virgo cluster, plus a differential expansion effect. The character of the velocity field has recently been discussed again by Nicoll and Segal,[28] who claim that the linear law is rejected at a level $p < 0.001$ in the LSG, while Tammann and Sandage,[29] on the basis of new criteria for sample selection, did not discover any significant local variation of H. They infer that H is isotropic, constant, and equal to 55 ± 5 km sec^{-1} Mpc^{-1}.

However, various results obtained earlier point to discrepancies between nearby and distant values of H and its variation with the position in the sky.

First, using close groups of galaxies (most of which evidently belong to the LSG), de Vaucouleurs has obtained values of H slightly over 100. In his latest paper, the value $H = 100 \pm 10$ km sec^{-1} Mpc^{-1} was adopted. This agrees well with the value $95 - 12 < H < 95 + 16$ obtained by van den Bergh,[30] using nine different methods and considering galaxies mostly belonging to the LSG, where the majority are in the direction of the central areas of the LSG. A supergalactic value $H \sim 100$ was also obtained by Heidmann,[31] utilizing a method based on a luminosity–diameter relation.

On the other hand, Sandage,[32] using ScI galaxies with radial velocities larger than 4000 km/sec and mostly out of the central direction of the LSG,[1] obtained $H = 55 \pm 7$. From the luminosity function of galaxies in clusters, Abell and Eastmond[33] arrived at the value $H = 47 \pm 5$ for the Coma and the Corona Borealis clusters. On the basis of these results, the metagalactic value of the parameter H is remarkably smaller than the supergalactic one.

In this context, a recent discovery by Rubin *et al.*[34] should be emphasized. Observing a sample of 74 ScI galaxies with $14.0 \leqslant m \leqslant 15.0$, they have shown that in one area the mean velocity is 4966 ± 122 km/sec and in the other (including the central area of LSG) it is 6431 ± 160 km/sec; 38 other galaxies displayed a similar redshift anisotropy.

The evolution of the discussion (Tammann and Sandage[29] and Karoji and Nottale[35]) on possible interpretations of the angular anisotropy of the redshift distribution on the sky map first discovered by Rubin *et al.*[34] and later confirmed on various types of sources (see Jaakkola *et al.*[36]) is evidently modified by the observational fact that a new, statistically complete, sample of distant faint radiogalaxies collected by Colla *et al.*[37] does not present the Rubin–Ford anisotropy but contains instead an even stranger anomaly in its radial velocity distribution.

We have utilized in this list all galaxies within the magnitude interval $13 \leqslant m_{corr} \leqslant 15.5$ within a velocity range $4800 \leqslant V \leqslant 20\,000$ km/sec. All apparent magnitudes have been corrected for galactic absorption by the usual formula

$$m_{corr} = m + 0.25 \text{ cosec } |b^{II}|$$

For this subsample one obtains

$$\langle HM \rangle_{II} - \langle HM \rangle_{I} = 0.02 \pm 0.045$$

for the difference of HM values $(HM = \log V - 0.2 \, m)$ between the two sky map regions proposed by Rubin *et al.*[34] This is an essentially negative result, for it is not compatible with an interpretation of the RRF Sc I data in terms of (1) solar motions (Rubin *et al.*[39]), (2) differential expansion (Haggerty and Wertz[40]), or (3) differential absorption (Hartwick[41]), since all these interpretations would act in the same way on all possible homogeneous samples of distant sources.

At first sight, of course, this strengthens the suggestion made by Tammann and Sandage[29] that this type of anisotropy results only from a random statistical fluctuation on the sky of absolute magnitude distribution of all samples considered until now.[36]

A closer examination following a proposal of Karoji and Nottale[35] shows, however, that this sample presents an even more remarkable anisotropy in the radial velocity distribution. If one subdivides it into two subsamples corresponding to two space regions, i.e., region B, where light crosses intervening clusters of galaxies, and region A, where light does not pass through any cluster, as represented geometrically in Fig. 1 and as projected on the sky map of Fig. 2 (where region B is limited by closed curves), one gets (since each region I and II of RRF now contain part of regions B and A and vice versa):

$$\langle HM \rangle_{B} - \langle HM \rangle_{A} = 0.16 \pm 0.036$$

for 41 objects. This yields a highly significant distribution (having a real effect probability $P > 99.9\%$) with a Student coefficient $t = 4.4$, i.e., a $>4\sigma$ deviation.

The following points now can be made:

1. This anisotropy is apparently completely independent of the RRF regions. We find, indeed, that, in region II (23 objects),

$$\langle HM \rangle_{B} - \langle HM \rangle_{A} = 0.15 \pm 0.039$$

and in region I (18 objects),

$$\langle HM \rangle_{B} - \langle HM \rangle_{A} = 0.19 \pm 0.055$$

which also yields $>99\%$ P values.

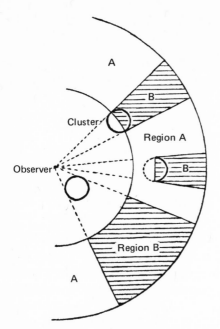

Fig. 1. Definition of Regions A and B.

2. It appears related to an excess redshift of region B since we have (41 objects)

$$\langle V \rangle_B - \langle V \rangle_A = 2412 \pm 1327 \text{ km/sec}$$

3. It does not appear related to any anomaly in the magnitude distribution, since we get

$$\langle m_{corr} \rangle_B - \langle m_{corr} \rangle_A = -0.26 \pm 0.218$$

$$\langle m \rangle_B - \langle m \rangle_A = -0.31 \pm 0.212$$

for the corrected and the noncorrected values. The observation window $13.0 < m_{corr} < 15.5$ is too large to support the bias suggested by Tammann and Sandage.[29]

4. It does not result either[35] from any significant difference in the average difference of the mean geometrical distances $\langle D \rangle$ of regions B and A, since we get

$$\langle D \rangle_B - \langle D \rangle_A = 200 \text{ km/sec}$$

in symbolic radial velocities.

5. It can explain the RRF effect, since the differences

$$\langle HM_B \rangle_{II} - \langle HM_B \rangle_I = -0.04 \pm 0.040 \text{ (20 objects)}$$

$$\langle HM_A \rangle_{II} - \langle HM_A \rangle_I = -0.00 \pm 0.055 \text{ (21 objects)}$$

are not statistically significant. This is also the case for all samples known to us, with the exception of the RRF ScI sample, which is not yet available.

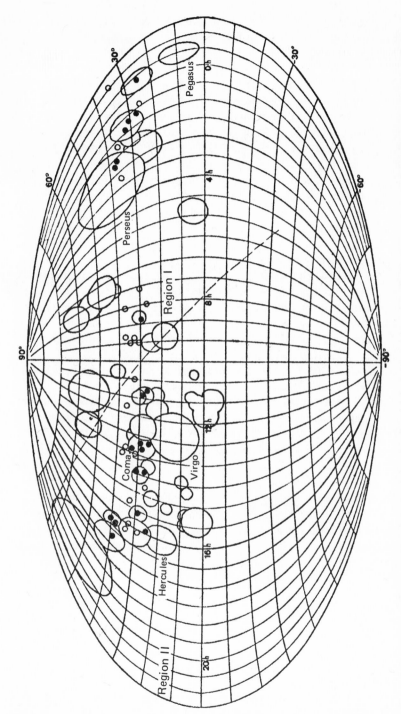

Fig. 2. Distribution on the map of the sky of the clusters of galaxies and of our sample of galaxies, in equatorial coordinates. The clusters are represented by their outlines. Open circles are radio galaxies in region A, filled circles are those in region B. The dashed line is the RFR border.

We conclude that the preceding results, which have been summarized by the statement,[35] "Everything goes as if light emitted from distant sources is redshifted when it travels through clusters of galaxies," evidently favors the existence of a "tired-light" mechanism first discussed by Hubble and Tolman[38] in the literature and recently revived by the observations of Arp,[1] Burbidge,[18] and others.[11] This result, together with the differences in the H values discussed above, suggests that the parameter H is not constant but fluctuates spatially, depending on the different kinds of mass concentrations. According to the aforementioned results, the metagalactic value of H is smaller than the supergalactic one. If this is interpreted in terms of expansion, the LSG is expanding faster than the less dense regions of the metagalaxy, and it can be noted that the gravitation in the LSG should lead to the contrary state of affairs.

One can evidently attempt to explain away the difference between $H \sim 50$ and $H \sim 100$ (in the LSG) by differences in zero-point calibration between different authors, so that this difference of ~ 50 between distant and close values of H would have no real physical existence. This is, of course, possible in principle, but (in the author's opinion) it is not supported by the data at the present stage of observational evidence. We propose the following points.

A zero-point error by Tammann and Sandage or by de Vaucouleurs explains only partly the difference between them. This is seen in the different values adopted for the distance of M101, i.e., 712 and 4.6 Mpc, respectively. The change in H of Tammann and Sandage would then be from 55 to 76, i.e., a difference of 20–30 remains, which can be compared with the difference (~ 20) observed by Rubin et al.[34] between their regions I and II.

The result of van den Bergh is based upon the generally adopted distances of the Local Group galaxies. Most of his nine methods utilize objects belonging to the LSG and mainly are located in the sky within Rubin's region II, which contains the center of the LSG. The same applies to Heidmann's result.[31]

The Abell and Eastmond result depends on the calibration of M87 in the Virgo cluster. This calibration is founded on the assumption that the brightest globular cluster of M87 is comparable to the brightest globular cluster of the Milky Way and M31. This means that a possible change in the distance scale—if M87's brightest globular cluster is brighter than the others—increases the difference in H and depresses H still further. As a consequence, the difference between van den Bergh and Abell–Eastmond cannot be explained as being simply due to different assumptions concerning the zero point of calibration.

The preceding analysis suggests that one should attach a physical significance to the distribution, over the sky and in depth, of the galaxies used in the studies yielding different H values—a fact which evidently conflicts with the extreme isotropy of the 2.7°K field. If this is true, we seem to be left with the only alternative (to explain Rubin's AR) of redshift cause other than expansion, namely, we must rule out the customary concept which links the isotropy of the 2.7°K radiation with an isotropic expansion. This view accords with various models developed previously by Burbidge[18] and other authors.

We are thus led to the idea that an explanation by a new physical property of light could explicate the ARs, on the one hand, and form the basis of a new comprehensive cosmological point of view on the other. This was already implied in our preliminary papers[17] and has been recently criticized by Puget and Schatzman,[42] who assumed mistakenly that we were giving the 2.7°K field's origin by distant sources with $z \geqslant 0.5$, as was recently attempted by Rowan-Robinson.[43] This view is definitely wrong and we shall now demonstrate that is is possible to construct a model of photon interactions which links the H values and their anisotropy with the local structure of the Universe around us.

POSSIBLE MODEL OF PHOTON-BOSON SCATTERING

Our interpretation rests essentially on the idea that the AR energy loss, associated with anomalous redshifts, is carried away by some new, very light boson φ which interacts strongly with the incident photon γ^T. Indeed, if one then accepts the de Broglie phase-correlation principle[44] (see Moles and Vigier[45]), one obtains a "massive" photon γ (with a longitudinal component γ^L) and a new, light, "massive," neutral "pseudoscalar" photon φ, with $m_e \gg m_\varphi \gg m_\gamma$, where m_e denotes the electron mass. This opens the possibility for introducing a new possible elastic interaction ($\gamma + \varphi \to \gamma + \varphi$) Hamiltonian which satisfies the main restrictions imposed by experiment, i.e., which implies:

a. A strong forward γ-scattering necessary to explain the pointlike character of distant sources.

b. A constant fractional energy loss $\langle \delta z \rangle_\gamma$ per collision for $10^{10} < \nu < 10^{15}$ Hz (since the observed AR is identical for optical and radio waves).

c. Compatibility with quantum electrodynamics. This means, for example, that the coupling $\gamma + \varphi \to \gamma + \varphi$, which implies $\gamma + \gamma \to \varphi + \varphi$, does not yield unwanted reactions of the type $e^+ + e^- \to \varphi + \varphi$.

d. Compatibility with the observational fact that no line shift has yet been observed when light is passed through strong radiation fields in present laboratory conditions (Weiss and Grodzins[46]).

In order to satisfy condition (a) and obtain a strong forward peak in the scattering cross section, one is immediately tempted to consider interactions of the electromagnetic type, since the exchange of a massless photon results in the familiar drastic infinite forward peak. This has been exploited, for example, in ν-e interactions with a finite neutrino magnetic dipole moment, first considered by Bethe[47] and later discussed by various authors.[48] Let us first recall their result. The general matrix element for such a ν-e interaction is given by

$$M = -\frac{e^2 \kappa}{2m_\nu} (p'_\nu) i\sigma^{\mu\rho} q_\rho (p_\nu) \frac{1}{t - m_\gamma^2} (p') \gamma_\mu | p_e) \tag{1}$$

where κ is a possible neutrino magnetic moment, m_ν and m_γ represent the neutrino and photon mass, respectively, p_ν and p_e (p'_ν and p'_e) are the initial (final) neutrino and electron four-momenta, and $q_\mu = p_\nu - p'_\nu$, where spinors and matrices follow the usual convention of Bjorken and Drell. The resulting differential cross section has the form

$$\frac{d\sigma}{dT}(T,E) = 4\pi \left(\frac{\alpha\kappa}{2m_\nu}\right)^2 \frac{E - (T - m_\gamma^2)}{E(T - m_\gamma^2)} \tag{2}$$

where α is the fine structure constant, $E(E')$ is the initial (final) neutrino energy, and $t = -Zm_e T = -Zm_e(E - E')$, m_e representing the electron mass.

Of course, this is only valid in principle for nonzero-mass fermions. However, one can evidently generalize this result if one considers spin-0 or spin-1 bosons as even combinations ("fusion" in de Broglie's vocabulary—de Broglie, *La Mécanique ondulatoire du photon*, 1940) of antiparallel or parallel basic spin-1/2 components, which for leptons behave as the heuristic fermionic free quarks (or partons) in the semiheuristic hadron models of the current literature.

More especially, following de Broglie, we shall construct nonzero-mass, spin-0 and spin-1 bosons with the fusion of two nonzero-mass, spin-1/2 fermions, represented by two four-component Dirac spinors u_α and v_α. This means that we introduce 16-component wave functions $\Phi = u_\alpha v_\beta$, with α, $\beta = 1, 2, 3, 4$. They satisfy the relations

$$[p_\mu \Gamma^\mu i m_0 c] \Phi = 0$$

where the p_μ represent the usual four-momentum operators and the Γ^μ are 16-component matrices that satisfy the relations

$$\Gamma^\mu \Gamma^\nu \Gamma^\rho + \Gamma^\rho \Gamma^\nu \Gamma^\mu = \delta^{\mu\nu}\Gamma^\rho + \delta^{\nu\rho}\Gamma^\mu$$

in de Broglie's representation, i.e., with the matrix elements

$$(\Gamma^\mu)_{i_1 i_2 ; l_1 l_2} = [(\gamma^\mu)_{i_1 l_1}\delta_{i_2 l_2} + \delta_{i_1 l_1}(\gamma^\mu)_{i_2 l_2}] \equiv \tfrac{1}{2}[\gamma^\mu_{(1)}\delta_{(2)} + \gamma^\mu_{(2)}\delta_{(1)}]$$

where $\gamma^\mu_{(1)}$ and $\gamma^\mu_{(2)}$ are two independent systems of Dirac matrices satisfying

$$\gamma^\mu_{(r)}\gamma^\nu_{(s)} = \gamma^\nu_{(s)}\gamma^\mu_{(r)}; \quad r,s = 1,2, \ r \neq s$$

One can easily show that de Broglie's representation yields wave equations for $aJ = 0$ pseudoscalar (φ) and $aJ = 1$ vector boson (γ) simultaneously. Indeed, the matrix elements of the square of the angular momentum operator $S_{\mu\nu}$ can be written in the form

$$(S^2)_{i_1,i_2;m_1,m_2} = [\delta_{i_1 m_1}\delta_{i_2 m_2} + \delta_{i_1 m_2} + \delta_{i_2 m_1}]$$

and yields

$$S^2\Phi = \begin{cases} 2\Phi & \text{for} \quad \Phi_{i_1,i_2} = \Phi_{(i_1,i_2)}\text{(symmetric)} \equiv \Phi_S \\ 0 & \text{for} \quad \Phi_{i_1,i_2} = \Phi_{[i_1,i_2]}\text{(antisymmetric)} = \Phi_A \end{cases}$$

This means that φ corresponds to the *antisymmetric* part of Φ and γ corresponds to the *symmetric* part of Φ.

One then defines the projection operators

$$\eta^{\pm}_{i_1, i_2; l_1, l_2} = \tfrac{1}{2} [\delta_{i_1 i_1} \delta_{i_2 i_2} \pm \delta_{i_2 l_1} \delta_{i_2 l_2}]$$

with $(\eta^{\pm})^2 = \eta^{\pm}$ and $\eta^{+}\eta^{-} = \eta^{-}\eta^{+} = 0$, which yields

$$\eta^{\pm}\Phi = \Phi_S \equiv \Phi_{\gamma} \quad \text{and} \quad \eta^{-}\Phi = \Phi_A \equiv \Phi_{\varphi}$$

which satisfy two independent wave equations for $J = 0$ and $J = 1$:

$$[\tfrac{1}{2} p_{\mu}(\gamma^{\mu}_{(1)} + \gamma^{\mu}_{(2)}) + iMc] \Phi_A = 0 \quad \text{for } \varphi \text{ bosons}$$

$$[\tfrac{1}{2} p_{\mu}(\gamma^{\mu}_{(1)} + \gamma^{\mu}_{(2)}) + imc] \Phi_S = 0 \quad \text{for } \gamma \text{ bosons}$$

For φ, the angular momentum can be written (Φ_{φ} representing the wave function of the φ particle)

$$\overline{\Phi}_{\varphi}(S_{\mu\nu})_{\varphi}\Phi_{\varphi} = \overline{\Phi}_A S_{\mu\nu}\overline{\Phi}_A = \overline{\Phi}\overline{\eta}^{-} S_{\mu\nu}\eta^{-}\Phi$$

with $\eta^{-}S_{\mu\nu}\eta^{-} = S_{\mu\nu}\eta^{-} \equiv S^{-}_{\mu\nu}$, i.e.,

$$\overline{\Phi}_{\gamma}(S_{\mu\nu})_{\varphi} = \overline{\Phi}S^{-}_{\mu\nu}\Phi$$

In the same way we get, with $(\Gamma^{\mu})^{+} = \Gamma^{\mu}\eta^{+}$, the relation

$$\overline{\Phi}_{\gamma}\Gamma_{\mu}\Phi_{\gamma} = \Phi\Gamma^{+}_{\mu}\Phi$$

where Φ_{γ} represents the γ particle.

Let us now introduce a γ–φ interaction represented by the interaction Lagrangian

$$\mathcal{L}_{\text{int}} = -g_0 : (\Phi_{\varphi}(S^{\mu\nu})_{\varphi}\Phi_{\gamma}) (\overline{U}(S_{\mu\nu})_W W) : -g_1 : (\overline{\Phi}_{\gamma}(\Gamma_{\alpha})_{\gamma}\Phi_{\gamma}) (\overline{U}(\Gamma_{\alpha})_W) : + \text{h.c.}$$

where W stands for an intermediate particle and $U = W(0)$ for its annihilation state.

Its propagator can be written

$$\overline{U}[S^{\mu\nu} T_F(q)\Gamma_{\alpha} + T_F(-q)S^{\mu\nu}] U \equiv A^{\mu\nu}_{\alpha}$$

where

$$T_F(q) = \frac{1}{\Gamma^{\mu}q_{\mu} - M_W + i\epsilon}$$

After a rather simple calculation (Popovic, 1975), one obtains

$$A_{\mu\nu\alpha} = \frac{1}{t - M^2_W} (g_{\mu\nu}q_{\nu} - g_{\alpha\nu}q_{\mu})$$

The effective Hamiltonian corresponding to $\mathcal{L}_{\text{int}}$ is

$$H_{\text{eff}} = \tfrac{1}{2} g_0 g_1 (\overline{\Phi}S^{-}_{\mu\nu}\Phi)A^{\mu\nu\alpha}(\overline{\Phi}\Gamma^{+}_{\alpha}\Phi) \tag{3}$$

Writing, as before,

$$\Phi_\varphi = u_1 v_1, \quad \overline{\Phi}_\varphi = \bar{v}_1 \bar{u}_1; \quad \Phi_\gamma = u_2 v_2, \quad \overline{\Phi}_\gamma = \bar{v}_2 \bar{u}_2$$

we get

$$\overline{\Phi} S_{\mu\nu}^- \Phi = \bar{v}_1 \sigma_{\mu\nu} v_1 + \bar{u}_1 \sigma_{\mu\nu} u_1$$

where u_1 and v_1 $\frac{1}{2}$ spins are antiparallel. For γ one obtains, in a similar way,

$$\overline{\Phi} \Gamma_\alpha^+ \Phi = \bar{u}_2 \gamma_\alpha u_2 + \bar{v}_2 \gamma_\alpha v_2$$

with parallel u_2 and v_2 $\frac{1}{2}$ spins. The total Hamiltonian then becomes

$$H_{\text{eff}} = g_0 g_1 (\bar{u}_1 \sigma_{\mu\nu} u_1 + \bar{v}_1 \sigma_{\mu\nu} v_1) \frac{q^\nu}{t - M_W^2} (\bar{u}_2 \gamma_\mu u_2 + \bar{v}_2 \gamma_\mu v_2) \qquad (4)$$

which, as expected, reduces to a sum of two independent interactions of type (1). The q_ν in (4) is twice the energy of one of these individual processes. The spin of W has been taken to be 1 (in each process), but since we have two exchanges with opposite W spins, the total interaction process amounts to the exchange of a scalar particle of the φ type, i.e., it can be written in the symbolic form $\varphi + \gamma \to \varphi \to \varphi + \gamma$, which is not built from current–current interactions.

We now turn to C conservation. If the exchanged particle is a φ, then C must be $+1$. This forbids S and U channels,

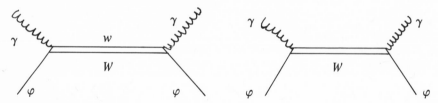

but allows the t channel,

and a very strong $\Phi\Phi\Phi$ coupling. This coupling is necessary in our scheme in order to isotropize the 2.7°K radiation field.

Now the result for the cross section reads (see Clark and Pedigo[48])

$$\frac{d\sigma}{dT} = \kappa \frac{E - T}{ET} \qquad (5)$$

where κ is a constant. This satisfies condition (a), since it enhances tremendously the forward scattering for $T \to 0$. Indeed, if one works in the φ rest system, the

γ^T are still relativistic for the frequency band $10^{10} \leqslant \nu \leqslant 10^{15}$ Hz. This can be justified by the fact that the known experimental limit[51] for m_γ is $m_\gamma \ll 10^{-48}$ g and from the assumption $m_\varphi \gg m_\gamma$. We find that in the rest frame of the φ, $R'(V_\varphi = 0)$, the γ velocity can be written ($h = c = 1$) in the form

$$V' \cong 1 - 2 \left(\frac{m_\gamma E_\varphi}{m_\varphi E_\gamma} \right)^2$$

i.e., the γ is still relativistic if $(E_\varphi m_\gamma / E_\gamma m_\varphi)^2 \ll \frac{1}{2}$. In this case, denoting by θ the deflection angle per collision, one gets

$$\langle \delta z \rangle = \langle \delta \nu / \nu \rangle \cong E_\gamma \theta^2 / 2 m_\varphi$$

with $z_t = N \delta z$ and $\theta_t = \theta \sqrt{N}$ for N collisions, i.e., $z_t = E \theta_t^2 / 2m$. This yields a numerical limit on m_φ, since Wardle and Miley[52] have measured an upper limit on the angular dimension ($\sim 10^{-5}$) for various quasars with $z_t = 2\text{-}3$ at a wavelength of 3 cm, i.e., $E_\gamma \cong 6 \times 10^{-17}$ erg. As a consequence, $m_\varphi \leqslant 0.5 \times 10^{-47}$ g. If one accepts $\langle \delta z \rangle \cong 2 \times 10^{-9}$, as necessary for the interpretation of the Pioneer-6 data,[53] one finds $\langle \delta \theta \rangle_\gamma \leqslant 10^{-11}$ rad per collision.

On the other hand, (5) satisfies condition (b), since we get for the average fractional energy $\langle T \rangle / E$ per collision the value

$$\frac{\langle T \rangle}{E} = \frac{1}{\sigma_t} \int \frac{T}{E} \frac{d\sigma}{dT} dT = \frac{\kappa}{2\sigma_t} \tag{6}$$

i.e.,

$$(\langle T \rangle / E) \sigma_t = \kappa / 2 \tag{7}$$

From the Pioneer-6 data, $\langle T \rangle / E \cong 2 \times 10^{-9}$, $\kappa / 2 \sim 2 \times 10^{-30}$ cm, and $\sigma_t \cong 10$.

The energy loss per unit of length for a γ^T crossing a φ "bath" can thus be written in the form

$$dE/E = -p_\varphi(r) \, dr \, \sigma_t \langle T \rangle / E \tag{8}$$

where $p_\varphi(r)$ represents the φ-particle density along the incident γ^T path.

From (7) and (8), one obtains

$$dE/E = \delta \nu / \nu = -\tfrac{1}{2} \kappa p_\varphi(r) \, dr \tag{9}$$

which yields the redshift law

$$1 + z = \exp \left[\tfrac{1}{2} \kappa \int_0^r p_\varphi(r) \, dr \right] \tag{10}$$

which is usually approximated by its linear development, i.e., by Hubble's law.

Relation (9) shows that, in this model, the fractional energy loss is practically independent of the initial frequency along the path of the γ^T.

Conditions (c) and (d) can now be shown to be satisfied. First, the new φ's do not interact with charged matter directly. Moreover, the discrete quantum character of the γ–φ collisions would not show up[11] in the Weiss–Grodzin type of experiment.

ASTROPHYSICAL AND COSMOLOGICAL CONSEQUENCES

Four astrophysical and cosmological consequences follow from the information given above.

1. When an incident γ crosses a γ-radiation field (with a density n_γ) during a time t, this γ density is necessarily linked with a φ density n_φ as a possible inelastic γ–φ coupling such as $\varphi \to \gamma + \gamma + \gamma$ or $\gamma + \varphi \to n_1\gamma + n_2\varphi$. We thus observe an anomalous redshift $\delta\nu/\nu = tH$, where H is the constant $H = n_\varphi\sigma_{\gamma-\varphi}\langle\delta z\rangle_\gamma c$. In a quasiequilibrium state (which we can assume around strong γ-radiating sources), where $n_\varphi \sim \frac{1}{2}n_\gamma$, we get $H = AT^3 c$; T is the "equivalent γ-blackbody" temperature and $A \simeq 1.85 \times 10^{-29}$ deg^{-3} cm^{-1}, if we want to interpret the Pioneer-6 data[11] as a typically accurate AR instance.

2. In the vicinity of any extended γ-field source (such as our galactic cluster), we obtain an isotropic φ distribution provided we assume that $\langle\delta\theta\rangle_\varphi$ associated with $\varphi + \varphi \to \varphi + \varphi$ is greater than $\langle\delta\theta\rangle \leqslant 10^{-11}$ rad in $\gamma + \varphi \to \gamma + \varphi$. Of course, if we focus our attention only on making the γ's angularly isotropic around the 3°K radiation field via the γ–φ interaction, this would only occur on a distance L satisfying $L\sigma_{\gamma-\varphi}(l/z)n_\gamma\langle\theta\rangle_\gamma \sim l$, i.e., $L \sim 10^{45}$ cm. But if we bear in mind that $\langle\delta\theta\rangle_{\varphi-\varphi}$ is $\sim\pi/2$ and utilize the known result[47] that if one injects a thermodynamic system into another system in thermodynamic equilibrium, the compound system reaches the latter's equilibrium, then we realize that— provided we can make the φ's locally isotropic around 3°K—the γ's will also be made isotropic in the same domain.

Let us now consider an interaction of the type $H = \Lambda\Phi_\varphi^4\Delta_\phi$ corresponding to $\varphi + \varphi \to \varphi + \varphi + \varphi$; it can be shown[48] that for $\langle\delta\theta\rangle_{\varphi-\varphi} \sim 10^{-3}$ rad[49] we have $L \sim 10$ Mpc. We thus obtain an isotropic φ distribution as a consequence of the strong $\langle\delta\theta\rangle_{\varphi-\varphi}$, i.e., an isotropic distribution of the associated γ's, which are in quasiequilibrium with this φ distribution. We assume here that this essentially local γ bath is responsible for the $T_3 = 2.7$°K radiation, which now results essentially from our own galaxy or possibly from the local cluster. An attempted detailed theoretical demonstration will be published subsequently; we wish to emphasize here that the observation of Collins et al.[50] provides experimental support for this assumption.

The γ "jets" observed (with the astonishingly high experimental cross section $\sigma \sim 3 \times 10^{-25}$ cm^2) can indeed be interpreted as reflecting possible inelastic γ–φ interactions of the type $\gamma + \varphi \to \gamma + \varphi + \gamma + \gamma$. A combination of this type of process would soon equilibrate the γ and φ distributions.

3. Supposing that the 2.7°K γ-temperature distribution (labeled T_3) is essentially a local fluctuation in the overall T distribution in space, we shall now try to evaluate T_2 in the local supercluster and T_1 between superclusters from

measured values of the Hubble constant. Indeed, we are now in the position to write (see consequence 1 above) the Hubble law in the general form

$$v/c = \overline{H}L/c = \sum_i AT_i^3 L_i$$

which implies a clustered and hierarchical type of universe.

In consequence, we can now readily explain Rubin's result. The difference of velocities between the two regions is $\Delta v = v_2 - v_1 = 1465 \pm 282$ km/sec for galaxies at a maximum distance $L = v/H_1 = 96$ Mpc. Our assumptions, applied to the observed directions, then yield $v_2 = H_1(L - l_z) + H_2 l_z$, where l_z is the characteristic size of the LSC (see Fig. 3), i.e., $l_z \sim 30$ Mpc. This gives $\Delta v \cong 1350$ km/sec, in excellent agreement with the value measured by Rubin.

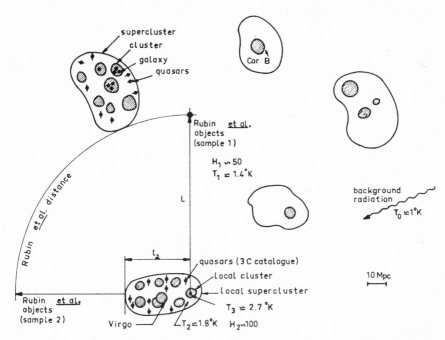

Fig. 3. The local aspect of the hierarchical universe (schematic). We distinguish, from large to small structures, superclusters of galaxies, clusters of galaxies, and finally galaxies and quasars. The background radiation, corresponding possibly to a still larger type of structure, is $T_0 \cong 1°$K. Outside superclusters, one has $T_1 \cong 1.4°$K and $H_1 \cong 50$. Within superclusters (such as our own Local Supercluster, LSC) one has $T_2 \cong 1.8°$K and $H_2 = 100$. Inside the local cluster, or near our own galaxy, $T_3 = 2.7°$K, and H escapes measurement, because of local motions. The Coma and Cor B clusters have led to the measurement of H_1, together with Sc moderately distant galaxies. H_2 has been measured within the LSC. These values are compatible with the measurements by Rubin of different redshifts for galaxies of the same magnitude, some in the direction of the LSC center, some in a perpendicular direction. The observed quasars are assumed to be affected by anomalous redshifts. They are "local," i.e., mostly localized in our LSC.

The above model also yields immediately $1.6 \leqslant T_2 \leqslant 1.9$ for $H_2 \sim 100$ and $1.3 \leqslant T_1 \leqslant 1.7$ for $H_1 \sim 50$.

We observe locally, as stated earlier, a temperature of $2.7°K$. For the time being, this temperature T_3 may be regarded as a result of the local, isotropic concentration of φ particles. It is easy to demonstrate—with reasonable estimates of the length L_3 that appears in the preceding general form of Hubble's law—that this local field of photons does not affect the interpretation of Rubin's results, since the contribution of the term $AT_3^3 L_3$ is less than 3% of the contribution of the two other terms. We further note that internal dispersion of velocities within the local cluster forbids, as is well known, the determination of any local value H_3 of H, which should, in principle at least, correspond to the $2.7°K$ radiation.

4. A background-blackbody radiation field $T_0 \sim 1°K$ in deep space can now be justified independently of any expansion assumption. Our AR mechanism—the "tired light" version of our theory—leads to a maximum distance D_b at which the galaxies have such a large apparent angular diameter (due to successive photon deflections on the φ bath corresponding to the T_0 distribution) that they constitute a uniform background. We start from the customary estimate that matter contained in galaxies corresponds to a density $\sim 5 \times 10^{-31}$ g/cm^3 and that galactic masses vary in the range $(10^{10}\text{-}10^{11})M_\odot$, i.e., from 2×10^{43} to 2×10^{44} g.

This implies an average of one galaxy in a volume of 0.3×10^{74} to 10^{75} cm^3 and corresponds to an average intergalactic distance of $R = 3 \times 10^{25}$ cm, i.e., $R \cong 10$ Mpc. The average angular separation on our continuous background becomes $\alpha \cong 10/D_b$, if D_b is expressed in Mpc. Using the value $\langle \delta\theta \rangle_\gamma \sim 10^{-13}$ rad, we see, on the other hand, that the angular diameter of such a background galaxy blurred by our γ collisions is equal to $\beta \leqslant 10^{-13}\sqrt{n}$, where n denotes the average number of collisions in the distance D_b, i.e., $n = D_b H/c\langle \delta z \rangle = 50/3 \times 10^5 \times 2 \times 10^{-9} D_b$. We thus obtain $\beta \leqslant 3 \times 10^{-11} D_b^{1/2}$, and can safely consider that, when $\beta = \alpha$, the blurring of galaxies creates a uniform isotropic background of radiation diluted (by a factor s) at a distance D_b. This background radiates like an average galaxy of infinite optical depth. Of course, this is only an approximation, and it will be necessary to obtain a complete solution of the transfer problem of the radiation coming from distances $>D_b$ and to utilize a more realistic distribution of the light sources in this computation. However, writing $\alpha = \beta$ already yields $D_b \cong 6 \times 10^7$ Mpc, which corresponds to a background redshift $z \cong 10^4$ far beyond our present possibilities of observation. Assuming that the "average" galactic sources are blackbodies with $T \sim 10^4$ °K, their displaced spectrum creates (as a consequence of the value $s \sim 10^4$ deduced from the actual average size and separation of galaxies) in the observable Universe a background quasiblackbody radiation of $T_0 \sim 1°K$ (see Fig. 4). Of course, if we take into account the importance of possible local fluctuations of this background temperature and the rough nature of this calculation, this rather

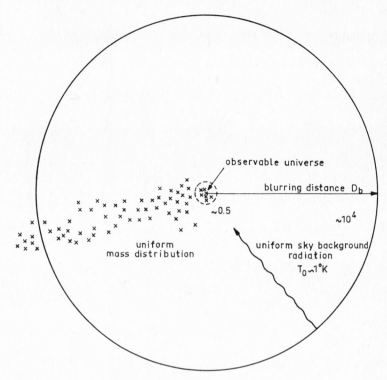

Fig. 4. General concept of the Universe (a working hypothesis). The Universe is infinite in time, locally fluctuating, and statistically stable. The background radiation comes from red-shifted light of galaxies that, because of their blurring by the cumulative effect of photon interactions, appear like a radiating continuous wall at about $z = 10^4$, or 5×10^7 Mpc. The Universe is, at large, uniform and homogeneous. Evolutionary processes, statistically, do not affect the structure or the observed parameters.

satisfactory result is only indicative of a possible reasonable agreement of our γ-φ process with the values deduced from the observed H values. It does not completely rule out some additional expansion of a much smaller amplitude than usually accepted, though this is no longer necessary in our model.

CONCLUSION

To summarize, our model implies the following results and hypotheses.

(a) The radius of the universe (if it exists!) is $\gg D_b$, which implies a very low large-scale density ($\sim 10^{-35}$ g/cm^3) despite the observed small-scale density $\sim 5 \times 10^{-31}$ g/cm^3. Our photon interactions blur (for any observer) all galaxies more distant than $D_b \sim 5 \times 10^7$ Mpc; this is responsible for the reddening. As in the expansion theory, the large-scale value of D_b of our model resolves Olbers' paradox when we introduce a finite γ mean free path in the universe, $d \gg D_b$, i.e., it explains why "the visible sky is dark at night." It further explains the minimum possible value of $H \cong 50$ km sec^{-1} Mpc^{-1}.

(b) Relatively strong density fluctuations in time and space and a statistical steady state obtain. The hierarchy of fluctuations (Fig. 4) explains the average value of H and its fluctuations, as well as the relatively high local value $T_3 \sim 2.7°$K. Note that T could be much higher near a QSS. There is indeed a close correlation between the mass density distribution (see Fig. 3) and the γ-density distribution; it is understood that the latter is smoothed out by the strong scattering angles of the associated φ's.

(c) Objects which radiate a large number of γ's (and thus of φ's), such as the QSSs, display strong ARs as source effects. This would justify Arp's assumption[1] that some 3C QSSs belong to the local supercluster.

Of course, we are quite conscious of the fact that these working hypotheses raise many unsolved questions. First, we need much more experimental support and less ambiguity in the interaction procedure than is permissible at present. Second, we must account for the chemical composition of the observed Universe—a fact which fits relatively well into the so-called Big-Bang theory. We notice, however, that there are also many points of obscurity, and that there is a need for refinement, in the classical expanding Universe model: How is it possible to explain within its frame the observed anisotropy of the Hubble constant?

REFERENCES

1. H. Arp, *Science* **174**, 1189 (1971).
2. H. Arp, Invited summary paper, *IAU Symp*. No. 58 (1973).
3. S. Collin–Souffrin, J. C. Pecker, and H. Tovmassian, *Astron. Astrophys.* **30**(2), 351 (1974); L. Botinelli and L. Gougenheim, *Astron. Astrophys.* **26**, 85 (1971).
4. H. Arp, *Astr. Lett.* **7**, 221 (1971).
5. R. Walker, Private communication (1974).
6. H. Arp, Am. Astron. Soc. meeting, review paper, according to P. W. Hodge, *Sky Telesc.* **44**, 23 (1972).
7. C. Balkowski, L. Botinelli, P. Chamaraux, L. Gougenheim, and J. Heidmann, *Astron. Astrophys.* **25** 319 (1973).
8. J. Heard and J. D. Fernie, *J. Roy. Astron. Soc. Can.* **62**, 99 (1968); J. D. Fernie, *Astrophys. J.* **183**, 583 (1973).
9. L. V. Kuhi, J. C. Pecker, and J. P. Vigier, *Astron. Astrophys.* **32**, 111 (1974).
10. F. Roddier, *Ann. Astrophys.* **28**, 478 (1964); D. Sadeh, S. H. Knowles, and B. S. Yapice, *Science* **159**, 307 (1968); D. Sadeh, S. H. Knowles, and B. Au, *Science* **161**, 567 (1968); D. F. Dickinson, A. E. Lilley, H. Penfield, and I. I. Shapiro, *Science* **167**, 1755 (1970); R. M. Goldstein, *Science* **166**, 598 (1969).
11. P. Merat, J. C. Pecker, and J.-P. Vigier, *Astron. Astrophys.* **30**, 167 (1974).
12. P. Merat, J. C. Pecker, J.-P. Vigier, and W. Yourgrau, *Astron. Astrophys.* **32**, 471 (1974).
13. A. Chastel and J. F. Heyvaerts, *Nature* **249**, 21 (1974); R. Aldrovandi, S. Caser, and R. Omnès, *Nature* **241**, 340 (1973).

14. R. L. Cohen and G. K. Wertheim, *Nature* **241**, 109 (1973).
15. H. Chew, *Nature (Phys. Sci.)* **242**, 5 (1973).
16. D. H. Weinstein and T. Keeney, *Nature* **247**, 140 (1974).
17. J. C. Pecker, A. P. Roberts, and J.-P. Vigier, *Nature* **237**, 227 (1972).
18. G. Burbidge, *Nature* **233**, 36 (1971).
19. E. P. Hubble, *Proc. Nat. Acad. Sci. U.S.* **15**, 168 (1929).
20. E. P. Hubble, *Proc. Nat. Acad. Sci. U.S.* **22**, 621 (1936).
21. G. de Vaucouleurs, *Ap. J.* **63**, 253 (1958).
22. G. de Vaucouleurs, *Ap. J.* **69**, 737 (1964).
23. G. de Vaucouleurs, in *Proc. of Galileo Conference*, ed. by L. Rosino (G. Barbera, Florence; Pub. Dept. Astron. Univ. Texas 1, No. 10, 1966), Vol. 2, T. 3, p. 37.
24. G. de Vaucouleurs, in *IAU Symp. No. 44*, ed. by D. S. Evans (D. Reidel, New York, 1972), p. 353.
25. H. Shapley and A. Ames, *Harvard Ann.* **88**, No. 2 (1932).
26. A. Reiz, *Lund. Obs. Ann.* **1941**, No. 9.
27. G. de Vaucouleurs, *Vistas in Astronomy* **2**, 1584 (1956).
28. J. F. Nicoll and I. E. Segal, Massachusetts Institute of Technology Preprint.
29. G. A. Tammann and A. Sandage, Hale Observatories, Carnegie Institution of Washington, California Institute of Technology Preprint.
30. S. van den Bergh, *Nature* **225**, 503 (1970).
31. J. Heidmann, *C. R. Acad. Sci. Paris* **271B**, 658 (1970).
32. A. Sandage, *Quart. J. RAS* **13**, 282 (1972).
33. G. O. Abell and S. Eastmond, *Ap. J.* **75s**, 161 (1968).
34. V. C. Rubin, W. K. Ford, Jr., and J. S. Rubin *Ap. J.* **183L**, 111 (1973).
35. H. Karoji and L. Nottale, unpublished.
36. T. Jaakkola, H. Karoji, M. Moles, and J.-P. Vigier, *Nature* **256**, 24 (1975).
37. G. Colla, C. Fanti, F. Fanti, I. Gioia, C. Lari, J. Lequeux, R. Lucas, M. H. Ulrich, *Astron. Astrophys. Suppl.* **20**, 1 (1975).
38. E. Hubble and R. C. Tolman, *Astrophys. J.* **82**, 302 (1935).
39. V. C. Rubin, W. K. Ford, J. S. Rubin, unpublished.
40. M. J. Haggerty and J. R. Wertz, *Monthly Not. Roy. Astron. Soc.* **155**, 495 (1972).
41. F. D. A. Hartwick, *Ap. J. Lett.* **195L**, 7 (1975).
42. J. L. Puget and E. Schatzman, *Astron. Astrophys.* **32**, 477 (1974).
43. M. Rowan-Robinson, *Monthly Not. Roy. Astron. Soc.* **168**, 45 (1974).
44. L. de Broglie, *C. R. Acad. Sci. Paris* **277B**, 71 (1973).
45. M. Moles and J.-P. Vigier, *C. R. Acad. Sci. Paris* **278B**, 969 (1974).
46. R. Weiss and L. Grodzins, *Phy. Lett.* **1**, 342 (1962).
47. H. A. Bethe, *Proc. Camb. Phil. Soc.* **31**, 108 (1935).
48. R. B. Clark and R. D. Pedigo, *Phys. Rev. D* **8**(7), 2261 (1973).
49. W. Band and J. L. Park, *Int. J. Theor. Phys.* **9**(6) 415 (1964).
50. G. B. Collins, J. R. Ficenek, D. M. Stevens, W. P. Trower, and J. Fisher, *Phys. Rev.* **8**, 982 (1973).
51. S. Goldhaber and M. M. Nieto, *Rev. Mod. Phys.* **43**, 277 (1971).
52. J. F. C. Wardle and G. K. Miley, *Astron. Astrophys.* **30**, 305 (1974).
53. P. Merat, J. C. Pecker, and J.-P. Vigier, *Astron. Astrophys.* **30**, 167 (1974).

Chapter XII

The Role of Time in Cosmology

Gerald J. Whitrow

Imperial College of Science and Technology

The history of cosmology since the time of the ancient Greeks has been successively dominated by three fundamental analogies concerning the general nature
of the Universe:

(1) The analogy between macrocosm and microcosm, which originated in
antiquity and influenced thought down to the sixteenth century, based on the
belief that 'mind' is the source of the orderliness in nature which makes natural
science possible, the world being a kind of organism, both alive and intelligent.

(2) The mechanistic analogy, which originated in the later middle ages and
renaissance and was only finally discarded in the course of the present century,
that the Universe is like a machine, being itself neither alive nor intelligent but
created by an external intelligence, just as machines have been created by the
agency of man.

(3) The historical analogy, which originated in the eighteenth century, that
the world is neither an organism nor a machine but more like human society,
being an aggregate of individual constituents or processes with a history.

The first analogy gave rise to the general idea of *the teleological Universe*,
primarily associated with the name of Aristotle, the second to *the clocklike
Newtonian Universe*, and the third to *the evolutionary Universe* of modern
science. Each of these analogies can be associated with a different view of the
nature and significance of *time*. It is the object of the present paper to examine
the role of time in the general development of cosmology, as an aid to understanding how our present world-picture has come about.

THE TELEOLOGICAL UNIVERSE

In Greek cosmology the concept of time was not of primary importance. Instead, the tendency was to regard the temporal aspects of phenomena as sub-

ordinate to the permanent. Even Heraclitus, who regarded change as the very essence of reality, adopted a concept of transmutation which involved simultaneous processes of creation and decay that maintained a permanent and not a progressive order of the cosmos. Similarly, Anaximander, who has been described as the author of "the first scientific cosmogony"[1] and as responsible for "the first theory of organic evolution,"[2] appears to have believed in the continual generation and destruction of worlds as a permanent process without beginning or end. Indeed, his conception of the cosmic process was not evolutionary at all. Instead, it involved the cyclic alternation of opposites, and the concept of time was associated with the idea of "justice" or balance rather than progress and with the theory of primary opposites rather than a single linear variable. The whole conception was no doubt suggested by the cycle of the seasons with its alternating conflict of the hot and the cold, the wet and the dry. Each of these advances in 'unjust' agression at the expense of its opposite and then pays the penalty, retreating before the counterattack of the latter, the object of the whole cycle being to maintain the balance of justice.

This concept of nature as a continual strife of opposite powers subject to the ordinance of time was submitted to penetrating criticism by Parmenides, who had a considerable influence on Plato, in whose cosmological dialogue the *Timaeus* time was merely a feature of the visible order of things based on an archetype ('Eternity') of which it is the 'moving image.' The essentially cyclical nature of change was stressed by Plato's pupil Aristotle, but he did not believe in Plato's conception of a transcendental timeless world of ideal forms. For him the proper object of scientific investigation was the empirical world revealed to us by the senses. Although he regarded motion and change as the essential characteristics of the physical world, his primary concern was with the states before and after motion rather than with the dynamic process itself.[3] For him, form and place were more fundamental than time. Scientific investigation was the quest for 'essences,' that is, for the underlying permanent qualities, or essential natures, of things. Aristotle's teleological outlook, with its emphasis on purpose-conditions, meant that time played a far more subsidiary role than in modern science, where we think instead of invariable *sequences* rather than essences and investigate laws of nature which determine the development of systems in time from given *initial conditions*.

The theory of final causes was the Aristotelian doctrine principally selected for attack by the advocates of the new experimental philosophy in the seventeenth century. Teleology, in Bacon's famous gibe, came to be regarded like a virgin consecrated to God, since it produces no offspring. For it was increasingly felt that an attempt to account for the production of a particular effect by saying that there was a natural tendency for it to be produced told one nothing. Instead, attention was directed to the role of efficient causes. A major influence on this revolution in scientific thought was a new attitude toward the concept of time and a greater awareness of its significance in daily life that can be traced back to the later middle ages.

THE CLOCKLIKE UNIVERSE

Until the fourteenth century A.D. the most reliable way to tell the time was by means of a sundial. The only mechanical time-recorders in antiquity were water-clocks, but although often of considerable complexity they were not at all accurate. This lack of precise measurement must have been an important factor in the failure of the Greeks to realize the potential significance of the metrical concept of time.

The fundamental distinction between water-clocks and mechanical clocks, in the strict sense of the term, is that the former involve a continuous process (the flow of water through an orifice), whereas the latter are governed by a mechanical motion that continually repeats itself. The mechanical clock, in this sense, appears to have been a European invention of the late thirteenth century or early fourteenth century.

Until recently, it had been generally supposed that the first truly mechanical escapement was the verge and foliot type found in various church clocks throughout Europe, but North[4] has shown that the earliest such escapement of which we have certain knowledge, namely that of the St. Albans clock designed by Richard of Wallingford, ca. 1328, was different. It was an oscillating mechanism involving an extra wheel, as compared with the common verge and crown wheel escapement. North, who has brilliantly reconstructed the St. Albans escapement from the purely verbal description given in the surviving manuscript, has since found that a similar escapement was known more than a century and half later to Leonardo da Vinci, but Leonardo can no longer be regarded as its inventor. (Edward III blamed Richard of Wallingford for spending more money on his clock than on the care of his church!)

During the fourteenth century a tremendous craze developed for the construction of elaborate astronomical clocks. As the historian Lynn White has remarked,[5] "No European community felt able to hold up its head unless in its midst the planets wheeled in cycles and epicycles, while angels trumpeted, cocks crew, and apostles, kings and prophets marched and countermarched at the booming of the hours." And he continues,[5] "These new great astronomical clocks were presented frankly as astronomical marvels, and the public delighted in them as such. This in itself indicates a shift in the values of European society."

Despite the practical difficulties of precise time measurement, the abstract framework of mathematically divided time gradually became the new medium of daily existence, so that, as Lewis Mumford has stressed, "Eternity ceased gradually to serve as the measure and focus of human actions" (Mumford,[6] p. 14). Mathematical time was also destined to become the medium for man's conception of the universe, but this only came about very slowly. For, despite the influence of Christianity, with its emphasis on the Incarnation as a unique event, not subject to repetition, the theory of cycles and of astral influences was accepted by most Christian thinkers until the seventeenth century. In their world-view, time was not conceived as a continuous mathematical parameter, but was split up into separate seasons, divisions of the zodiac, and so on, each exerting its specific influence. That was why, for example, down to the seventeenth century medical students were required to study astronomy and astrology.

One of the great mental innovations that were foreshadowed in the fourteenth century was the idea that the physical Universe is a kind of machine. The origins of this idea can be traced to the surprisingly elaborate astronomical clocks that were, as we have seen, designed and produced during that century. One of the most famous was the astrarium of Giovanni de' Dondi of Padua, designed between 1348 and 1364. De' Dondi's complex instrument was only incidentally a timepiece. Primarily it was a mechanical representation of the universe.[7] It was much more elaborate than the first of the famous series of astronomical clocks in Strasbourg Cathedral that was installed about the same time, 1350. The original Strasbourg clock probably contained, besides moving figures, an annual calendar dial, and possibly a lunar dial and an astrolabe; but the instrument designed by de' Dondi incorporated a perpetual calendar for all religious feasts both fixed and movable, and also indicated the celestial motions of the Sun, Moon, and planets, including even the motion of the nodes of the Moon's orbit, which take over eighteen years to make a complete revolution around the ecliptic.

The invention of clockwork and its application to mechanical models of the Universe, such as de' Dondi's, made a powerful impact on men's minds. Although, as a rule, medieval scholars were not concerned with machines, they became more and more interested in clocks, particularly because of their connection with astronomy, since it was generally believed that a correct knowledge of the relative positions of the heavenly bodies was necessary for the success of most earthly activities. A particularly interesting indication of the way in which the invention of clockwork began to influence thought occurs in a treatise by de' Dondi's contemporary, Nicole Oresme, on the question of whether the motions of the heavenly bodies are commensurable or incommensurable. Part of the treatise is in the form of an allegorical debate between Arithmetic, who favors commensurability, and Geometry, who upholds the opposite view. Arithmetic argues that incommensurability and irrational proportion would detract from the harmony of the universe. "For if someone should construct a material clock, would he not make all the motions and wheels as nearly commensurable as possible?"[8] This is the earliest instance known to me of the mechanical simulation of the Universe by clockwork suggesting the reciprocal idea that the Universe itself is a clocklike machine.

Nevertheless, it was not until the scientific revolution of the seventeenth century that this idea came to the fore. Early that century Kepler specifically rejected the old quasianimistic magical conception of the Universe and asserted that it was similar to a clock. Among others who drew the same analogy was Robert Boyle. In a passage in which he maintained that the existence of God is not revealed so much by miracles as by the exquisite structure and symmetry of the world—that is, by regularity rather than irregularity—he argued that the universe is not a puppet whose strings have to be pulled now and again but,[9] "it is like a rare clock, such as may be that at Strasbourg, where all things are so skillfully contrived, that the engine being once set a-moving, all things proceed according to the artificer's first design, and the motions . . . do not require the particular interposing of the artificer, or any intelligent agent employed by him, but perform their functions upon particular occasions, by virtue of the general and primitive contrivance of the whole engine."

The analogy between the Universe and the clock has two quite distinct aspects, which I will call 'mechanical' and 'mathematical.' As regards the mechanical analogy, Boyle's words clearly imply a conception of nature from which all traces of the animistic world-view, such as was still evident at the beginning of the seventeenth century in Gilbert's book on the magnet, have been banished. In the development of the mechanistic conception of nature in the course of that century the mechanical clock played a central role. It was surely no coincidence that the greatest practitioner of the mechanical philosophy in its formative period, Christiaan Huygens, who in the first chapter of his *Traité de la Lumière* declared that in true philosophy all natural phenomena are explained *par des raisons de mechanique*, was also responsible for converting the mechanical clock into a precision instrument.

This development was based on Galileo's discovery of a natural periodic process that could be conveniently adapted for the purposes of accurate timekeeping. As a result of much mathematical thinking on experiments with oscillating pendulums, Galileo arrived at the conclusion that each simple pendulum has its own type of vibration depending on its length. In his old age (he died in 1642) he contemplated applying the pendulum to clockwork which could record mechanically the number of swings. The first pendulum clock was constructed by Huygens in 1656, and was based on his discovery that theoretically perfect isochronism could be achieved by compelling the bob to describe a cycloidal arc. Great as was Huygens's achievement, particularly from the point of view of theory, the ultimate practical solution of the problem came only after the invention of a new type of escapement. Huygens' clock incorporated the verge type, but about 1670 a much improved type, the anchor type, was invented that interfered less with the pendulum's free motion.

The invention of a satisfactory mechanical clock had a tremendous influence on the concept of time itself. For, unlike the water-clocks that preceded it, the mechanical clock (if properly regulated) can tick away continually and accurately for years on end, and so must have greatly influenced belief in the homogeneity and continuity of time. The mechanical clock was therefore not only the prototype instrument for the mechanical conception of the Universe, but for the modern idea of time. An even more far-reaching influence has been claimed for it by Lewis Mumford, who has argued that "it dissociated time from human events and helped create belief in an independent world of mathematically measurable sequences: the special world of science" (Mumford,[6] p. 15).

There is a particular aspect of this influence to which I should like to draw attention. I have already suggested that besides the purely mechanical analogy between the Universe and the clock there is what I called a 'mathematical' analogy, which has proved to be of more permanent value. In referring to the famous Strasbourg clock, Boyle said "that the engine being once set a-moving, all things proceed according to the artificer's first design." In the case of a clock, 'design' refers to the action of its mechanism and has no teleological significance. This is in marked contrast to Aristotle's conception of nature. For, as I mentioned earlier, this was based on the importance he attached to *purpose-conditions*, that is, to the natural places, states, or fully developed forms to which, in his view, all things, inanimate as well as animate, aspire. Consequently, for him, substantial or formal *essences*, rather than temporal *sequences*, were the primary object of scientific inquiry. Similarly, it was commonly assumed by medieval scholastic philosophers that each of the properties of a body was due to a special quality inherent in the body. This way of thinking came under fire in the seventeenth century because, as I have said before, it was increasingly felt that it failed to explain anything. Instead of postualting *ad hoc* qualities, scientists who rejected Aristotle and his scholastic followers invoked hypothetical mechanical systems of corpuscles or vortices to elucidate natural phenomena. Insofar as such a system operates according to general laws from given initial conditions rather than teleologically, it has some similarity to a clock, irrespective of its structure. For, if a clock is to read the correct time, its mechanism must not only function properly but the hands must be set correctly beforehand. From this point of view, however, the clock analogy is mathematical rather than mechanical, since it puts primary emphasis on calculating the course which a physical system will follow in time from given initial conditions rather than on providing a complete explanation of how it operates.

This was the method adopted by Newton in the *Principia*, the full title of which specifically refers to the *Mathematical* rather than the *Mechanical* principles of natural philosophy. As he said in one of his letters to Bentley, "Gravity must be caused by an agent acting constantly according to certain laws; but whether this agent be material or immaterial I have left to the consideration of my readers." Unlike his main continental critics, Huygens and Leibniz, Newton was willing—at least in the *Principia*—to by-pass the problem of explaining gravitation mechanistically. Instead, taking time as the independent variable, he formulated mathematical laws of motion and gravitation in terms of which gravitational phenomena could be described and predicted.

The success of Newton's method came to be generally recognized during the course of the eighteenth and nineteenth centuries by its triumphs in celestial mechanics, and led ultimately to the conclusion that the clocklike behavior of the *Universe* is of less significance than that of the *laws* formulated by Newton to investigate it. For, with increasing precision of observation, it became clear that, strictly speaking, we cannot base our measurement of time on the observed motions of any of the heavenly bodies: The Moon's revolutions are not strictly uniform, but are subject to a small secular acceleration; minute irregularities occur in the diurnal rotation of the Earth; and so on. Instead, astronomers came to regard 'an invariable measure of time' simply as a measure that leads to no contradiction between the observations of celestial bodies and the rigorous theory of their motions.[10]

This idea that the ultimate standard of time must be based on our concept of universal laws of nature was explicitly recognized in one of the great classics of Newtonian mechanics, the treatise on *Natural Philosophy* by Thomson and Tait, published 1890. In discussing the law of inertia, they pointed out that it could be stated in the following form: The times during which any particular body not compelled by force to alter the speeds of its motions passes through equal spaces are equal. And in this form, they claimed, the law expresses our convention for measuring time.[11]

Historically, the law of inertia was closely related to the concept of time. It was no accident that the idea of what we may call circular inertia—that is to say, the primacy formerly assigned to circular motion—prevailed when the cyclic conception of time was dominant, and that the modern idea of inertial motion came in with the mathematical concept of time as a linear variable that can be represented geometrically by an infinitely extended straight line.

The idea of time as a mathematical concept that has many analogies with a line was first discussed explicitly in the *Geometrical Lectures* of Isaac Barrow, who was the first occupant of the Lucasian chair of mathematics at Cambridge, in which he was succeeded by Newton in 1669. He argued that,[12] "time has length alone, is similar in all its parts and can be looked upon as constituted from a simple addition of successive instants or as from a continuous flow of one instant; either a straight or a circular line." The reference here to "a circular line" shows that Barrow was not completely emancipated from traditional ideas. Nevertheless, his statement goes further than any of Galileo's, for Galileo only used straight line *segments* to denote particular intervals of time.[13] Barrow was, however, influenced by the success of the kinematic method in geometry that had been developed with great effect by Galileo's pupil Torricelli. Barrow's views on time influenced Newton, whose *Principia* was published in 1687. Newton's views, in turn, made a great impression on the philosopher John Locke. In his *Essay concerning Human Understanding*, published in 1690, we find the clearest statement of the scientific conception of time that was evolved in the seventeenth century:[14] "duration is but as it were the length of one straight line extended *in infinitum*, not capable of multiplicity, variation, or figure, but is one common measure of all existence whatsoever, wherein all things whilst they exist equally partake. For this present moment is common to all things that are now in being, and equally comprehends that part of their existence as much as if they were all but one single being; and we may truly say, they all exist in the same moment of time." And so by the end of the seventeenth century we have arrived at an explicit formulation of the idea of time that dominated physical science until the advent of Einstein's special theory of relativity—an idea that can be baldly summarized in the symbol t denoting the independent variable of classical dynamics.

THE EVOLUTIONARY UNIVERSE

Although the mechanistic clocklike conception of the Universe differed in so many respects from the teleological, and in particular assigned a far greater significance to the role of time, it, too, tended to be originally associated with the cyclical conception of change. The persistence of this doctrine is strikingly illustrated by the case of Newton. Although, as we have seen, he was not a thoroughgoing mechanist, he was far from being an evolutionist. For example, in a letter of 7 December 1675 to Oldenburg he explicitly stated his belief that "nature is a perpetual circulatory worker" (Turnbull,[15] Vol. I, p. 366). Later he used his theory of gravitation to explain how the orbital motions of the planets and satellites can be maintained but not how these motions may have originated. There is evidence that he believed that the true cause of gravity is the direct action of God.[16] In his correspondence with Bentley, he argued that the motions which the planets now have in their nearly circular orbits "could not spring from any natural cause alone but were imprest by an intelligent agent" (Turnbull,[15] Vol. III, p. 234). To support this conclusion he pointed out that without such intervention it was impossible for the planets to fall under gravity from some very remote region and acquire the appropriate velocities. The divine power would be required in a double respect: to turn the descending motion of the falling planets into a side motion and at the same time to double the attractive power of the Sun (Turnbull,[15] Vol. III, p. 240). The latter condition arises because at any given distance from the Sun the velocity of fall from infinity exceeds the velocity in a circular orbit at that distance unless the Sun's attractive power is suddenly doubled.

A different view was taken by Descartes. Like Newton, he abandoned Aristotle's dichotomy between the changing, corruptible Earth and the unchanging, incorruptible celestial regions and regarded all matter as subject to the same physical laws. But as a mechanical determinist he did not invoke divine intervention to explain the origin of the solar system. In his *Principia* of 1644, he tried to explain the uniformity of direction of the motions in the solar system and their approximation to the plane of the ecliptic by his theory of vortices. He assumed that originally the world was filled with matter distributed as uniformly as possible, and he sketched out qualitatively a theory of successive formation of the Sun and planets, including the Earth, which he regarded as composed of a series of different layers.

Consequently, despite being a more thorough mechanist than Newton—indeed, paradoxically, because he was this—Descartes became the founder of modern cosmogony. His idea of the Universe evolving by natural processes of separation and combination was the source of a succession of theories of cosmic evolution. Nearly a century later, Swedenborg, in his *Principia* of 1734, advocated a modified view of the Cartesian cosmogony. In particular, he suggested that planets were ejected from the Sun, but his idea of how this may have happened was rejected by Buffon, who, in 1745, put forward the first tidal theory of the origin of the solar system. Assuming that comets were far more massive than we believe today, Buffon suggested that a comet colliding with the Sun may have torn out sufficient material to form the planets.

Neither Swedenborg nor Buffon was responsible for applying Newtonian ideas to the problems of cosmogony. The first to do this was Immanuel Kant in his *Universal Natural History and Theory of the Heavens*, published in 1755. Kant began with the idea, which he was careful to point out was only a hypothesis, that in the beginning all matter was in a gaseous state and was spread more or less uniformly throughout the Universe. Consequently, he assumed— as many have done since—that we live in an evolutionary, or developing, Universe in the sense that the past was essentially simpler than the present. Kant had to postulate, however, that there were some primeval irregularities, regions of higher density acting as centers of condensation. One such center was the origin of the solar system. Assuming that gravitation alone was the primary agency giving rise to celestial bodies,[1] Kant supposed that rotational motions were generated through the collisions of particles whereby they were laterally deflected from their original gravitational motions. In this way Kant thought that eventually coplanar circular orbits about the Sun could arise, and that the motions would tend to be in the same sense. But this would mean that the solar system generated its own angular momentum by the interaction of its parts, in contradiction to the principle of the conversation of angular momentum and hence of Newton's laws of motion. (Kant should not be judged too harshly, however, for this principle does not appear to have been formulated in full generality before 1775, by Euler.) Laplace's nebular hypothesis, put forward in 1796 in his *Exposition du Système du Monde*, was free from this defect and his primeval solar nebula was assumed to rotate *ab initio*.

Laplace's most celebrated achievement in astronomy concerned the stability of the present state of the solar system. Perturbations in the orbits of the planets, which Newton had assumed would be cumulative, so that divine intervention would be needed to correct them, were shown instead to be periodic and self-correcting. Consequently, although Laplace believed that the solar system had evolved in the remote past from an entirely different original state, he thought that it had already attained its final state which would remain undis-

---

[1] Writing some sixty years later than Kant, the great pioneer of modern observational sidereal astronomy William Herschel suggested that "the state into which the incessant action of clustering power has brought the Milky Way at present is a kind of chronometer that may be used to measure the time of its past and future existence."[34]

turbed for ever. This conclusion, which contrasted with Newton's belief that the world was coming to an end, was also strikingly different from Kant's insight that the creation of new stars and systems is continuing.

Although the idea of evolution was in the air, one of the obstacles that it had to contend with was the widespread inherited conviction that the range of past time was severely limited. Archbishop Ussher in 1658 gave his assent to one of the various estimates of the age of the world made by the early patristic writers, obtained by adding up the generations since Adam. His conclusion that God created the world on Sunday, 23rd October, 4004 B.C.[17] became well known because it was incorporated in the margin of the Authorized Version of the Bible.

It is remarkable what a straitjacket this Bible-based chronology was for scientists in the seventeenth century studying the nature of fossils. Both Steno and Hooke realized that they were the petrified traces of former living organisms and were led to a dynamic theory of geological change, but were confronted with the difficulty of fitting this into the accepted time scale. Hooke declared in his *Discourse on Earthquakes*,[18] "that there have been in former times of the world, divers species of creatures, that are now quite lost." But the naturalist John Ray, who was at first inclined to accept Hooke's views, and also suggested that if Steno were right in asserting that mountains had not all existed from the beginning, then perhaps "the world is a great deal older than is imagined or believed,"[19] eventually under the influence of his theological beliefs altered his opinion on fossils in favor of an inorganic origin.[20]

An outstanding example of the limitations imposed in the seventeenth century on ideas about the past is Leibniz's *Protogaea* of 1691. As court historian at Hanover he was ordered to write a history of the Brunswick-Luneburg family. With typical teutonic thoroughness, he decided that "in order to show the remotest origin of our state, we must say something about the first configuration of the Earth, of the nature of the soil and what it contains" (Haber,[21] p. 84), but his ideas were severely constrained by his attempt to harmonize them with the book of *Genesis*. Nevertheless, he brought the idea of historical process into natural history and had a stimulating influence in the eighteenth century on Buffon (Haber,[21] p. 115).

During that century scientists and others began to discard the Bible-based chronology of nature. In 1721, Montesquieu wrote in his *Lettres Persanes*, "Is it possible for those who understand nature and have a reasonable idea of God to believe that matter and created things are only 6,000 years old?" In mid-century, Diderot thought in 'millions of years' and Kant suggested that the universe may be hundreds of millions of years old. Buffon, when writing his *Epoques de la Nature*, published in 1778, privately estimated that the first stages of the cooling of the Earth would have required at least a million years.[22] In print he was more cautious and estimated the Earth's age as being at least 75,000 years. Some of his ideas were condemned by the faculty of theology of the University of Paris (Taton,[23] pp. 572-573).

In 1788, the geologist James Hutton in his *Theory of the Earth* rejected the sudden catastrophic agencies that had been previously invoked to explain the stratification of rocks, the deposition of oceans, etc. He realised that the true scientific approach is not to invoke such *ad hoc* hypotheses but to test whether or not the same agents as are operating now could have operated all through the past. In his view, the world has evolved and is still evolving (in one place he actually likens it to an organism). He concluded that vast periods of time were required for the Earth to have reached its present state, and from his study of sedimentary and igneous rocks he concluded that "We find no vestige of a beginning—no prospect of an end."

The idea of using fossils to establish a chronology of the rocks was first suggested by Hooke, but was not acted on for over a hundred years. Toward the end of the eigtheenth century, William Smith, an English engineer and surveyor who collected fossils, realized that each geological stratum could be recognized by the fossils found in it, and that the same succession of strata occurred wherever the rocks concerned were found. He produced in 1815 the first large-scale geological map of any extensive area. Meanwhile the science of stratigraphical palaeontology was being founded independently in France by Giraud-Soulavie, who was the first to recognize that the stratigraphical ordering of rocks can be regarded as a chronological ordering. A member of a monastic order, he was violently attacked by the clergy and by Buffon and was forced to keep silent for many years until the French revolution. In 1793, he declared that traces of volcanic eruption had persisted for more than six million years (Taton,[23] p. 574).

During the nineteenth century the idea of time as linear advancement finally prevailed through the influence of the biological evolutionists, but the climate of thought that made it possible to contemplate the hundreds of millions of years required for the operation of natural selection to account for present and past species was prepared primarily by the geologists. It was therefore not surprising that Charles Darwin began his life's work as a geologist, as well as a naturalist. Nevertheless, Darwin's demands on the extent of past time came as a great shock to many, as Sir Archibald Geikie explained forty years later:[24] "Until Darwin took up the question, the necessity for vast periods of time, in order to explain the characters of the geological record, was very inadequately realized. Of course, in a general sense the great antiquity of the crust of the earth was everywhere admitted. But no one before his day had perceived how enormous must have been the periods required for the deposition of even some thin continuous groups of strata."

For measurements of geological time, as distinct from guesses, appeal must be made to physics, and here Darwin met what he believed to be one of the gravest objections to his theory. This came from the new science of thermodynamics, which introduced 'time's arrow' into physics. Newton's equations of motion are unaltered by a reversal of the direction of time, and Laplace's conclusion that the solar system could last forever had been based purely on dynamical considerations. In 1854, Helmholtz suggested that the Sun maintains its enormous outpouring of radiation by continually shrinking and thereby releasing gravitational energy which is converted into thermal energy of radiation. He calculated that the current rate of solar radiation could not have been maintained by the Sun for more than about twenty million years. This conclusion was supported by William Thomson (later Lord Kelvin), who thought that at most this estimate could be lengthened to fifty million years. "What then," he wrote, "are we to think of such geological estimates as 300,000,000 years for the 'denudation of the Weald'?"[25]

In confirmation of his view that the hundreds of millions of years demanded by the geologists and palaeontologists could not be allowed, Kelvin considered the flow of heat through the Earth's crust. He argued that this indicated that the Earth must be cooling and must therefore have been hotter in the past. He calculated the epoch at which the Earth was molten and found that this must have been between 20 and 40 million years ago.

While this controversy was raging and the physicists were compressing the extent of past time required by the geologists in just as severe a way as the theologians had done in the previous two centuries, the concept of evolution was being extended to the history of the Earth-Moon system.

The importance of tidal friction in this context had, in fact, already been realized as long ago as 1754 by Immanuel Kant in the most remarkable of his evolutionary speculations, the short essay that he wrote on the question, *Whether the Earth has undergone an Alteration of its Axial Rotation*. The frictional resistance of the Earth's surface to the tidal current in the seas and oceans, induced primarily by the gravitational action of the Moon, is very slow in its action but is irreversible and over long periods of time could give rise to great changes in the rotation of the Earth and the orbit of the Moon. Kant's discussion was not quantitatively correct, the effect he obtained being much too large, but it was the first indication that the time of celestial mechanics is not cyclic and eternal in the sense suggested by the investigations of Laplace.

Toward the end of the nineteenth century a more thorough and accurate analysis of the dissipative effects of tidal friction on the Earth-Moon system was made by Charles Darwin's son, Sir George Darwin, who tried to fit his results into the time scale allowed by Helmholtz and Kelvin. "If millions of millions of years were necessary," he wrote," the theory would have to be rejected."[26] He calculated that the minimum time required for the transformation of the Moon's orbit from its supposed initial condition to its present shape would be 50-60 million years. He realized that the actual period was probably a good deal longer, "yet I cannot think," he wrote, "that the applicability of the theory is negatived by the magnitude of the period required."

The resolution of these difficulties was possible only after the discovery of radioactivity and the investigation of nuclear transmutations by Rutherford and others. It is now known that there is a sufficient supply of radioactive elements in the crustal rocks to make the net loss of heat from them extremely small, and Kelvin's estimate of a few tens of millions of years has to be replaced by estimates of a few thousands of millions of years. Similarly, since the investigations of Bethe and Weizsäcker in the later 1930s it is generally accepted that the Sun's heat is maintained by thermonuclear processes that can continue steadily for thousands of millions of years.

The discovery of radioactivity not only resolved the difficulties faced by the evolutionists, but had important consequences for the concept of time. The t variable symbolizing the time scale of Newtonian mechanics is reversible, since the fundamental laws of dynamics would be unaffected if time were to run backward. On the other hand, dissipative phenomena in nature, such as frictional effects, are irreversible and therefore indicate time's arrow. The introduction of this latter concept into theoretical physics came, as I have already mentioned, with the rise of the science of thermodynamics in the middle of the nineteenth century. The famous Second Law is of quite a different character from the laws of dynamics and those governing natural forces such as gravitation. It is concerned with the temporal behavior of large numbers of particles and asserts the tendency for orderly arrangements of large numbers of molecules to break down into disorderly arrangements. This law expresses the general time-directional tendency of the physical world but provides no method of measuring this tendency and so, unlike the laws of dynamics and gravitation, cannot be regarded as defining theoretically a clock. On the other hand, the Rutherford–Soddy law of radioactive decay, formulated in 1902, not only indicates time's arrow but also provides means of measuring time. It can therefore be regarded as a noncyclic or evolutionary clock, but does it measure the same uniform time as that implied by the laws of mechanics? At first this question was not raised because of the firm belief that time is, in the words of Locke previously quoted, "one common measure of all existence whatsoever," a belief that was fostered by the growing tendency in advanced industrial civilization for men's lives to become increasingly regulated by the clock.

THE ROLE OF TIME IN MODERN COSMOLOGY

In view of the central role that time had come to play in modern life as well as in the scientific world-view, it was a tremendous shock when, in 1905, Einstein pointed out a previously unsuspected gap in the theory of time measurement. In pondering on the role of the velocity of light in physics he was led to think about the dependence of the measurement of motion on the measurement of time. It occurred to him that time measurement depends on the idea of simultaneity—for example, the coincidence in time of a particular event with a particular position of the hands on the face of a clock. The crucial stage in his thinking occurred when he realized that, although the idea of simultaneity is perfectly clear for two events at the same place, it is not equally clear for events at different places. As a result, he was prepared to discard the previously unquestioned assumption that time is universal and unique.

Einstein was not only ready to abandon the classical theory of time but also the mechanistic conception of nature which, despite the development of the evolutionary principle, was still a powerful influence in physics. He was thus led to discard all attempts to describe light and other forms of electromagnetic radiation as oscillations in a universal medium called the ether, for the mechanical properties that had to be assigned to the ether seemed to defy explanation. In place of classical mechanics and the universal clock as the foundation of physics, Einstein resorted to the principle of relativity, or equality of status, of all observers in uniform relative motion. In particular, since the properties of light must be the same for all of them, he claimed that each must assign the same velocity to it, although this condition is incompatible with the classical assumption of universal simultaneity. Einstein therefore rejected this assumption and explored instead the consequences of using the invariance of the velocity of light as a means of comparing the time readings of clocks in uniform relative motion in different places. He found that, on this basis, different observers would, in general, assign different times to the same event and that a moving clock would appear to run slow compared with an identical clock at rest with respect to the observer.

Einstein's special theory of relativity and the general theory that he formulated some ten years later imply that there is no universal time the same for all possible observers in relative motion. Since these theories were therefore incompatible with the classical presupposition that there is a unique sequence of temporal states of the Universe as a whole, the idea of a cosmic time scale might be thought to have no objective significance. Such a conclusion would, however, be mistaken, as we see from the remarkable developments that have since occurred in cosmology.

In 1924 the astonomer E. P. Hubble, using the recently installed 100-inch reflector on Mount Wilson, showed conclusively that the general background of the observable Universe is formed not by the stars but by the galaxies. Five years later he found that the galaxies appear to be systematically receding from each other. This discovery made as great a change in man's conception of the universe as the Copernican revolution four hundred years before. For, instead of an overall static picture of the cosmos, it seemed that the Universe must be regarded as expanding, the rate of the mutual recession of its parts increasing with their relative distance.

Hubble's discovery stimulated much work in theoretical cosmology and aroused great interest in the fundamental papers on expanding world-models by A. Friedmann and G. Lemaître that had been written several years before but had previously attracted little attention. Other models that were somewhat simpler were devised in 1932 by Einstein and de Sitter and also by E. A. Milne. As a result there was a general revival of the idea that there are successive states of the Universe defining a universal scale of time. This came about because in all the models considered there were certain privileged observers, namely those moving with the galaxies. In general, the local times of these observers fitted together into one universal time, called 'cosmic time.' (The only exception was Milne's model, but even in this a universal time was later obtained by a simple transformation of time scale.) This meant that the relativity of time need only be invoked when considering observers not associated with galaxies, that is to say, observers in motion with respect to the cosmic background in their immediate neighborhood.

The existence of cosmic time was challenged by the mathematical philosopher Kurt Gödel in a remarkable paper published in 1949.[27] He agreed that the relativity postulate that all observers are equivalent as regards the formulation of the laws of physics does not preclude the possibility that the particular arrangement of matter and motion in the actual Universe may offer a more 'natural' or 'simpler' aspect to some observers than to others. But he did not believe that the aggregate of local times associated with such a class of privileged observers must automatically constitute a universal time. His argument was based on his discovery of a homogeneous world-model in which the local times of the privileged observers who move with the galaxies cannot be fitted together into one universal time.

A key assumption in the construction of previous homogeneous models of the expanding Universe was that an observer associated with a galaxy would see himself to be at a center of isotropy or spherical symmetry, so that there would be no preferential directions in the Universe. Consequently, in these models there is no cosmic rotation, in other words, no cosmic motions transverse to the line of sight. The absence of cosmic rotation meant that at each point the directions of cosmic recession are like the spokes of a wheel (except that they form a three-dimensional pencil). These directions coincide with the directions of inertial motion—that, is with the directions of free motion not subject to local forces—and the set of these directions from any given point form a pencil of straight lines without relative rotation. This pencil is called the local 'compass of inertia.' In the world-model constructed by Gödel the system of galaxies is observed from each point to rotate relative to the local compass of inertia. The model was open to the objection that it was theoretically possible for a sufficiently fast-moving observer to travel into his past or future and back again, making a complete circuit in time. In 1962, Ozsvath and Schücking found a locally rotating expanding world-model free from this defect.[28]

Consequently, assuming that in a first rough approximation the Universe is effectively homogeneous in the large, the crucial factor in deciding whether the idea of universal cosmic time is likely to be well-founded is evidence bearing on cosmic rotation. In other words, empirical evidence for world isotropy, and hence for the absence of cosmic rotation, can also be regarded as evidence for the idea of cosmic time.

Impressive support for the assumption of world isotropy has come in recent years from the discovery of what has come to be called the 'primeval fireball.' In 1965, the radioastronomers A. A. Penzias and R. W. Wilson, at the Bell Telephone Laboratories in New Jersey, found that some unexpected radiation was leaking into the antenna of their apparatus. They soon discovered that the source of this radiation must be more or less isotropic and that at the wavelength at which they were working (7.4 cm) the radiation appeared to be equivalent to the intensity of a blackbody source at about 3.5° Kelvin. Subsequent work has corrected this estimate to approximately 2.7°. Meanwhile, at Princeton University, R. H. Dicke and his colleagues were arguing on purely theoretical grounds that a universal microwave radiation with similar properties should exist. Their work was done in ignorance of that carried out by George Gamow some fifteen years earlier, who, on the basis of an explosive origin of the expanding Universe, had predicted the existence of a primeval fireball with a present temperature of about 5°.

The primeval nature of the observed background radiation is indicated by the blackbody character of its spectrum, but the best confirmed feature of this radiation is its isotropy, the departures from which as indicated by a very small temperature fluctuation with direction of not more than about 0.1% can be ascribed to the proper motion of the solar system with respect to the local supercluster of galaxies.[29] This isotropy is sufficiently precise to exclude the possibility of any local origin for the radiation, since a source restricted to the solar system, our Galaxy, or even the Local Supercluster would not appear so nearly isotropic to an observer located, as we are, far from the centers of these systems. Moreover, any large-scale departures from homogeneity or isotropy in the Universe would affect the radiation and make it appear anisotropic to us. Consequently, we have powerful evidence that the Universe as a whole is predominantly homogeneous and isotropic, and this conclusion, as we have seen, is itself a strong argument for the existence of cosmic time.

The existence of universal scale of time gives objective meaning to the concept of the age of the Universe, but it still leaves open the question of whether there is a *unique* uniform time in nature. For, given a scale of time t, we can, by a monotonic transformation of the form $\tau = f(t)$, obtain another scale τ that keeps pace with t only if τ is a linear function of t. In other words, if τ is not a linear function of t, then natural processes that are periodic according to the t-scale would not be periodic according to the τ-scale. Similarly, if there are any natural processes which are periodic according to a τ-scale that is not a linear function of t, they will not be periodic on the t-scale. Two scales of time can therefore only be regarded as effectively the same (except for possible differences in the choice of time zero and time unit) if they are linearly related. Otherwise, they are essentially distinct. We are therefore faced with the type of question that I have mentioned previously. Is the uniform time defined, for example, by the Rutherford–Soddy law of radioactive decay effectively the same as that implied by the laws of dynamics or gravitation, or is each the basis of a different scale of cosmic time?

This type of question was first raised about 1935 by Milne in exploring the properties of his world-model. The usual way in which it has been discussed since is in terms of the various fundamental constants of physics and the possible secular variation of some of them in terms of the postulated invariance of others. The possibility of the secular variation of a universal physical 'constant' was first considered in the case of G, the constant of gravitation, by Milne, who found that in his uniformly expanding world-model G increases linearly with the time t. In 1937 Dirac suggested that the ratio of electric to gravitational forces may be changing.[30] His particular hypothesis was equivalent to assuming that, in terms of the charges and masses of protons and electrons, G decreases inversely with time. In 1961 Brans and Dicke were led to predict a much slower variation, namely that G varies as t^{-n}, where n is of the order of 0.1.

To test the possibility of a variation in G, Pochoda and Schwarzschild in 1964 calculated the effects on the evolution of the Sun.[31] They computed evolutionary models under various assumptions of the type G varies as t^{-n}. In accordance with our ideas of the age of the Earth and solar system, they assumed that the Sun was in its initial hydrogen transmutation, or 'Main Sequence,' stage about 4.5×10^9 years ago. For G varying inversely as t, i.e., $n = 1$, they found that the present state of the Sun could only be accounted for satisfactorily if the age of the Universe were 15×10^9 years or more. This limitation arose because they found that, if the age of the Universe were less, the initial Main Sequence stage of the Sun would have occurred relatively earlier in the history of the Universe when, under the assumption the G varies inversely as t, G would still be rather high. Since the luminosity of the Sun depends on the eighth power of G, its initial luminosity would have been so high that the rate of transmutation of hydrogen in the Sun's thermonuclear core would have been so fast that this hydrogen would have been exhausted and the Sun would have left the Main Sequence before now, i.e., before 4.5×10^9 years had elapsed, and would now presumably be a Red Giant. In other words, the Sun could not now be in the state in which it is.

In 1964 it seemed that this conclusion ruled out Dirac's hypothesis, because the age of the Universe was thought to be less than 15×10^9 years. However, in 1971 Sandage obtained a new value for the rate of expansion of the Universe implying a possible age of 15×10^9 years or more, so that on the basis of solar evolution calculations a variation of G of the type envisaged by Dirac cannot be ruled out. The much slower rate of variation predicted by Brans and Dicke is too insensitive to be tested in this way.

Although there has not been until recently any compelling reason for discarding the hypothesis of a unique, uniform cosmic time in terms of which the fundamental constants and laws of physics are secular invariants,[32] Van Flandern, of the U.S. Naval Observatory, claims that a careful comparison of atomic time and ephemeris time in the timing of occultations of stars by the Moon between 1955 and 1973 indicates that G is decreasing secularly by about one part in 10^{10} per annum. [33] It remains to be seen whether this preliminary result of Van Flandern's analysis will be confirmed. If it is, the theoretical consequences will be far-reaching, especially for our ideas of gravitation, since a varying gravitational constant is incompatible with general relativity.

Be that as it may, the discovery of the expansion of the universe has not only reinforced the tendency in recent centuries for time to become the dominant feature of the scientific world-view, but it has definitely thrown new light on the old problem of the total extent of past time. For the cylical Universe of the Greeks, time was unlimited, without beginning or end, but it was a concept of secondary significance. A very different view began to develop in the middle ages. Under the influence of Christianity, the idea of time as linear advancement without cyclical repetition began to emerge. Strict limits were placed on the total duration of the world, but these limits were believed to be due to divine intervention. Following the scientific revolution of the seventeenth century, the temporal restrictions imposed by an out-of-date biblical chronology gradually became intolerable. In place of a limited past of a few thousand years, a tremendous vista began to open up extending over hundreds of thousands, and later over hundreds of millions, of years. Increasing awareness of the extent of past time was accompanied by a growing realization that the world is not only old but has a history. Attempts to measure past time and to date the various stages of terrestrial and cosmic evolution became much more precise early in this century with the discovery of radioactivity, and later with the development of nuclear physics and astrophysics. We now believe that the Earth, Sun, and solar system are about five thousand million years old and that the age of the Milky Way and oldest stellar clusters is about ten thousand million years. In view of these results, it is surely of profound significance that the respective ages assigned to the expanding Universe according to observational evidence interpreted in terms of the simplest world-models obeying Hubble's law of recession, are of the order of ten to fifteen thousand million years. But whether this figure is to be taken as an indication of the total time that has elapsed since the whole physical world was created or only since it took the form which we now study is a question that takes us beyond the limits of scientific enquiry. In the light of our present knowledge, it would seem to be the longest stretch of past time over which we can extend the laws of nature as we know them.

REFERENCES

1. S. Sambursky, *The Physical World of the Greeks* (London, 1956), p. 185.
2. G. Sarton, *Introduction to the History of Science* (Baltimore, 1927), Vol 1, p. 72.
3. Aristotle, *Physics*, 225a.

4. J. D. North, Monasticism and the First Mechanical Clocks, paper read at the Second World Conference of the International Society for the Study of Time, Lake Yamanaka, Japan (1973).

5. Lynn White, *Medieval Technology and Social Change* (Oxford, 1962), pp. 124–5.

6. L. Mumford, *Technics and Civilisation* (London, 1934).

7. S. A. Bedini and F. R. Maddison, *Mechanical Universe: the Astrarium of Giovanni de' Dondi* (Philadelphia, 1966).

8. E. Grant, *Nicole Oresme and the Kinematics of Circular Motion* (Madison, Wisconsin, 1971), p. 295.

9. Robert Boyle, *The Works of the Honourable Robert Boyle*, ed. by Thomas Birch (London, 1772), Vol. 5, p. 163.

10. G. M. Clemence, *Am. Sci.* **40**, 267 (1952).

11. W. Thomson and P. G. Tait, *Natural Philosophy* (Cambridge, 1890), Part 1, p. 241.

12. I. Barrow, *Lectiones Geometricae*, transl. by E. Stone (London, 1735), Lect. 1, p. 35.

13. Galileo Galilei, *Dialogues Concerning Two New Sciences*, transl. by H. Crew and A. de Salvio (New York, 1933), pp. 176 *et seq.*

14. J. Locke, *Essay concerning Human Understanding* (1690), Book II, Chapter 15, Paragraph 11.

15. H. W. Turnbull (ed.), *The Correspondence of Isacc Newton* (Cambridge, 1959).

16. J. E. McGuire and P. M. Rattansi, *Notes and Records of the Royal Society* **21**, 112 (1966).

17. J. Ussher, *The Annals of the World Deduced from the Origin of Time* (London, 1658), p. 1.

18. R. Hook, *Discourse of Earthquakes* (London, 1705), p. 435.

19. W. N. Edwards, *The Early History of Palaeontology* (London, 1967), p. 33.

20. J. M. Eyles, *Nature* **175**, 103 (1955).

21. F. C. Haber, *The Age of the World* (Baltimore, 1959).

22. N. Hampson, *The Enlightenment* (London, 1968), p. 220.

23. R. Taton (ed.), *The Beginning of Modern Science*, transl. by A. J. Pomerans (London, 1964).

24. Sir Archibald Geikie, *The Founders of Geology* (London, 1897), p. 283.

25. Sir William Thomson, *Popular Lectures and Addresses* (London, 1889), Vol. 1, p. 361.

26. G. H. Darwin, *The Tides* London, 1898), p. 257.

27. K. Gödel, *Rev. Mod. Phys.* **21**, 447 (1949).

28. L. Ozsvath and E. Schücking, *Nature* **193**, 1168 (1962).

29. E. Conklin, *Nature* **222**, 971 (1969).

30. P. A. M. Dirac, *Nature* **139**, 323 (1937); *Proc. Roy. Soc. A* **165**, 199 (1938).

31. P. Pochoda and M. Schwarzschild, Variation of the Gravitational Constant and the Evolution of the Sun, *Astrophys. J.* **139**, 587 (1964).

32. F. J. Dyson, Fundamental Constants and Their Time Variation, in *Aspects of Quantum Theory*, ed. by A. Salam and E. P. Wigner (Cambridge, 1972), pp. 213–36.

33. T. C. Van Flandern, A Determination of the Rate of Change of *G*, (Abstract), *Bull. Am. Astron. Soc.* **6**, 206 (1974); *Sci. Am.* 1974 (October), 56–7.

34. William Herschel, *Phil. Trans. Roy. Soc.* **1814**, 284.

Chapter XIII

On Some Cosmological Theories and Constants

Wolfgang Yourgrau

University of Denver

It is certainly desirable, or perhaps even necessary, to review from time to time the state of the exact or applied sciences. This attitude definitely holds in the case of cosmology. Granted, cosmology and cosmogony—this separation of terms has become obsolete today—are, no doubt, two of the oldest examples of man's innate curiosity. Although thousands of years ago they already contained traces of scientific elements and sometimes even noteworthy attempts at more or less plausible theories—nevertheless, cosmology and cosmogony have retained, in the main, their speculative character.[1]

In contrast, astronomy (and later astrophysics), based almost exclusively on observation, could claim to be of an empirical nature. Unlike cosmology, which gradually absorbed cosmogony and was based on more and more diverse and often incompatible hypotheses, astronomical science was able to combine some fundamental premises with experiential data and thus to support cosmological theories. Because those theories had to be not only consistent but also testable in the empirical sense, astronomy joined the other physical sciences long before cosmology could be considered to be a genuine science employing astronomical results as its verifiable and falsifiable foundation.

These introductory remarks were made in order to do justice to modern cosmology, i.e., to the findings in this domain of intellectual curiosity.

With the developments of the last sixty years we are fully justified in considering cosmology as almost a valid natural physical science. True, physical sciences are supposed to fulfill the three following main criteria: observation, experimentation, and mathematization. Cosmology, which we interpret as the study of the origin, structure, and evolution of the Universe, will probably for a long time remain incapable of performing experiments to the same degree as the other sciences. Yet, with the rapid development of new techniques and the birth of novel hypotheses, we may optimistically expect that even in cosmology experimentation will gradually increase. In part at least we can already check critically not only many hypotheses and theories concerning our Earth and the solar system, but also whole varieties of cosmological models of the Universe. In consequence, many explanations, interpretations, and models have been eliminated as scientifically untenable.

Let us first make certain fundamental assumptions without which it would be difficult to construct any possible or even probable cosmological theory. I am referring to homogeneity and isotropy. These are vital suppositions. Nevertheless, it cannot be maintained with complete certainty (probability 1) that the so-called laws of nature applicable to the Earth and to the solar system and our current knowledge of the observable Universe must also hold true uniformly for the Universe at large. This means that we cannot entirely ignore the fact that there may be in the Universe regions where the laws of nature are not the same as those with which we are familiar. After all, this possibility cannot be entirely dismissed, since even our natural laws are not absolute, but only of statistical validity.[33]

We have now reached a point which should enable us to proceed, in a more or less methodical manner, to consider some cosmological theories and constants.

In his first version of general relativity theory, Einstein conceived of a spherical, static universe. As Friedmann later pointed out, the static structure of this universe was due to a mathematical error (division by zero). It has been sometimes stated that Einstein *had* to introduce into his general equations a new term which he called the "cosmologic constant." He gave it the symbol Λ. Yet, according to Einstein,[11] "The introduction of this constant constitutes a complication of the theory which seriously reduces its logical simplicity." Its introduction can only be justified by the difficulty produced by the almost unavoidable introduction of a finite average density of matter.

His equations of gravitation were therefore enlarged and yielded

$$(R_{ik} - \tfrac{1}{2} g_{ik} R) + \Lambda g_{ik} + \kappa T_{ik} = 0$$

where

$$\kappa = 8\pi G/c^2 = 1.86 \times 10^{-27}$$

It is immediately evident that already in the first approximation the physical structure of the gravitational field is drastically and fundamentally incompatible with Newton's theory of gravitation. The reason for this difference is that the gravitational potential is represented by a tensor and not by a scalar.

Friedmann demonstrated convincingly that one can obtain Einstein's equations of gravitation without having to inject the cosmologic, or cosmological, constant, as it was soon called.

In a later publication Einstein admitted that the cosmological constant "was not required by the theory as such, nor did it seem natural from a theoretical point of view."[12] Friedmann showed that Einstein's theory leads us directly to an expansion of space. In other words, Einstein's original field equations allow a solution in which the (ambiguous) "world radius" depends on time, and thus suggests space to be expanding. But Einstein departed even further from his original version of GRT and considered for a while the logical possibility of the Universe being infinite. He also approved of a proposal by Seeliger that, by a modification of the Newtonian law of gravitation, it would be possible for the mean density of matter to be constant everywhere, even to infinity and without resulting in infinitely large gravitational fields. It follows from the above that the Universe cannot possess a center.

Later Einstein insisted that an infinite universe is only possible, if the mean density of matter in the material universe vanishes. However, he still contended that though Friedmann's conception of an infinite universe with diminishing density of matter is *logically* possible, "it is less probable than the assumption that there is a finite mean density in the Universe." Friedmann's thesis of an expanding universe was soon confirmed by Hubble's famous discovery that redshift of the spectral lines does increase linearly with distance.[18]

Einstein also accepted the so-called "Cosmological Principle" ("Problem"). This principle means simply that the distribution of the galaxies seems to be the same in all spatial directions. It can now be claimed, at our present state of observational techniques, that the galaxies (previously and sometimes still called nebulae) are all receding from our own galaxy, the Milky Way. Our observations of many galaxies indicate that the light emitted from those galaxies shifts its spectral lines more to the red than the light emitted from galaxies that are closer to us. The recession of the galaxies corresponds to the expansion of the Universe. There is no logical contradiction between the cosmological principle and galactic recession. Hubble's law, which observationally confirmed the recession of galaxies and hence the expansion of the Universe, will be discussed at some length later.

In this connection it may be apposite to mention that Einstein refused to account for the Hubble shift of spectral lines by any other means than the Doppler effect. This interpretation has been challenged by several authors as not being the only plausible one.

Einstein questioned the assumption of a "beginning of the world" (i.e., start of the expansion) for empirical and theoretical reasons. In particular, he doubted that 10^9 years ago expansion of the Universe began.

Further, he stated without qualification that the "theory of evolution" of the stars is based on weaker foundations than his own relativistic field equations. Moreover, he refused to accept the current and generally accepted viewpoint that the beginning of the expansion is tantamount to a singularity in the mathematical sense. In other words, for Einstein the so-called beginning of the world was identical with the development of the now existing stars in systems of stars at a point when all those stars "did not yet exist as individual entities." He asked why space is not filled with radiation so that the nocturnal sky would appear to us like a "glowing surface" (Olbers' paradox). Yet the answer is quite simple: If it is correct that the Universe is infinitely expanding, then we should not expect it to be drowned in, or inundated by, a lasting luminosity, i.e., by a "flood of light," even from the most *remote* regions of the Universe. According to this explanation, Olbers' paradox can be considered to be resolved.[4],[1] Of course, such a resolution of the time-honored paradox requires that to be consistent with the postulate of uniformity in the Universe, the "motion of expansion" has to be such that the "velocity of recession" is proportional to distance (Bondi). And this means that Hubble's law (redshift) is assumed to be valid.

Let us repeat: The sky which we observe at night is dark due to the fact that the Universe is expanding. We have incontrovertible empirical evidence that radiation is absorbed by the Universe at large. It is an interesting afterthought that Friedmann was able, by his version of Einstein's general relativistic field equations alone, to predict theoretically the expansion of the Universe. Hubble's law, based on observations, requires only some reformulation in order to become adequately consistent with general relativity.

Discussing, though only in brief, de Sitter's relativistic model of the Universe, one has first of all to take, however perfunctorily, a glimpse as to how he interpreted the cosmological constant Λ. He stresses that Einstein's constant "is a name without any meaning. ... We have, in fact, not the slightest inkling of what its real significance is. It is put in the equations in order to give them the greatest possible degree of mathematical generality."[6] And, with regard to its mathematical function, it is completely undetermined. It could be positive, negative, or zero.

He reminds us that in Einstein's paper of November 1915 the Λ term did not appear at all. Hence, Λ had obviously the value zero. Cosmologically speaking, this implies that the curvature of the Universe is proportional to Λ.

In 1917, two solutions of Einstein's field equations for a homogeneous, isotropic universe were offered. If we call with de Sitter the solutions A and B, we arrive at the conclusion that in A the universe has a finite density, whereas in B the average density is zero, i.e., the universe is empty. De Sitter was aware of the fact that A and B were no more than approximations to the actual Universe, since the Universe, treated not as a model but examined by observational astronomy, can be neither A nor B. If one regards A as a static universe, then B is an empty universe. We are thus confronted with a static universe containing matter and having no expansion on the one hand, and on the other an

---

[1] Redshift entails a diminishing of cosmic illumination. The total background radiation in the actual or a model Universe is numerically very small ($2.7°K$). Astrophysical investigations show the intergalactic starlight density to be $\sim 10^{-2}$ eV cm^{-3}; our own galaxy's starlight density is only 1 eV cm^{-3}. At any arbitrary point within the Milky Way, all the other galaxies contribute only 1% of the background starlight.

empty universe void of matter and expanding. Since the Universe in reality contains matter and is expanding, neither solution A nor B is correct. We must bear in mind that, when these two solutions were constructed, the expansion of the Universe had not yet been discovered.

De Sitter assumed that the *only* reasonable physical interpretation[2] of Λ was that it denoted "the curvature of the world." Moreover, the square root of its reciprocal (the radius of curvature) *seemed* to furnish us with a natural unit of length. In the nonstatic case, the mysterious constant Λ as well as the curvature did not have to be positive but could also be negative or zero. He proved that in solutions A and B the curvature of three-dimensional space (static) was by necessity positive. Perhaps one should emphasize that both A and B assumed a static (finite) universe, which was originally considered (by Einstein) as the most plausible or possible one.

De Sitter proposed three types of nonstatic universes: the oscillating universes and the expanding universes of the first or second kind. The main characteristic of the expanding "family" of the first kind is that the radius is continually increasing from a definite initial time when it had the value zero. This universe becomes infinitely large after an infinite time. In the second kind of nonstatic universe, the radius possesses at the initial time a definite minimum value and also increases so that it becomes infinite after an infinite time.

Needless to say, the above assertions were sheer speculations. I may add here that in the Einstein model (in all three static universes assumed by GRT) the cosmological constant is supposed to be equal to the reciprocal of R^2, whereas de Sitter computed for his interpretation the constant to be equal to $3/R^2$. Whitrow correctly points out the significant fact that in special relativity the cosmological constant is omitted—for obvious reasons. We are only interested in this connection in the fact that the cosmological constant was given an explicit physical interpretation.

It is a strange state of affairs that although Friedmann demonstrated rigorously that the field equations of GRT do not require at all the introduction of the Λ term, he employed this constant on more than one occasion when dealing with the *actual* physical Universe. When he discarded later the original Einsteinian model of a closed or finite universe, he advocated an ever-expanding type of universe and introduced for it the expression of a "monotonic universe." But let us return to de Sitter's relativistic hypotheses. If Λ is negative, "only oscillating universes are possible," independent as to whether the curvature is positive, negative, or zero. If Λ is zero and the curvature positive, we still obtain the model of an oscillating universe, but if the curvature is zero or negative, we arrive at the model of expanding universes. Finally, if the cosmological constant is positive, then in the case of positive curvatures, the three types of families of nonstatic universes are possible. If we assume negative and zero curvatures, then

---

[2] Yet he had earlier stated that Λ "is a name without any meaning" (!).

the first so-called family of expanding universes alone is feasible. Should the curvature be small, then the Λ term must be automatically small and fundamental; in the case that the curvature is very small, the cosmological constant will be very small.

Later on de Sitter admits that in view of the above, and other considerations, it becomes extremely difficult to interpret Λ as providing us with a natural unit of length. Like Einstein, de Sitter is extremely cautious when he considers the beginning of the expansion of the Universe and does not offer any definite initial, numerical value of t_0. He reasons that it is, of course, possible "to relegate the epoch of the starting of the expansion to minus infinity, e.g., by using instead of the ordinary time, the logarithm of the time elapsed since the beginning." However, he immediately realizes that his approach amounts to no more than to a "mathematical trick." (We are here reminded how Schrödinger applied a similar trick when he derived mathematically the concept of negative entropy.)

De Sitter's model of the universe is void of matter but contains motion. In contrast, Einstein's model of the universe contains the largest possible concentration of matter, yet it is entirely without any motion. I agree with Whitrow when he regards the models of the universe as created by Einstein and de Sitter merely as a limited image of our actual Universe. The model of Einstein as well as that of de Sitter are both interesting solutions of the modified gravitational equations. Orginally, they were models of static universes. Today we know that the "de Sitter model is best regarded as a limiting form of an expanding universe in which the mean density is everywhere zero." Einstein introduced Λ into the equations of GRT because he was convinced that all stellar velocities are, and will be always, very small compared to that of light. Here we have another explanation for the occurrence of the cosmological constant in the field equations!

It was not until 1932 that Einstein, as well as de Sitter, finally realized that Einstein's assumption was no longer necessary for the general theory of relativity. They coauthored a memoir in which they stated, "There is no direct observational evidence for the curvature, the only directly observed data being the mean density and the expansion, which later proves that the actual universe corresponds to the non-statical case. . . . Historically the term containing the 'cosmological constant Λ' was introduced into the field equations in order to enable us to account theoretically for the existence of a finite mean density in a static universe. It now appears that in the dynamical case this end can be reached without the introduction of Λ."[31]

It was not before 1931 that Heckmann found that the curvature as well as Λ could be positive, zero, or negative.[14,15] It seems to be evident that de Sitter was undoubtedly influenced by Heckmann's contention.

We shall later see that the somewhat "mysterious" cosmological constant has never entirely ceased to haunt cosmological theories or hypotheses, not only as a term necessary for a physical interpretation of the Universe, but also in theoretical models.

It was my original intention to treat wherever possible some cosmological theories or hypotheses and cosmological constants separately. I soon began to realize that this is an almost impossible task, because most cosmological theories or hypotheses are evolutionary and this means that they are relativistic. One can only show that the cosmological constant has been interpreted in different ways, depending on the hypothesis or theory which contains this term.

Lemaître's hypothesis of the primeval atom can be considered a purely cosmogonic attempt to account for the origin and development of the Universe. I never understood the nature of the so-called primeval atom. The initial conditions for the so-called cosmic repulsion, which constitutes the contrast to the gravitational force (attraction), have never been clear to me. To quote Lemaître, "A sort of gas must have resulted from the splitting of the primeval atom which was broken into smaller and smaller pieces, a gas that was, of course, not much like present-day gases which are in statistical equilibrium and perfectly homogeneous."[22]

This statement, which presupposes initial conditions responsible for the evolution of the Universe, is regarded as sufficient and necessary. We must, however, realize that Lemaître does not inform us as to the nature of the primeval atom, but only tells us that a sort of gas was the *result* from the splitting of the primeval atom. He does not inform us as to the material constituents of this original atom. We only learn that a gas resulted *after* the primeval atom was split! Once the inhomogeneous gas was produced, a mixture of more homogeneous gaseous clouds formed, and these clouds caused expansion, later slowed down so that "repulsion and attraction balanced one another." Whenever the gravitational force overcomes the repulsive force, gaseous nebulae (galaxies) are formed and with progressive condensation toward the center of these gaseous galaxies (nebulae), stars and possibly also planets finally evolve.

We are here very much reminded of the Laplace hypothesis.

Lemaître's hypothesis suggests that at the time when the primeval atom disintegrated, matter was strongly condensed. This whole hypothesis entails an interplay of cosmic gravitation and repulsion and allows for a model that is theoretically compatible with an Einsteinian universe, with the exception that Einstein never envisaged a primeval atom as the origin of the Universe.

Observational verification of Lemaître's hypothesis is only in part available. For example, according to current empirical data, repulsion in the Universe very much exceeds attraction. We are thus able to resort to the universe model of de Sitter, i.e., we can apply his formula for an empty universe. Consequently, from the velocity of expansion one can derive the value of Einstein's cosmological constant. And this means in our case that Λ is positive. If Λ were negative, it would represent attraction. Whenever gravitational and repulsive forces neutralize each other, we again arrive at an Einsteinian universe.

Gamow, whose own hypothesis of the Big Squeeze is to a great extent based upon Lemaître's hypothesis, analyzed and sometimes criticized the intellectual "father" of his own theory. Examining the hypothesis of Lemaître's primeval atom, Gamow expressed his disagreement with the expression "primeval atom." In its place he proposes the expression "primeval nucleus." The reason for this is that, according to Gamow, Lemaître proposed that before the expansion began, "all the matter of the universe was in the state of dense nuclear fluid, and formed an enormously large nucleus."[13] Gamow ignored the following fact: Lemaître stated explicitly that a kind of gas *resulted* from the splitting of the primeval atom. He nowhere suggested that this atom *itself* was either a gas, liquid, or solid. He leaves the whole issue an open question. But Gamow interpreted Lemaître's view regarding initial expansion in such a way that "the original fluid became mechanically unstable and began to break up into fragments of all possible sizes."[13]

He then continued his interpretation by asserting that "as the result of spontaneous [proton → neutron + electron]"[3] transformations, Lemaître's original "polyneutrons" had to become very soon positively charged and were "surrounded by thin electronic atmospheres." The "primeval atoms" (?) could easily be compared to the atoms with which we deal at present.

I think that Gamow's choice of terminology for Lemaître's hypothesis was not a very felicitous one. First of all, it would have been an improvement if he would not have used an expression like "such fragments were formed by the mechanical breakup process of the original nuclear fluid. . . ." Also, the expression "spontaneous transformations" does not seem to me suitable. Instead of speaking of the result of "spontaneous transformations," it would have been in agreement with modern terminology to refer to neutron or β^- decay under the emission of an antineutrino ($n \to p + e^- + \bar{v}$).

Since Lemaître employed throughout his work the terms gas and gaseous clouds, he committed himself to a definite viewpoint. In contradistinction, Gamow always used for Lemaître's gas the term fluid and thus left it open whether he had a liquid or a gas in mind. And by using the term "fluid" he seemed to have ignored the fact that (only too well known to him) a fluid can be a liquid or a gas.

We recall that when Einstein made the assumption of the "cosmological problem," he tried to show that in contrast to Newton's reasoning, there could be no center of the Universe. It follows that at any given time an observer on any other galaxy (or nebula) would see (roughly) the identical picture of the Universe as does an observer on the Earth or any other point in space. The requirement of a cosmological problem or principle can be, claimed Lemaître, fully satisfied by his primeval atom hypothesis. Unfortunately, the distinction between the expressions "Cosmological Problem" and "Cosmological Principle" can lead to some confusion. There is undoubtedly a close relationship between homogeneity and the Cosmological Problem or Principle.

However, the most interesting part of Lemaître's contribution to cosmology is for me not his hypothesis of the primeval atom, which was more successfully

[3]Of course, this reaction should read [neutron → proton + electron].

superseded and enlarged by Gamow's Big Squeeze hypothesis—I consider his almost fanatic defense of Einstein's cosmological constant rather fascinating. Lemaître insisted on the logical convenience, or even further, on the theoretical necessity, of that fateful constant. In the very early stages of cosmic evolution, Λ was naturally rather a negligible quantity. But when cosmical repulsion became the dominating force, expansion started to increase with tremendous momentum until the de Sitter condition $\rho \ll \rho_0$ was attained. If Λ is negative, it becomes an attractive force and combines easily with the gravitational force. Further, Λ was a necessary constant to render acceptable "the short scale of time which is imposed by the value of the red-shift of the nebulae .... it might account for the formation of stars and nebulae."[21]

Again, velocities of recession are assumed to be proportional to the distance; this constant ratio depends on Λ by the equation

$$v/r = 1/T_H = \sqrt{\tfrac{1}{3}\Lambda}$$

T_H determines the time scale for an adequate description of the expanding Universe. It seems to be unwarranted to assume that the development of the Universe did not last more than T_H; this term stands also for the duration of the geological ages. It follows that the value of Λ (positive) must decrease. In addition, let us suppose, as an approximation (not a Newtonian one), not merely spherical symmetry, but "complete equivalence of each point and each direction, i.e., homogeneity and isotropy." As a result, the behavior pattern of the whole Universe is "completely described" by the successive values that the variable radius of space R assumes. And we know that there is a mutual relation between the cosmological constant Λ and the cosmical density ρ_0. We also know that within the limits imposed upon the astronomer "by the precision of planetary observations," ρ_0 or Λ is still entirely unknown in cosmological theory. For Lemaître, it was "a happy accident" when Einstein in 1917 introduced the cosmological constant into the equations of gravitation.

It is ironical that despite everything mentioned above, Lemaître made the following statements in obvious contradiction to his ardent defense of the theoretical and physical necessity for the Λ constant: "The history of science provides many instances of discoveries which have been made for reasons which are no longer considered satisfactory. It may be that the discovery of the cosmological constant is such a case."[21]

Before concluding this analysis of Lemaître's hypothesis, one important point deserves mention. In the paper published in 1949 he talks about a single atom nucleus and does not see any sense in referring to space and time in connection with this atom. Space and time, he contends, are statistical notions; they "apply [only] to an assembly of a great number of individual elements." Hence, they did not possess any meaning at the instant of the disintegration of the primeval atom.

The Big Squeeze cosmological hypothesis due to Gamow deserves special attention because, in spite of some serious defects, it rests not only on an interesting presupposition, but also on a plethora of valuable information. This information is in no way entirely original, but it has the advantage that it is presented in a coherent and often plausible manner. We recall that already Friedmann had convincingly demonstrated that the modified version of Einstein's field equations must lead to the conception of expanding and contracting universes. While physical forces on the whole decrease with distance, cosmological repulsion was supposed to be extremely weak over short distances, but significantly strong in the case of intergalactic distances. Gamow joins those cosmologists for whom the Λ constant (universal repulsion) was a mathematical generalization of Einstein's general relativity equations. Granted, when we became aware of the fact that the Universe was not static, but expanding, Λ became redundant. Still, this constant may nevertheless be of some value in cosmology even though the primary reason for its introduction has become obsolete.[13]

To illustrate that Λ has at least physical meaning, we may express the repulsive force (positive Λ) that acts on a particle with mass m by

$$F = -\tfrac{1}{3}c^2 \Lambda md$$

where d is the distance to the particle.

Before we discuss Gamow's explanation for the expansion of the Universe and his hypothesis of the Ylem, we should realize that he resorted, more than the "axiomatic" cosmologists in our time, to the findings of astronomers and astrophysicists. He devoted much of his attention to the characteristics, origin, structure, and development of galaxies; we may classify galaxies thus: spiral galaxies (ordinary and barred spirals), elliptical, spherical, and irregular galaxies.[4] He referred without exact specification to the Great Spiral Nebula of Andromeda, which is an unorthodox galaxy since it is not expanding, but moving toward us, and therefore displays a violet-shift in the lines of its spectrum. Baade resolved photographically the central body (i.e., the core) of Andromeda.

There was no doubt in Gamow's mind that the whole system of galaxies is expanding. With Hubble, he concluded that when the source of light is receding from an observer, light waves are lengthened and thus all colors shift toward the red end of the spectrum. In contrast, whenever the light source approaches an observer, light waves are shortened owing to the motion of the source, of course, and all colors shift toward the blue end of the spectrum. This is the so-called Doppler effect and is numerically rather small. It should be stressed that ordi-

---

[4] Hubble denotes Gamow's "ordinary" spirals by the expression "normal spirals." Spherical galaxies have also been called "circular galaxies." Gamow omits in his classification to mention irregular galaxies.

nary nebulae often consist of very large clouds of dust floating in interstellar space. For instance, Orion represents such a nebula. Further, we can divide nebulae into two groups of observation: galaxies proper, and diffused material within a galaxy. The term "nebula" has now been replaced by the term "galaxy," yet we still talk about the nebulous mass in Orion; here we mean by "nebulous" nothing else but diffused matter.

Gamow asserted that the galaxies were never in a state of rest but are moving away from one another (with high initial velocities) so that the distances between the neighboring galaxies will never cease to increase, and as a final result, the whole system (the Universe) expands.

Obviously, if the Universe is expanding, according to Gamow, there must have been a remote time in the past when the Universe was in a state of high compression. At that time the temperatures must have been extremely high, of the order of magnitude of 10^9. We do not know much about the maximum density of the compressed primordial state of matter. Perhaps the preexpansion density of the Universe resembled that of a nuclear fluid. Again, Gamow does not inform us whether this fluid was a gas or a liquid in that precompression state. But since the Universe is supposed to be infinite, the space outside the compressed primordial state of matter must have been, in concordance with Gamow's reasoning, entirely populated by matter, too. And, if we assume that matter occupies an infinite space (infinite Universe), and that it can be either compressed or expanded and still fill the same infinite space, then we are confronted with a paradox of infinity. Such a paradox was plausibly illustrated by Hilbert. It is still neither resolved nor dissolved.

Gamow provided a physically impressive explanation of why and how the expansion of the Universe started. He postulated the so-called "Big Squeeze," which occurred sometime in the very early history of the Universe. This Squeeze was due to the collapse which took place at an even earlier state in the development of the Universe. The expansion which we observe today is no more than an "elastic rebound." It began when the maximum density of primordial mass had been reached. We do not know anything positive about the presqueeze phase of the Universe. But we postulate, not unreasonably, that almost immediately after the maximum compression had been reached, expansion began. Perhaps it is also probable that in the precollapse state there existed an ocean of neutrons, protons, and electrons. But all this is sheer speculation and certainly not a theory based on any cogent arguments or empirical evidence.

Gamow proposed 1.7×10^9 yr as the optimal "astronomical" value for the occurrence of the Big Squeeze. With Lemaître, he agreed that expansion is accelerated. This implies that the recession velocities of all the neighboring galaxies were much smaller in the past than now. Hence, the date for the origin, i.e., the age of the Universe, had to be shifted back deeper in time. It was a comparatively simple step to treat nebulae and galaxies as synonymous. However, like many other authors, he used the term intergalactic and interstellar in a very ambiguous manner. When we talk about "extragalactic" we know exactly what the term means. But when we talk about "intergalactic" or "interstellar," then those expressions may connote either that we refer to distances between galaxies or stars, or that we refer to the internal or inner content of a galaxy or star. Some continental workers use the term "intra" when they refer to what happens within these celestial bodies.

Behr computed that the distances between the galaxies are about twice as great as had been hitherto estimated. Consequently, Gamow had to raise the time of the start of the Big Squeeze to roughly 3.4×10^9 yr and postulate an open and infinite universe.

In contrast to Einstein's closed model, he suggested an open, limitless, infinite universe. Einstein's closed universe can undergo expansion up to a certain limit. Beyond that limit, expansion will stop and contraction will take place. A universe that is closed and periodic in space had to be, of course, also periodic in time. This would lead to a so-called oscillating or pulsating universe. Gamow claimed that it would be physically more defendable to assume the limitless "expansion" of the Universe, for it also entails its limitless "extension." After warning the reader not to confuse distance with age in cosmology, he suggested the present age of the Universe to be $t = 10^{17}$ sec. He was obviously wrong when he accepted the results of some workers that roughly 55% of matter in the Universe is hydrogen and c. 44% helium. We are left with 1%, which was to account for all heavier elements.

The most original contribution of Gamow to cosmology is probably his hypothesis of the "Ylem." This almost unknown term means, according to Webster's dictionary, "the first substance from which the elements were supposed to be formed." He finally postulated, together with several collaborators, the original state of matter to be a *hot nuclear gas* (not a liquid). During the first minutes when expansion started and the temperature of the Universe was of the order of magnitude 10^9 (at least 3×10^9 deg), obviously no composite nuclei could have been formed. Thus, the state of matter was envisioned to be the formulation of a hot gas by nuclear particles such as neutrons, protons, and electrons. We may recall that free neutrons are unstable and decay into protons and electrons, on the average, within about 12 min—of course after having been pushed or kicked out of the nucleus.

But successive decay of neutrons, β^- decay $(n \to p + e^- + \bar{\nu})$ will be made up for by the formation of new neutrons by means of a reverse process. That is, collisions of protons and electrons result in a neutron-producing reaction $(p + e^- \to n + \nu)$. Already during the first hour, the temperature decreased (at the end of the first hour after the expansion started) and this dropping of the temperature was conducive to the "aggregation process" in which the still remaining neutrons attached themselves to protons. After that aggregation hour, no neutrons were left and the temperature of the Ylem fell below the limit which is necessary to get proton reactions.

Indeed, it sounds rather bizarre that according to the Ylem hypothesis the entire formation of atom-building required less than one hour. Gamow's answer was that the nuclear chain reaction in an exploding atom bomb takes no more than a few microseconds. And yet, some products of radioactive fission are still present even several years after the explosion.

Perhaps not to the same extent as Lemaître, Gamow was inclined to resort to the equations of GRT in order to account for the universal expansion and try to reconcile those equations with empirical data obtained by divers nuclear reactions. His criterion for the legitimacy of this relation was based on the circumstance as to whether or not the computations led to results that were compatible with, and similar to, the observed abundances of the various known atomic species.

Whereas Gamow's theory, like that of Lemaître, may furnish a feasible account for the formation of heavier elements, his attempt at explaining the formation of light elements (apart from hydrogen and helium) was unsuccessful, and so were the hypotheses offered by Fermi and Turkevich.

When the totality of adequate species (chemical constitution of the elements) had been completed during the first hour of expansion, the next 3×10^7 yr was without any significant events. The primordial hot gas expanded continuously, with the result that the temperature dropped more and more. Let us assume that the temperature decreased from the initial magnitude of several 10^9 deg to that of only a few 10^3 deg. Then the part of the gas produced by vapors of diverse elements (having high melting points) condensed into very fine dust (molecules). This dust kept on floating in the already existing mixture of hydrogen and helium. Gamow, for unknown reasons, omitted to classify the primordial hot gas mixture as a plasma which transformed, by very drastic cooling, into hydrogen and helium. The mixture of dust and gas still prevails in interstellar space and causes the reddening (not Doppler redshift) of the very distant stars.

As soon as this interstellar (?) material came into being, it accumulated into enormous large clouds of irregular shapes. We designate today such clouds as either luminous or dark nebulae, depending entirely on whether or not these nebulae are illuminated by stars in close proximity. However, this process did not go on forever, because if it had, the Universe would today be no more than a highly diluted mixture of gas and dust and the cosmic temperature would be almost absolute zero. Such a state would be incompatible with the fact that we know for certain that matter throughout the Universe is differentiated into galaxies, stars, and planets. Nonetheless, apart from galaxies, stars, and planets, we have reliable evidence that the Universe throughout contains a gas-dust mixture.

Jeans had contended that a gas subjected to gravitational forces and occupying an unlimited space will be unstable and fragmentize into giant gas clouds. But later computations showed that this applies exclusively to a nonexpanding gas. It is not quite clear why the expansion of the Universe was responsible for the formation of matter as we know it today. One explanation would be that as soon as the mass density of radiation dropped below the density of ordinary matter, it became the prominent factor in the evolutionary process. Jeans' theory is only in part valid and can no longer be upheld. But neither does Gamow's account of the formation of gaseous protogalaxies answer some very important questions. For instance, we assume today that space between the various galaxies is, for all practical purposes, vacuous. And this could imply *realiter* that the galaxies existing in their present state are in no way altogether subjected to mutual gravitational attraction.

In consonance with Gamow's Big Squeeze hypothesis, one can maintain that when the galaxies were separated, their mutual distances amounted merely to 1% of what we assume they are now; it follows that their gravitational interaction was at least 100 times as great in the past as it is now. It makes sense, therefore, to contend that the mutual recession of galaxies was initially still considerably retarded by their mutual attraction due to gravitational forces. To quote Gamow, *"The original break-up of the uniformly expanding material of the universe took place at the time when this material ceased to be 'gravitationally coherent.' "*[13]

Since Gamow's whole hypothesis is based upon the assumption of gradual evolution, we have to accept the conjecture that the protogalaxies, produced completely by cooled gas, did not yet contain stars. His own explanation is rather unsatisfactory because it is not original and often very vague. He must have realized that himself because he has recourse to von Weizsäcker's quite persuasively proposed turbulence motion theory. If one combines this turbulence motion with the rotation theory, then one arrives at a condensation. This condensation of the initially cool, dark, gaseous protogalaxies led finally to the huge number of shining stars. And this entire process may have required a period of no more than some 10^8 years. It is highly probable that the cool and dark gas of the initial protogalaxies condensed quite rapidly and thus may account for a whole avalanche of stars appearing in the respective galaxies. According to Gamow, however, the dust cloud became eventually so large and heavy that it increased gravitational attraction and pulled interstellar gas as well as greater quantities of dust into itself and thus ultimately became a nucleus of a new growing star. Assuming the existence of giant intergalactic nebulae, we can easily imagine that new stars would be created. But Gamow considered such a development an exception rather than a normal process.

Two analogous evolutions occurred during the formation of our Universe. First, we recall the initial fragmentation or breakup of the uniformly (?) expanding primordial gas, which resulted in billions of individual, i.e., separated galaxies. Second, the material within each galaxy condensed and produced many billions of individual stars. To the unsophisticated observer, it seems that the galaxies are scattered in space in a rather random fashion. It has been accepted today, however, that very often galaxies are clustered into groups and that these groups frequently consist of a few hundred singular galaxies. Thus, our own galaxy belongs to the so-called "Local Group." This Local Group is made up of three spiral (e.g., Andromeda), six elliptical, and four irregular galaxies.[5] As noted above, Gamow, in his classification of galactic shapes, mentioned that we distinguish between an ordinary and a barred spiral.

But let us return to von Weizsäcker's theory, which can boast of many followers in Europe and in this country, too. He postulated that the primordial gas, when expansion began, was not as homogeneous as Gamow thought. Of course, this viewpoint has a certain affinity with Lemaître's claim that the primordial gas was definitely inhomogeneous. Von Weizsäcker held that the process of expansion (probably homogeneous) was interfered with by a kind of turbulence. Further, he assumed that the motion (regular, or uniform) of the expanding gas became fragmented into an enormous number of turbulent eddies of different sizes and shapes.

Gamow was not satisfied with this account because it would presuppose turbulence in the primordial state, whereas he regarded turbulence as a "natural consequence of the expansion process." Maxwell argued that rotational forces would destroy any condensation the very moment it began to consolidate. From his premises he showed that gravitational condensations could never have taken place unless one assumes that the amount of material in the rotating, gaseous disk (that allegedly surrounded the still very young Sun) was at least 100 times greater than the entire mass of all the planets.

On those grounds, the Kant-Laplace hypothesis was relinquished. Cosmologists (or cosmogonists) reverted again to the collision hypothesis of Buffon. But it is a fact, known to all cosmologists, that the "rejuvenated" collision hypothesis, presented by Jeans, Moulton, and Chamberlin, respectively, did not succeed in convincing critical workers in this domain. The whole issue—Gamow regarded it perhaps too zealously as a paradox—was treated in a plausible manner by von Weizsäcker; he maintained correctly that Maxwell's objection cannot be considered as valid because of our advanced knowledge with respect to the chemical constitution of cosmic matter. It is indubitably a very original aspect of von Weizsäcker's theory that he recognized the significant contribution to cosmology of introducing turbulent motion and rotation as the main factors in producing condensation. His theory provided an intelligent explanation for the formation of stars within the gaseous protogalaxies. Moreover, it was explicated in a reasonable manner why the galactic clouds produced billions of stars in contrast to the solar disk that produced merely a very small number of planets. In short, von Weizsäcker and later Kuiper enabled us to appreciate the crucial role turbulent motion played "within the solar nebula." I think one cannot question the fact that, as far as planetary formation is concerned, von Weizsäcker and

[5] In 1960, ten ellipticals were counted, but the number for the seven other types of galaxies was not revised.

Kuiper have succeeded in presenting a reasonable, credible, and not merely speculative theory of planetary formation. The question as to how stars developed in the galaxies has been less satisfactorily accounted for than von Weizsäcker's explanation of planetary formation. Still, even his hypothesis concerning the birth of stars makes more sense than most of the attempts made by other cosmologists. Von Weizsäcker's theory seems also, on the whole, to coincide with some of Gamow's ideas, although he pointed out that some aspects of von Weizsäcker's hypothesis are not compatible with his own views.

We are still ignorant concerning the density distribution in a star that is *in statu nascendi*. But we know that whenever a star is created by gravitational attraction, its luminosity and surface temperature will grow continuously and very rapidly with time. Eddington was the first to present a scientifically plausible theory of stellar structure. For instance, he demonstrated that a certain kind of nuclear reaction will be able to arrest contraction. He examined the physical conditions that have to exist in the central domains of the stars constituting the "Main Sequence." The ingenious Hertzsprung–Russell diagram informs us about the logarithms of the luminosities of stars plotted against the various colors of the stars determined by their surface temperatures. This never-challenged diagram permits us to recognize that the majority of stars do exist along a rather narrow strip or path: It is called the "Main Sequence" and traverses the "tracks" of evolution due to contraction. Hence, Eddington's theory and the Hertzsprung diagram made it possible to obtain information about the structure of stars, which was scientifically a tremendous advance as far as stellar structure is concerned.

Unfortunately, we are still not able to gather enough statistical data to arrive at a sensible hypothesis about the early contractive phases of stellar evolution.

Astronomers have provided evidence that there are categories of stars prone to violent periodic explosions. As pulsating stars display an enormous range of periods, the time intervals between explosions on or in the stars differ within very wide limits. Gamow and Longmire have convincingly demonstrated that a star may change to a state where its radius will undergo very slow periodic variations. I think there is no doubt whatsoever among astronomers or cosmologists that belief in the existence of pulsating stars is not based any longer on a hypothesis but rather on hard empirical evidence.

The so-called "novae" do not explode more than once. These nova explosions can be observed in our galaxy in rather large numbers. In contrast, the "supernovae" explosions differ from the novae explosions in that whenever a supernova explodes, a star increases its luminosity at least a billion times. Supernova explosions are rather rare, compared with nova explosions. It has been found that, on the whole, only one supernova explosion occurs in four centuries. Yet, Tycho Brahe (1572) and Kepler (1604) reported two supernova explosions in the Milky Way. We have no idea as to why these nova and supernova explosions take place. We can only hope that astrophysical researches will soon inform us about the reasons for these spontaneous explosion processes. Gamow was inclined to interpret stellar explosions as successive phases in the evolution of an aging star. Yet he was forced to admit that there are aging stars which do not undergo spontaneous explosions. As a matter of fact, stars which originated much more recently than older ones aged more rapidly than older stars. We know that the so-called "White Dwarfs" have been considered dead stars, despite their enormous density and high temperature. For instance, the Companion of Sirius is a characteristic example of a White Dwarf. White Dwarfs form the ultimate phase of stellar contraction. And this means that they do not possess any hydrogen at all.

Gamow, as we have seen, proposed in his theory that the matter occupying the Universe must have been the result of the Big Squeeze, which later broke up and formed the primordial Ylem of neutrons, protons, and electrons. The Ylem cooled down very rapidly owing to expansion. The above elementary particles started to closely adhere to one another and produced aggregates of various degrees of complexity, and these became the prototypes of our present atomic nuclei. That this early process of "nuclear cooking" should have lasted no more than an hour seems to me to be an entirely arbitrary assumption. It is difficult to conceive that the primordial gas was homogeneous. Lemaître convincingly claimed that the primordial "substance" was inhomogeneous; it only became homogeneous later. The so-called protogalaxies arise due to Newtonian gravitation. But this would logically imply, if we accept Gamow's hypothesis, that these protogalaxies were inhomogeneous. How all this is reconcilable with our fundamental presupposition of homogeneity, valid for the whole Universe, is difficult to grasp. It may be parenthetically pointed out that in current cosmological literature Gamow's hypotheses of the Big Squeeze and of the Ylem are often called the "Big-Bang" theory. I can assure the reader that he never in his writings, lectures, or in personal conversations with me ever employed this expression. Unconfirmed rumor has it that Hoyle coined this term rather facetiously.

Perhaps at this point we may sum up by saying that the models of Einstein, de Sitter, Friedmann, Lemaître, and Gamow stemmed from GRT. They are cosmological hypotheses based on some more or less plausible extrapolations. Whereas one can derive from Einstein, de Sitter, and Friedmann a variety of models from each of their multifaceted hypotheses, Lemaître and Gamow advocated specific theories, i.e., each advanced only one type of cosmology. We have already seen in the foregoing that Gamow's theory is, to a great extent, a further development of Lemaître's original model.

Eddington as well as Friedmann realized that Einstein's spherical universe was dynamically unstable. Eddington discovered independently that the actual Universe is best pictured as an unstable equilibrium between two opposing forces: cosmic gravitation and cosmic repulsion. But he, too, identified this repulsion with a positive Λ constant. According to him, the Universe is now expanding with much greater velocity than in the past. Hence, it must be much older than what is given by the relativistic equations for the universal expansion. As far as the rate of expansion is concerned, Gamow, in contrast to Eddington, was extremely imprecise. First, he seemed to propose, like Lemaître, an ever-increasing velocity, that is, the acceleration of the expansion process. But later on, without any supporting arguments, he pleaded for a universal (constant) expansion. Eddington's model, in contrast to Gamow and his cited precursors, is nonrelativistic. Yet he as well as the cosmologists espousing classical, if not dogmatic, GRT considered space (not galaxies) to be itself expanding and "carrying with it the individual galaxies and clusters like straws in a stream." In view of our later discussion of some constants in cosmology, it might be important to emphasize already here that Eddington was firmly convinced that the expansion of the Universe in no way affects fundamental physical constants. Thus, G is a genuine constant, comparable to a mathematical invariant, whereas ρ and T (the reciprocal of Hubble's constant α) vary with the arrow of time. Eddington made some very imaginative contributions to the structure of stars and to cosmology in general, but his highly controversial *Fundamental Theory*[10] and his cosmological hypothesis, though they contain some valuable points, are no longer regarded as of significant value among modern cosmologists. Besides, Slater discovered some rather serious mistakes in Eddington's computations.

Before we take a brief glance at Milne's model, we ought to bear in mind that the rate of expansion of the Universe seems to be still an open question. Hubble's observations appear to support the view that the galaxies scattered throughout the space of the Universe are moving away from us and, moreover, that the more distant they are, the more rapidly they move. Yet, about 20 years ago, Humason, Mayall, and Sandage[19] maintained to have discovered reliable evidence for their claim that the expansion of the Universe has probably been more rapid in the past than today (as already anticipated by Eddington). It follows that the age of the Universe could be less than the age calculated upon the uniform expansion hypothesis. According to Robertson, the so-called "age of the universe" t_0, is not accessible to "direct observation." In this connection Whitrow correctly pointed out that the *Einstein–de Sitter* model of the Universe contained already the notion that the rate of expansion was greater in the past than today. However, it was much smaller than the expansion velocity suggested by Humason, Mayall, and Sandage. When Gamow accepted the relativistic equations for universal expansion, he must have had in mind one particular phase in Einstein's cosmology, but not the *Einstein–de Sitter* model.

Milne's cosmology[25,26] based the expansion of the Universe on the Second Law of Thermodynamics. The relativistic hypotheses of the models of the Universe assumed, as necessary premises, homogeneity and isotropy. Milne started with a postulate which he called the *Cosmological Principle*. He accepted Hubble's law in such a way that it would hold at all periods of time: His so-called kinematic relativity was in no way similar or related to Einstein's relativity theory. His cosmology attempted to show that Newtonian "analogs" could easily be substituted for the diverse expanding hypotheses of the Universe based on relativistic cosmology. His uniformly expanding model of the Universe (which had no dynamical properties) represents a closed system in the sense very well summarized by Milne's former student Whitrow.[6] This closed model is nevertheless infinite as far as material content is concerned. The Universe is closed, that is, "Any event occurring anywhere at any time is in principle communicable to the observer if he waits long enough to receive an optical or electro-magnetic signal." Of course, Milne's model rests upon observations and experiments that cannot be actually performed. But since these observations or experiments are possible in principle, we cannot call them imaginary or mental, i.e., they are not *Gedankenexperimente.*

In view of our future reference to the steady-state theory, it may be appropriate to give here for his uniformly expanding model of the Universe the formula

$$\tfrac{4}{3}\,\pi G\rho t^2 = 1$$

where ρ is the average density at a time t. It is relevant to note that in this equation, G is not a genuine constant but increases proportionally with the time t.

I think it can be claimed that relativistic cosmology is generally regarded to be evolutionary and assumes a definite beginning or creation of the Universe in the finite past. One may agree with Bonnor that in GRT "space itself" expands and carries "the nebulae [galaxies] with it—like leaves in the wind—and not all nebulae" move "away from each other through . . . emptiness."[5]

Cosmology deals mainly with two kinds of probable models. For instance, we may conceive that expansion of the Universe will never cease. Consequently, the galaxies one can observe are bound to become more and more faint. Furthermore, the average density of matter will continually decrease. Alternatively, we may envisage that expansion is retarding and will, at some point in time, change over into contraction. Both alternatives lead to the conclusion that expansion began about 8×10^9 yr ago.

Bonnor does not agree with this interpretation of GRT cosmology. He does not deny that, at the creation of the expansion of the Universe in time, the differential equations appearing in evolutionary theories, which assume a definite beginning of the expansion in time, will become infinite and therefore lead to a mathematical singularity. And, as we all know, a singularity occurring in mathematical equations applying to a physical event or problem is tantamount to a breakdown of the physical theory in question. Bonnor therefore suggests that we change many of the cosmological models that accept a definite numeri-

---

[6]Whitrow has given an excellent exposition and analysis of Eddington's and Milne's cosmological theories.

cal estimate for the start of the expansion. According to him, those cosmological theories or models have simply to be changed so that no singularities occur when we extrapolate back into the remote past. Models of expansion, changing either gradually or rapidly to contraction, contain some highly unsatisfactory aspects. I do not mean that contraction implies that the galaxies approach each other closer and closer and finally arrive at infinite density. Because if this condition is reached, we are confronted again with a singularity as in the case of expansion or creation. In other words, the Universe becomes an infinite series of pulsations or oscillations, yet it still contains singularities which we are most anxious to eliminate from any physical theory. Heckmann (1930) maintained that matter throughout the Universe may have a slight, hitherto undetected rotation.[14,15] And this would remove the possibility of infinite density. To put it more precisely: Owing to the centrifugal force of the rotation, contraction may become reversed as soon as the Universe has reached not an infinite density, but a finite density at the completion of one of its never-ending oscillations.

Bonnor, despite his occasional criticism, is today, no doubt, one of the most ardent protagonists not only of relativistic cosmology, but of GRT as a gravitational theory. According to him, Einstein's field equations hold for "all states of the universe." He realizes that the Second Law of Thermodynamics is obviously incompatible with the hypothesis of a never-ending sequence of expansions and contractions. We are reminded of Milne, who asserted that the second law considered the expansion of the Universe as a "supreme manifestation" of the trend of entropy to reach a maximum. In its classical form, however, the entropy law was definitely supposed to apply only to closed or rather thermally insulated systems. Entropy was originally assumed to increase continually with the flux or growth of time. Bonnor does not seem to be convinced, however, that the Second Law applies to the Universe at large. Of course, it depends whether one considers the Universe as a closed or open system, quite apart of the condition of being thermally isolated.

His contention that the Universe is in its past and future unlimited is certainly not in agreement with orthodox GRT. Since the model of an expanding Universe makes more sense than any other interpretation of GRT, it follows that recession of galaxies necessarily implies that the Universe will become more and more thinned out and that the average density of matter will decrease. And thus, we have to conclude that the present state of the Universe must differ significantly from that in the cosmic past and future. It is true that relativistic cosmology does not automatically entail the conception of a Universe with a finite history or creation in time. Bonnor contends—and this seems to be one of his main arguments in favor of relativistic cosmological models—that GRT has been confirmed as far as terrestrial confirmation is concerned; it also applies successfully to the solar system. But he ignores the fact that Einstein interpreted the evolution of the presently existing system of stars as some kind of beginning of a Universe. After all, it has become common practice in science to extrapolate laws of nature holding on Earth to the Universe at large. These laws, interpreted by GRT, do not vary with time. Bonnor—and I think here his "faith" is somewhat naive—maintains rather boldly that "the field equations of general relativity apply at all times."[5] It is the state of the Universe that undergoes changes with the passage of time. He differs from many relativistic cosmologists in that he avoids expressions like the "age of the Universe," and "creation in the past." He considers such views obvious defects of evolutionary cosmologies, i.e., theories based on GRT. Still, the notion of an unlimited past and future of the Universe would have certainly not been accepted by Einstein himself.

There cannot be any doubt that the steady-state theory, as conceived and elaborated at great detail by Bondi and Gold (1948), who were soon joined by Hoyle, caused a tremendous sensation among cosmologists. According to their theory,[2-4,16,17] the Universe is uniform, in space as well as in time. In the Universe there is a never-ending process of continual creation of matter. The space between the galaxies does not remain empty, because matter is constantly created. There are sufficient convincing grounds to assume that stars as well as galaxies age and finally burn out, i.e., die. The observed recession of the galaxies implies a diminishing density in an expanding Universe. But an unchanging Universe can only be maintained if fresh matter is created out of nothing so that the mean density remains constant and the Universe uniform. Thus, new galaxies are created in intergalactic space with the result that the average distance between galaxies remains constant despite the fact that, owing to the expansion of the Universe, the distance between the still-existing galaxies is continually increasing. It is a main tenet of the steady-state theory that there is no contradiction between continuous expansion, the growing intergalactic distance, and the creation of new matter, forming new galaxies. Of course, all this applies only to a large-scale model of the Universe, uniform and unchanging (Bondi).[4]

According to Bonnor, creation of matter out of nothing cannot be upheld because it clearly violates the principle of the conservation of energy as well as other conservation laws, like the conservation of linear momentum, angular momentum, and so on. The steady-state theorists (in contradistinction to the argument of the alleged violation of some conservation laws) stress the fact that we arrived at those conservation laws by observation; they emphasize that experiments performed in the laboratory support the view that matter and energy have been shown to be never precisely conserved. In other words, conservation of energy cannot be considered to be an exact law.

It seems to me that the whole discussion of conservation laws by the steady-state exponents is not quite satisfactory. Let us enumerate what we call "absolute conservations laws":

1. mass-energy
2. linear momentum
3. angular momentum
4. electron-family number

5. meson-family number
6. baryon-family number
7. T ⎫
8. CP ⎬ CPT
9. charge.

Further, there are certain relationships among the four types of interaction:

1. strong
2. weak
3. electromagnetic
4. gravitational

We have plausible reason to assume these absolute laws to be valid for the microcosmos as well as for the macrocosmos. The introduction of the neutrino by Pauli (1930) is only one, though highly significant, example of the attempt to save conservation of energy. Since then there have been many discoveries which strengthen our conviction that these absolute conservation laws are not merely idle speculations but represent true and valid states of affairs, empirically and theoretically.

If we accept the steady-state theory, then time is of no avail. In other words, the Universe in the past looked "just the same as it does now." Near and distant galaxies cannot be distinguished—they do not vary with distance. In contrast, the relativistic models of the Universe require variations with distance. Consequently, should we ever discover such variations, then the steady-state theory would have to be relinquished. Only very numerous and reliable observations can conclusively decide the issue.

Gamow was a very eloquent opponent of the steady-state hypothesis. He pointed out that it was unable to account for the beginning of the Biq Squeeze and therefore it failed to inform us about the origin of the various atomic species. Moreover, the steady-state theory was unable to explain the actually observed abundances of chemical elements. He stressed that observational evidence seemed to be in contradiction with the theory, which he called "artificial and unreal."

Relativistic cosmologies imply that at the early stages of cosmic evolution the density of the Universe was extremely high. Therefore the origin of the heavy elements coincided with the so-called primeval phase of the Universe. Bondi and his whole school of thought does not agree with this viewpoint. If it was possible some time in the past to synthesize heavy elements from hydrogen, then it should not only be possible but very probable that such things can take place also today. Well, it looks that a crucial part of Gamow's criticism of the steady-state theory was wrong. It is a fact that not too long ago we found evidence that heavy elements are synthesized in numerous stars. These stars later explode or disintegrate so that the observed abundant heavy elements produced in the center of the stars are dispersed throughout space.

One of the most enticing features of the steady-state theory was undoubtedly the fact that no singularities blemished it from a purely mathematical standpoint —in contrast to relativistic cosmologies.

It was Hoyle who felt very strongly about the fact that the steady-state theory did have nothing to say about gravitation. His awareness of this shortcoming led him to a rather impressive formula, namely,

$$\tfrac{4}{3}\,\pi\,G\rho T^2 = \tfrac{1}{2}$$

where ρ is the local average density, T the reciprocal of Hubble's constant α, and G the gravitational constant. We are here reminded of Milne's very similar formula, but there is an essential difference: In the formula of Hoyle, G, ρ, and T are genuine constants, while in Milne's formula the constants vary with the arrow of time.

In contrast to Bonnor, Bondi and his co-workers questioned the idea that we are fully justified in extrapolating from GRT to the whole Universe. Although Hoyle tried to save some aspects of GRT for the steady-state theory, he and later Bondi as well as Gold withdrew their brainchild from the scientific scene. Some observations turned out to be incompatible with the theory. I think that it is within the realm of possibility that some major modifications and radical revisions may suggest to us to consider this model of the Universe as worthwhile to retain, in a drastically changed form. It was not only one of many conceptual contributions to modern cosmology, but it inspired astronomers and astrophysicists in their empirical research. Even today one encounters all over the world quite a number of cosmologists who do not believe that the steady-state theory is dead and 'burned out.' For instance, the distinction between the ordinary Cosmological Principle, which supposes that the Universe is uniform in space, and the Perfect Cosmological Principle, which was introduced by Bondi and Gold and maintains that the Universe was uniform in space as well as time, seems to me fruitful enough to be always considered as a valid contribution worthwhile to be retained in future cosmological theories or models. Furthermore, the steady-state theory demands from us a thorough investigation as to the nature of the laws of physics. Do they vary with time? After all, to satisfy our intrinsic and aesthetic 'ideal' belief, time should not appear as an explicit quantity in physical laws! And finally, the steady-state theory has forced us to examine again the question of whether or not the extrapolation from terrestrial physical laws to cosmological phenomena has to be accepted as necessary to form models or theories or whether such an extrapolation is compatible with the state of the actual Universe. I personally am inclined to be very cautious with respect to this issue. One has always to consider possible pitfalls or errors when the *extrapolative distance* becomes too great. I introduce the expression 'extrapolative distance' in order to avoid all possible misinterpretation.

I should now like to discuss some of the novel aspects of a theory which, for some odd reason, never attracted the attention of cosmologists which it merited.

Lyttleton accepted the expansion of the Universe as a hard datum (to use Russell's favorite expression), that is, as an undeniable physical fact. This does not mean that we know for certain the way in which this expansion actually progresses. For instance, the receding galaxies could be accelerated in such a way that their velocities keep constantly the same ratio with respect to their distances. Alternatively, the velocities might remain constant in such a manner that the most rapidly moving have gone farthest away. Moreover, in order to understand expansion, we cannot rely merely on observation, but have to resort to a theory, hypothesis, model, or guiding principle. Lyttleton suggested a very interesting, if not fascinating, so-called electrical model which in no way takes gravitational forces into account.[23]

Already Einstein had stressed that "matter consists of electrically charged particles"—a firmly established fact. We know that protons and electrons, when interlocked, produce a hydrogen atom. It has been generally assumed that the charges of an electron and proton are exactly equal and opposite, i.e., negative and positive. We can measure those charges indirectly in the laboratory with considerable accuracy. Yet we realize that there does not exist any empirical evidence for precise equality. Lyttleton proposed on these grounds the possibility of a small charge difference. Let the proton have minimal greater charge than the electron. It follows that instead of those charges cancelling out to zero total charge, the proton and electron will produce a hydrogen atom with a slight positive charge excess. And this means that the charge excess, i.e., interpreted as electrical repulsion, will be greater than gravitational attraction. It is only necessary that the numerical value of the excess be one part in 10^{18}. It is an established fact that electrical forces are extremely strong compared with gravitational forces, which are very weak indeed. Thus, even a minute charge excess results in a significant repulsion. This electrical repulsion is sufficiently strong to result in an expansion of the Universe in a manner consistent with our observations. Let us assume, for the sake of argument, that matter is created in space in the form of hydrogen atoms or as equal numbers of free protons and electrons. Of course, electrical charge will be automatically produced at the same time. In this case, charge cannot be conserved, but we remember that the classical Maxwell equations inform us that the charge is always conserved. Lyttleton suggested that we modify these equations by introducing a few very minute terms which would justify the creation of charge. He found that those necessary terms are analogous to Einstein's cosmological terms which he inserted in his field equations and which are closely related to the size of the Universe.

Then Lyttleton raised the most relevant question: Assuming that the background material has charge excess, it will certainly not condense into dense clouds, since the charge excess dominates or outweighs gravitation—How, under these circumstances, can galaxies ever come into being? The answer is relatively simple. There is a condition that would allow us to distinguish one region of space from another: Namely, if cosmic material anywhere would become ionized. And this means that the electrical bonds closely linking protons and electrons to form hydrogen atoms should become severed or completely broken. It is common knowledge that such a process always takes place at sufficiently high temperature. Owing to this process, condensations come about and finally develop into galaxies (with their stars and planets) entirely by reason of gravitational forces. Throughout any ionized parts of space one will always encounter an excess number of electrons which will compensate for the somewhat greater charge of the proton. The result would be that every region of volume would consequently become electrically neutral. It follows that gravitational forces become dominant because the charge excess vanishes to zero, so that condensation into galactic bodies will continually take place.

Because the charge excess is numerically extremely small, cosmic material requires not more than that for every 10^{18} protons there should exist no more than $10^{18} + 2$ electrons.

Another relevant aspect of Lyttleton's theory may tell us how cosmic rays originate and traverse or penetrate the Universe with a speed approximating that of light. These rays owe their existence to the protons which are "expelled from the huge ionized regions as they rid themselves of the continually-arriving slight charge-excess." The ratio of the electrical to the gravitational force between a photon and an electron amounts to $\sim 10^{39}$. If we consider the ratio of two known lengths, namely, the size of the Universe (i.e., the limiting distance which permits possible physical observation) and the radius of the electron, then we obtain again $\sim 10^{39}$. Like Eddington, Lyttleton combines pure numbers with those like our customary empirical units of mass, length, and time. In this vein, he divides the whole mass of the Universe by the mass of the hydrogen atom and he gets 10^{78} to within a very small numerical factor; thus one arrives at the square of 10^{39}. Lyttleton seems to doubt whether these rather amazing coincidences are due to pure chance. Like many other physicists and cosmologists, he sees in those numerical coincidences something rather profound with regard to the relation of the atom to the (observable) Universe. Finally, if we assume the electrical theory to account for the expansion of the Universe and if it should turn out to prove correct in every respect, this could only mean that 10^{39} is the inverse square of the charge-excess number, which is a "moderate" multiple of 10^{-18}.[23]

In conclusion, we have to inform the reader that repeated sound tests have shown that Lyttleton's postulate of a very minute charge excess has not been supported by empirical evidence. Hence, the theory is untenable and has therefore been discarded. If I am not mistaken, Bondi showed a certain sympathy for Lyttleton's theory. One should perhaps mention here that Lyttleton favored the steady-state theory, which is certainly, in many respects, compatible with his own cosmological hypothesis. The advantage of Lyttleton's postulate was that it was testable and not simply axiomatic. Had Lyttleton's electrical theory been proved correct, the dream of combining cosmology, quantum theory, and general relativity would have come closer to physical realization.

Unfortunately, shortage of space and time allows only for very brief mention

of the Robertson-Walker models of the Universe.[27,28,30] Sciama investigated
the relation between the models conceived by Robertson and Walker and the
actual Universe, which contains a huge amount of irregularities.[29] These irregularities reach a scale of at least 1 Mpc (cluster of galaxies) and may perhaps reach
out to a scale of 50 Mpc (in the case of superclustering) or perhaps even further.
Sciama supports the view that the Universe as a whole was in the past very much
denser than the material found at present in galaxies. The formation process of
galaxies is still not adequately explained, nor do we have any clear idea of how
the galaxies could originate in an expanding Universe. Furthermore, assuming
that the Universe was uniform some time in the past, then the rather small irregularities due to statistical fluctuations could not have evolved into galaxies
during the available time. Therefore, one is compelled to suppose that large density fluctuations occurred already in the early stages of the Universe. We have no
knowledge as to the cause of large fluctuations. Some authors have suggested
that they were the result of a very early state of collapse of the Universe, preceding the present phase of cosmic expansion. But nobody knows whether this
hypothesis has more merit than other explanations for large density fluctuations.

All these points are crucially related to another most vexing problem,
namely, that of the singular moment at $t = 0$ in the Robertson-Walker models
at a time when the density of the Universe was allegedly infinite. Some authors
were of the opinion that this singularity is actually no cause for any investigation, since it could easily be the result of the "artificially exact symmetry assumed in these models."[29] It has been maintained that in an irregular universe
no singularity need to have ever existed. But this assertion is definitely wrong, as
was demonstrated by Hawking, Ellis, and Penrose. If one appeals to quantum
mechanics, one might be able to ignore a "literal singularity," yet quantum
mechanics cannot ignore the probability that the density will become very high,
no less than 10^{59} g cm^{-3}. And one cannot deny that such a high density could
legitimately be considered as representing a singularity. However, we are still
quite ignorant concerning the significant characteristics of the singularities in
our irregular Universe. Robertson himself seems to have flirted with the idea of
rejecting the controversial cosmological constant Λ. Yet he still invoked the field
equations of GRT. Further, like Einstein, he realized the danger of regarding
t_0 as the age of the Universe. It is certainly beyond the limits imposed upon us
by direct observation. None the less, he regarded t_0 as a still useful and significant parameter. The Robertson-Walker model is, despite all of its modifications,
basically a relativistic hypothesis or theory. Robertson expected that "any
universe which expands without limit [i.e., continual acceleration] will approach the empty de Sitter case and that its ultimate fate is a state in which
each physical unit . . . is the only thing which exists within its own observable
universe."[37] Historically speaking, we should not forget that he—a mathematician like Friedmann—was the first to elaborate significantly some aspects of
the relativistic hypotheses of Friedmann and Lemaître. And we should also re-

mind the reader that the contributions of Friedmann and Lemaître, produced in the twenties, were almost entirely ignored by the majority of astronomers and cosmologists until Hubble's law appeared in print.[18]

It is customary to describe model universes of GRT as evolutionary. McVittie objects to this term because, according to him, evolution should be only ascribed to individual objects in the observed Universe. For instance, one may state that a star evolves and during a passage of time its chemical composition changes, it loses or gains mass, or its luminosity may vary. Also, an interstellar gas cloud may experience a sequence of phases so that it becomes ultimately transformed into a star. But all these individual objects are no more than discrete astronomical entities or bodies within our observed Universe. If we postulate a gas that fills a model universe, then we generalize in order to obtain an idealization. Consequently, the concept of evolution does not apply any longer and has to be discarded. An "evolutionary" model of the Universe is merely one in which density and pressure of the assumed, all-penetrating gas vary with the time. McVittie does not consider this state of affairs as compatible with the evolution attributed to observed individual objects in the universe. GRT cannot at present treat a so-called evolving universe, because it would require the solution of enormous mathematical problems that occur whenever we focus our attention on the discreteness of astronomical objects.[24]

Although Hubble's law and Hubble's constant have become established physical constituents of modern astronomy and cosmology, so that it may appear to be unwarranted to examine them as if they were in need of critical scrutiny, recent observations, together with ensuing theoretical considerations, suggest that it might be apposite to have another look at them. Suppose λ is the laboratory wavelength of a line that in the stellar spectrum appears as having wavelength $\lambda + d\lambda$, where $d\lambda$ is, of course, the change in λ. It follows that we can measure the displacement of the line in question by positing $\delta = d\lambda/\lambda$. This displacement does not depend on the chosen line, provided the observed datum is due to stellar motion. Thus we are able, by resorting to the displacement of the lines in spectrum of a star, to determine the stellar radial velocity. We know, from empirical evidence, that the shift (displacement) can be pointing to the long-wavelength, i.e., the red, end or the short-wavelength, i.e., the violet, end of the spectrum. There exists a very close relationship between the magnitude of the radial velocity denoted by v_R, the displacement δ, and c, the velocity of light. The so-called Doppler formula expressing this relation is

$$v_R = c\delta$$

We notice that there is no mention made as yet of Hubble's law or Hubble's constant.

Hubble's law is customarily formulated by $v = \alpha r$, where α denotes the constant of proportionality, r the radial distance, and v the radial velocity; α is Hubble's constant. To state it explicitly: Observational data lead us to propound an intimate relationship between redshift and distance; the relevant factor in this relation is supposedly α, the proportionality constant. But if we adhere for a while to the aforementioned Doppler formula and extrapolate from stellar radial velocity to applying $\lambda + d\lambda$ to the galaxy spectrum, we obtain—as McVittie put it—"a convenient measure of the redshift for the [displacement of] the line,"[24] viz. again the relation

$$\delta = d\lambda/\lambda$$

exactly the same connection as found with regard to the stellar spectrum. And now we have to decide, not merely by pure theoretical postulates, but by the results obtained by reliable observational data, whether or not δ remains constant or varies from line to line. One could not question the constancy of δ in a galactic spectrum if it were possible to measure redshift not only at optical wavelength, but at radio wavelength, too. Empirical evidence tells us that any increase of distance does incontrovertibly entail increase of velocity of recession. But we have to keep in mind that the relationship between distance and redshift is not tantamount to simple proportionality. All we can say is that redshift provides a plausible account for the recession of galaxies, that is, for an expanding Universe.

McVittie was the first to reject the notion of Hubble's constant being a true constant of nature. He regarded it merely as a parameter. Moreover, he introduced this distinction between the Hubble parameter and the acceleration parameter, viz.

$$h_1 = R_0'/R_0, \qquad h_2 = R_0''/R_0$$

Hubble's parameter h_1 can also be denoted by H in order to make it clear that we refer to Hubble's misinterpreted quantity, his wrongly named "constant." Hubble's parameter was viewed as a constant, since it is apparently proportional to the square root of Einstein's cosmological constant λ. H varies with time, for it is inversely proportional to the exact instant t_0 when we make our observations. McVittie evaluated the Hubble parameter as basically the reciprocal of *a* time, not of *the* time. It should also be recalled that Hubble's "constant" α was conceived to represent a direct measure of the age of the Universe. However, since the notion of "age" automatically invokes the variable "time," it follows cogently that α (or δ) cannot be regarded a true constant of nature, because by definition a *genuine* physical constant is never allowed to be a function of time.

Parenthetically one should remind the reader that for a long time Hubble refused to accept the Doppler-effect interpretation of observed redshifts. It was due to the insight of Heckmann and McVittie that an at least plausible exegesis of Hubble's findings, namely, the Doppler effect, was finally attained. Yet we must never forget that we can only equate the velocity (of recession) with the product of the Hubble parameter and the corresponding distance if the redshift of the source is relatively small. In the case of larger redshifts the relationship becomes nonlinear. And when we deal with small redshifts, *all* recessional velocities reduce automatically to the Doppler formula, i.e., to $c\delta$. McVittie found that the Doppler relation is in no way a unique one—there are several ways of correlating an observed displacement δ with a recessional velocity or rather with stellar or galactic velocities in general. Fundamentally, any adequate analysis of the problems arising from Hubble's law and Hubble's parameter depend ultimately upon an explicit, not only operational, definition of distance. I am convinced that resolving all the questions arising from a sensible interpretation of the whole 'Hubble complex' will enable us to make significant progress in our search for a model of the Universe that in the most reasonable and consistent manner dovetails with our astronomical (astrophysical) observations.

Recent research has shown that there exist undeniable redshift anomalies. Further work, already in progress, might lead to a drastic revision of Hubble's law or even to a new conception of the redshift phenomenon.[20,33]

So far our discussion has focused upon Λ, the force of repulsion ($\Lambda > 0$), and on Hubble's parameter H or the so-called constant $\alpha(\delta)$. We might even be forced to reexamine the scientific justification for our postulates of homogeneity and isotropy—principles that are meant to apply to large-scale cosmological models of the Universe alone. To be more precise: Only if we treat linear dimensions exceeding the magnitude of 10^9 parsecs can our cosmological studies be of any relevance. Dirac had already questioned the constancy of the large-scale dimensionless constants; he regarded them rather as variables changing their numerical values during cosmological phases, i.e., with the lapse of time.[7] Jordan, Dicke, and others have offered some very cogent arguments for their contention that the gravitational constant G (and consequently also Einstein's relativistic κ) is a function of time, viz., of the age of the Universe.

Dirac, utilizing an *ad hoc* cosmological principle, investigated the significance of Eddington's two celebrated (or notorious?) numbers 10^{39} and 10^{78}. He related them to the age of the Universe A and, propounding the thesis that units of time are given by atomic constants—for instance, $e^2/m_e c^3$—he presented the following combinations of diverse atomic constants and depicted them as units of time:

$$\frac{e^2}{m_e c^3}, \quad \frac{e^2}{m_N c^3}, \quad \frac{h}{m_e c^2}, \quad \frac{h}{m_N c^2}, \quad \frac{\hbar}{m_e c^2}, \quad \frac{\hbar}{m_N c^2}$$

Dirac concluded that large numbers of the order of magnitude 10^{39} and 10^{78} approximate very closely A and A^2. In other words, numbers such as 10^{39} and 10^{78} are not true or genuine constants of nature at all; they are functions of A and vary with respect to time.[7,32] The consequences of Dirac's cosmological principle are manifold. Let us cite only two of the most crucial ones: The principle of conservation of energy is incompatible with the hypothesis of permanent creation of matter (steady-state theory); Eddington's cosmic number N can no longer be regarded an invariant, because the number of particles (protons and electrons) in the Universe will increase with respect to A^2. Dirac's conjecture is, therefore, incompatible with Eddington's rather simple model of a Universe containing a finite, constant number of particles (protons and electrons). The steady-state theory seems to tend to concur with Dirac's views, at least as far as this issue is concerned. I never quite understood as to whether or not he assumed the atomic constants themselves, treated as singular entities as it were, to be genuine constants—"world-invariants" in Eddington's interpretation.[7-9]

Dicke and Brans envisaged field theories in which the parameter (not constant!) G satisfies a wave equation; their conjecture has the advantage that it can be tested observationally. And thus we may be able to decide eventually as to whether classical GRT or the Brans–Dicke modification, i.e., a 'patched up' GRT, carries more convincing power. It seems as if all fundamental constants in the atomic as well as in the cosmological realm are no true constants at all, but parameters or blunt variables. This would entail that our time-independent laws of nature are no more than wishful thinking—an aesthetic 'dream' rather than hard scientific truth, where theory accounts for the observational (empirical) data. It is needless to stress that conjectures such as constants of nature, homogeneity, isotropy, etc. are simpler and easier to deal with than parameters, variables, inhomogeneity, or anisotropy. We may have to relinquish our intrinsic tendency of formulating time-independent laws of nature and rely more and more upon statistics and the flow of time. The mathematician will then become the only scientist for whom elegance, namely utmost simplicity, is the ultimate aim. The cosmologist, like the theoretical physicist, could perhaps be made to realize that concepts such as symmetry, simplicity, and uniformity are inconsistent with empirical observations. Again, the *extrapolative distance* has to be considered when we try to formulate universally valid physical laws. The tools of modern cosmology no longer consist of speculations, hypotheses, and *a priori* assumptions. The astronomer and astrophysicist have at their disposal techniques such as optical telescopy, spectroscopy, radio and radar astronomy, infrared spectroscopy, and neutrino astronomy, and, moreover, they have discovered cosmic rays, pulsars (neutron stars), black holes, quasars, gravitons, stellar collapse, and other phenomena or concepts, which make it somewhat difficult to support any of the hitherto proposed models of the actual Universe as the most adequate and correct one.

Regrettably, it is impossible, owing to the restraint imposed upon the author, to discuss concepts such as ultraviolet catastrophe, the apparent symmetry of matter and antimatter, the primordial fireball, etc. In the solar system there are grounds to suppose that strong asymmetry prevails such that matter gains supremacy over antimatter. Does this asymmetry also apply to the whole Universe? It could be argued that the Andromeda nebula consists of antimatter instead of matter. But a decision with regard to this issue depends obviously on the information we hope to obtain from neutrino astronomy. Also, the discovery of magnetic monopoles—if it becomes an established fact—will certainly make an impact on the search for *the* cosmological model corresponding to observational data.

Indeed, cosmology is on the way to becoming a true physical science, despite the still speculative nature of most cosmological models.

REFERENCES

1. P. G. Bergmann, Cosmology as a Science, *Found. Phys.* **1**, 17 (1970).
2. H. Bondi, *Monthly Not. Roy. Astron. Soc. (London)* **cviii**, 252 (1948).
3. H. Bondi, *Cosmology* (Cambridge Univ. Press, 1952).
4. H. Bondi, in *Rival Theories of Cosmology*, ed. by R. A. Lyttleton (Oxford Univ. Press, London, 1960).
5. W. B. Bonnor, in *Rival Theories of Cosmology*, ed. by R. A. Lyttleton (Oxford Univ. Press, London, 1960).
6. W. de Sitter, *Kosmos* (Harvard Univ. Press, Cambridge, 1932).
7. P. A. M. Dirac, *Nature* **139**, 323 (1937); *Proc. Roy. Soc.*, **165**, 199 (1938).
8. A. S. Eddington, *Monthly Not. Roy. Astron. Soc. (London)* **xc**, 668, 678 (1930).
9. A. S. Eddington, *The Expanding Universe* (Cambridge Univ. Press, 1933).
10. A. S. Eddington, *Fundamental Theory* (Cambridge Univ. Press, 1946).
11. A. Einstein, *The Meaning of Relativity* (Methuen, London, 1946, 1950).
12. A. Einstein, *Relativity: The Special and General Theory* (Crown Publishers, New York, 1961).
13. G. Gamow, *The Creation of the Universe* (Viking, New York, 1952).
14. O. Heckmann, *Göttinger Nachrichten* **1932**, 97.
15. O. Heckmann, *Theorien der Kosmologie* (Springer, Berlin, 1968).
16. F Hoyle, *The Nature of the Universe* (Blackwell, Oxford, 1951).
17. F. Hoyle, *Frontiers of Astronomy* (Heinemann, London, 1955).
18. E. Hubble, *The Realm of the Nebulae* (Dover, New York, 1958).
19. M. L. Humason, N. U. Mayall, and A. R. Sandage, *Astron. J.* **61**, 97 (1956).
20. T. Jaakkola, M. Moles, J.-P. Vigier, J. C. Pecker, and W. Yourgrau, Cosmological Implications of Anomalous Redshifts—A Possible Working Hypothesis, *Found. Phys.* **5**, 257 (1975).
21. G. Lemaître, The Cosmological Constant, in *Albert Einstein: Philosopher–Scientist*, ed. by P. A. Schilpp (The Library of Living Philosophers, Evanston, Ill., 1949).
22. G. Lemaître, *The Primeval Atom* (Van Nostrand, New York, 1950).
23. R. A. Lyttleton, in *Rival Theories of Cosmology* (Oxford Univ. Press, London, 1960).
24. G. C. McVittie, *Fact and Theory in Cosmology* (Eyre & Spottiswoode, London, 1961).
25. E. A. Milne, *Relativity, Gravitation and World Structure* (Oxford Univ. Press, 1935).
26. E. A. Milne, *Kinematic Relativity* (Oxford Univ. Press, 1948).
27. H. P. Robertson, *Proc. Nat. Acad. Sci. (U.S.)* **xv**, 822 (1929).
28. H. P. Robertson, Cosmological Theory, in *Jubilee of Relativity Theory* (Birkhäuser, Basel, 1956).
29. D. W. Sciama, *Modern Cosmology* (Cambridge Univ. Press, 1972).
30. A. G. Walker, *Quart. J. Math.* **vi**, 89 (1935).
31. G. J. Whitrow, in *The Structure and Evolution of the Universe* (Harper, New York, 1959).
32. W. Yourgrau, Some Problems concerning Fundamental Constants in Physics, in *Current Issues in the Philosophy of Science*, ed. by H. Feigl and G. Maxwell (Holt, Rinehart and Winston, New York, 1961).
33. L. Nottale, J.-C. Pecker, J.-P. Vigier, and W. Yourgrau, *La Recherche* **7**(68), 529 (1976).

GENERAL SOURCES

P. Jordan, *Naturwiss.* **xxv**, 513 (1937); *Schwerkraft und Weltall* (Vieweg, Braunschweig, 1955).

G. C. McVittie, *Cosmological Theory* (Methuen, London, 1937).

G. C. McVittie, *General Relativity and Cosmology* (Chapman & Hall, London, 1956).

M. K. Munitz, *Space, Time and Creation* (The Free Press, Glencoe, Ill., 1957).

M. K. Munitz, *Theories of the Universe*, ed. by M. K. Munitz (The Free Press, Glencoe, Ill., 1957).

J. D. North, *The Measure of the Universe* (Oxford Univ. Press, 1967).

P. J. E. Peebles, *Physical Cosmology* (Princeton Univ. Press, 1971).

G. J. Whitrow, *Z. Astrophys. Ap.* **xiii**, 113 (1937).

G. J. Whitrow, in *Rival Theories of Cosmology*, ed. by R. A. Lyttleton (Oxford Univ. Press, London, 1960).

Chapter XIV

John Wyclyf on Time

Allen D. Breck

University of Denver

In considering the relationships between cosmology, history, and theology, let us go back in our thinking to the Middle Ages, particularly to the late medieval period, when the unity of all knowledge was still strongly defended, and the nature and origins of that unity hotly debated. Let us limit our focus to England, to the University of Oxford, and to the fourteenth century in particular. There the most forceful proponent of the unity of cosmology, history, and theology, and of the ability of man to perceive eternal truths concerning them, was the philosophical realist, John Wyclyf (1320-1384).

Our questions concern his dominant world views, the intellectual and literary tradition from which they stemmed (to the degree that they were not original with him), the nature of the attack on them, and the successive audiences which considered them. Time and history have served Wyclyf in very strange ways. In the six centuries since his death, almost none of his philosophical and theological writings have been translated into other languages from the original Latin. Despite the fact that in his own time he was a bold and in many ways an original, enthralling speaker and lecturer, no original manuscripts survive, and those of the second or later generations remained in collections scattered throughout Europe.[1]

In systematic medieval scholastic style he attacked the great problems concerning the nature of the Universe, its origins and operation, and the relationships of man with nature and the Divine. In time, these lectures were collected into a substantial *Summa De Ente* (sometimes called the *Summa Intellectualium*). Another collection, the *Summa Theologica*, is rather a series of important tracts on the Commandments, civil dominion, the truth of Scripture, the nature of the church, of the king, and the like,[2] all stemming from the first assumptions with

---

[1] See Manning[1] for an excellent introduction showing Wyclyf's place in medieval throught. A master work is Workman's *John Wyclif: A Study of the Medieval Church.*[2]

[2] *Shirley's Catalogue of the Latin Works of John Wyclif*, revised by Johann Loserth,[3] is in need of much revision and enlargement.

which we are concerned, viz., those dealing with the nature of universals and the eternity of time. It would be easy to say that Wyclyf was condemned solely for his treatises on the church, its hierarchy and mission, but those were only his conclusions, deriving from deep philosophical convictions. Consequently, the copying and printing of the works of such a thoroughgoing heretic were proscribed in England and on the continent. The fact that in 1417 the Council of Constance ordered his body disinterred, the bones burned, and the ashes poured into the river Avon is an adequate indication of the virulence of the opposition. The whole *corpus* of his writings was equally suspect, regardless of aim, content, or degree of orthodoxy—but in a real sense they were all of one piece.

Of course, manuscripts continued to be copied and circulated, though mostly in secret, as fires were ready both for the body and for the parchment. Archbishop Zbynek of Prague, to give only one example, committed to a public bonfire in 1410 all the Wyclyfite manuscripts he could find. Small wonder, then, that these writings were widely dispersed through the succeeding decades, and that, when printing became common in the 1450s, these major Latin writings were not included as suitable subjects. In substance and in detail, however, his influence had already been mediated to reformers through the Lollards in England and through Jan Hus and the Hussites in Bohemia, who in turn influenced Martin Luther and the continental theologian-philosophers.[4]

It was not until the late nineteenth century that Wyclyf's major works began to appear in print, and even then only in Latin, though with complete analyses, copious explanatory marginalia and introductions. This *corpus* was the work of the Wyclif Society of London, which from 1883 to 1922 issued 12 major philosophical and polemical writings in 36 volumes. Unfortunately for our greater understanding of his cosmology and 'anthropology,' substantial parts of his writings remain in manuscript. The *De Ente*, including two books, the first of seven, the second of six substantial treatises, has important parts yet to be published. S. Harrison Thomson, the dean of Wyclyf studies, has published the first and second tractates of book one,[5] and the present writer an edition of the *De Personarum Distinccione sive De Trinitate.*[6]

It is with one of those unpublished works that we are now concerned, the *De Tempore.* In the "grand design" it appears as tractate seven of book one, following logically treatises on Being in general, and on truths and universals in general. In sum, they give us answers to questions which start from the viewpoint of man toward the cosmos. Book two of the *De Ente* moves on to state and solve theological propositions concerning God's intellection, His will, the Trinity, 'ideas,' and His ability to create outside Himself. From references to earlier works in the manuscripts, the *De Tempore* has been dated between 1365 and 1370, before Wyclyf commenced his major works in theology.[3]

Divided into a preface and thirteen chapters, the *De Tempore* is a considerable expansion of Wyclyf's third treatise, *De Logica,* a work of his early

[3] Thomson's[7] is still the best dating of the manuscripts.

productive years, and one which needed much amplification. We shall have a look at its content later, but we first need to come to grips with an important problem, namely that of the authenticity of the text itself. What precisely *were* the conclusions of a fourteenth century Oxford professor in the ultrarealist tradition, when he turned his attention to the problem of time in its relation to cosmology on the one hand, and man in the world on the other? Let us look at the several manuscripts so far as they are known today.

One says 'known', for there are doubtless as many more copies somewhere as we have before us, awaiting identification in collections throughout the continent. In the absence of thorough catalogs and the superabundant presence of manuscripts without heading or ascription to any author, identification is rendered difficult. To give a single example: There are a number of known Wyclyf works in Poland ascribed to Thomas Aquinas as a means of preventing ready identification and consequent destruction. Parenthetically, I thought I had found one such work on 'time,' but it turned out, on close examination, to be a genuine Thomist work.

Despite these difficulties, it is now possible to identify positively 15 separate copies (in varying states of completeness and accuracy), scattered through Ireland (1), England (3), Czechoslovakia (6), Austria (1), Poland (1), Sweden (1), and Italy (2). They can be identified briefly as follows:

A. Trinity College, Dublin C.I.23 (350^a-387^a). This is a quite reliable text, but not the best; it survived the slaughter of the siege of Drogheda in 1649 and was subsequently sent to England. It had originally been part of the library of the famous Anglican Archbishop James Ussher, who in his *Annales Veteris et Novi Testamenti* of 1596 had come to the restful conclusion (based on Scripture) that the world had been created *in noctis illius initium, quae XXIII Octobris praecessit, in anno periodi Julianae 710*. Hence the famous date, 4004 B.C. Cromwell purchased the manuscript, among others, to be given "to Ireland", but Charles II at the Restoration gave it to Trinity College. It is a fifteenth century parchment manuscript, with rather nice rubrication of headings.

B. Trinity College, Cambridge B.16.2 (37^r-46 bis^r/2). This is a sumptuous manuscript with beautiful, though rough, initials and borders, some in bright gold, blue, and other colors. Dating from the end of the fourteenth century, it seems to be the best of the surviving manuscripts, having been copied by an English scribe for some wealthy owner.

C. Lincoln Cathedral C.I.15. This is a vellum manuscript of the late fourteenth or early fifteenth century, in a clear rendition in several late medieval book hands.

D. Gonville and Caius College, Cambridge 337/575 fo. 48*v*. It is bound, appropriately, with 47 folios of the *De Universalibus*, but all we have here is a single page, containing the chapter headings for each section. But that folio alone is enough to show us a somewhat different version of certain lines, and one which makes good sense.

E. Oesterr. Nationalbibliothek 4316, fos. 85^a-125^a. This is a paper manuscript of the fifteenth century, with rough red initials and some other rubrication. We know it was in the library at the time of the Emperor Charles VI (1685-1740), presumably brought here from Bohemia at the time of the imperial takeover.

F. Státní Knihova, Prague 535 (III.G.11), fos. 28^a-69^b. This is a paper manuscript, datable 1397, part of the vast copying effort of Czech scribes, working both in England and in Bohemia.

G. Státní Knihova, Prague 733 (IV.H.9), fos. 94^a-113^b. This is an early fifteenth century codex on parchment.

H. Státní Knihova, Prague 1553, fos. 87^a-113^b. This is a parchment of the late fourteenth century. We know this manuscript, together with others, to have been given to the Jesuits of the Clementinum by a certain Frau von Platenstein on the occasion of her conversion.

I. Prague, Lubkovice 153, fos. 75^a-109^a. This is a paper codex of the first quarter of the fifteenth century, by a scribe who signs himself *"Petrus de Wrbka, Bacc. Arcium."*

J. Prague Metropolitan Chapter Library 1410, fos. 170^a-210^b. This is paper, without rubrication, first half of the fifteenth century, a simple medieval book hand, deposited in the Archbishop's library on the Hradčany.

K. Prague, Metropolitan Chapter Library 1543, fos. 72^c-110^a. This is parchment of the late fourteenth century, with bold rough red capitals; a late medieval book hand.

L. Uniwersitet Jagiellonski, Krakow 848, fos. 72^c-96^d. Here we have a clearly identifiable Bohemian Latin script, signed by Mgr. Stephan Palecz, at the beginning of the fifteenth century.

M. Royal Library, Stockholm 22, fos. 1-33, in the handwriting of Jan Hus. Unhappily, this is not a good text: it must have been transcribed for some immediate purpose other than the transmission of a complete readable and accurate copy. Such codices were part of the booty brought back by Swedish troops during and after the Thirty Years' War.

N. Pavia, Universitá 311 (139.G.46), fos. 38^a-59^d, but bound with the order of folios badly mixed. A clear text, parchment of the fourteenth century.

O. Venezia, San Marco MS Lat. 172, fos. 1^a-27^c. It is by the hand of one "Michael Theotonicus de Prusia," perhaps a copyist working in Ferrara early in the fifteenth century. The manuscript was purchased at Padua in 1440 by Giovanni Marchanova. Other parts of the complete codex have been lost. It appears to be closer to the English than to the Bohemian tradition.

All the immediate foregoing is in aid of the conclusion that the recreation, the reassembling of the text on time, as it left the hands of John Wyclyf, is a difficult task. But it is also a rewarding one, for it enables us to rely somewhat less on what enemies and friends alike, as well as historians of medieval science, history, and philosophy, have said about his ideas, and the way they put them together. We should bear in mind that these surviving copies of copies often differ widely one from another, and that the production of something resembling the original is tantamount to assembling an ancient mammal from an incomplete collection of bones—only worse. We have no manuscript of the first, or, in all probability, of the second generation, although Trinity Cambridge B.16.2 and the fragment of the Caius codex are certainly early recensions. Nevertheless, despite all the difficulties (haste in copying, scribes whose Czech or German was better than the Latin of a man who thought in English, the wanderings, disappearances and misattributions of manuscripts), it is possible, indeed necessary, to produce a clear (even if not definitive) rendering of Wyclyf's contribution to the subject of time.

We may begin by noting his attributions to, and quotations from, the intellectual and literary tradition from which he derived his inspiration. C. S. Lewis in *The Discarded Image* goes too far when he says that medieval thinkers were reluctant to reject anything that had been written.[4] Still, his remark emphasizes the fact that such philosopher-theologians as Wyclyf expropriated the whole classical–Jewish–Christian *corpus* of writings to support their concept of the nature of God and the created world. What one authority said could (and indeed must) be reconciled with the *dicta* of another, or shown to be false with respect to some given standard.

For Wyclyf that standard is obviously Scripture; was he not the *Doctor Evangelicus*, "the Morning Star of the Reformation," in his concern that all assumptions about the faith, the church, man's actions, and his world-view (including time) be subordinated to another and higher order of things, as found in God's presence *in* Scripture? We have thus in the *De Tempore* 86 citations from the Bible—25 from the Old Testament, 56 from the New Testament, and five from the Apocrypha. "Does God predestinate?"; see Scripture. "What is the sin against the Holy Ghost?"; a chapter of the *De Tempore* is devoted to answering this question, again from a Scriptural point of view.

But nothing could be allowed to rest here on the grounds of faith alone. Medieval philosophers were sure that the pagan and Christian philosophers as well as the theologians were supportive of the search for 'Truth' and its manifestations. Hence we see a total of 29 references to Aristotle (read with the commentaries of Averroës): *The Metaphysics* (8), *The Physics* (11), *On Generation and Corruption* (8), and *On the Heavens* (2). Augustine comes next with 17 citations: his *Epistles* (4), the *Confessions* (3), the *Retractions* (3), three references to his sermons, two to *The City of God*, and one each to the *De Vera Religione* and to the *De Sermone Domini in Monte*.

Other authorities are quoted (read often in compilations, or *florilegia*, collections of "flowers"), There is a single reference each to Plato and Euclid, though

---

[4]Lewis,[8] p. 11: "They are indeed credulous of books. They find it hard to believe that anything an old *auctour* has said is simply untrue."

the latter receives much attention. Heraclitus appears twice, as does Boëthius (sixth century). We have one comment on Pelagius, and another on Arian, as well as John of Damascus. Writers from an age nearer to that of Wyclyf include Robert Grosseteste (5), Richard Fitzralph (2), Albertus Magnus commenting on Aristotle's *Physics* (2), Anselm (1), Hugh of St. Victor (1), and Gilbert de la Porrée (1). Thomas Aquinas is quoted only once, from the *Summa Contra Gentiles*. Wyclyf has 14 citations from his own works, including parts of the *De Ente*, the *De Accione*, or the *De Actibus Anime*. We have already seen this treatise in the embryonic *De Logica*, III.

So much, then, for Wyclyf's sources. Of some, such as Augustine, he had much available. Some he read through the mediation of other writers—quotations, handbooks, even stylized outlines of the proper organization of knowledge were the appropriation of every medieval scholar. But these (except for Scripture, and here he was a fundamentalist) were only supportive of his dogged pursuit of the Truth, wherever it might lead him. He thrived on controversy, opposition, and denunciation, but he came at them through hard scholarship. Goethe centuries later put it: *Es bildet ein Talent sich in der Stille, Doch ein Charakter in dem Strom der Welt.* "Talent develops in solitude, but character in the stream of life." He took his stand on almost all points (the nature of the Divine, for example) with a startling insistence that proofs are fundamentally not theological but philosophical, and available to us. Such calm certitude and rigorous dialectic made Wyclyf the most notable English thinker since William of Ockham.

The prevailing philosophical climate of Wyclyf's day was closer to what one might call 'conceptualism,' a sort of middle ground or folding-in of the two extremes of nominalism and realism. For him, however, the Christianized Platonism of Augustine and the Augustinian tradition in England constituted the one substantial way of looking at reality. Whereas Plato was silent or less precise about the structure of the world, Aristotle was found to be quite usable. In the *De Tempore* we have large sections devoted to the atomic character not only of nature but of time, and these devolve from Aristotelian concepts directly, and by quotation.

Wyclyf's theory of knowledge takes us to the heart of his thinking about God, man, and history.[5] By what means is it possible really to 'know'? Pure inspiration, the mystical way, unaided intuition, continuing tradition, and authority? No, rather in pure philosophy: *Deum esse probari potest per infallibilem demonstracionem a puro philosopho.* But how? We can—he continually contended against the logicians of his day—know certain propositions as unalterably true and without subject to modification, once we have our minds directed toward a right understanding of 'Being' and of its implications. His opponents charged him with being led by his inexorable logic to say of the subject 'Being' and its predicate 'to be' that there is one continuum of 'being,' and that particulars share in their universals. For, if God and man participate in 'being,' then

[5] Robson[9] gives a splendid analysis of Wyclyf's ideas in their context.

they are *to that degree* identical. Here we are in the midst of a vast medieval argument, or rather disputation, and tempted to contend, with some of Wyclyf's biographers, that he was a pantheist! But no, a vast gulf, of which the Scripture bears witness, exists, the one perfect, the other imperfect—the creator in contrast to the created.

Are there therefore real universals, types laid up in the mind of the Maker, distinct and coeternal with the Creator? Wyclyf finds this *factum* at the heart of his elaborate system. Universals do exist *ex parte rei*, hence such dogmas as the Incarnation and the Trinity are explicable. Of the latter, he says, *Deum esse trinum sicut oportet fideles quibus solus est nobis sermo concedere.*

Here we are at the heart of a great "chain of being," a flow of ideas (*ideas* in the Platonic sense). These included all Scripture as an emanation of the divine, God Himself, hence the proof-value of the Bible in disputation. We 'know,' for God implants in us the means of knowing and of realizing, as much as we can, the nontemporal, the universal beyond the singular which our senses can perceive. The truth we know exists "in being," and being is itself a radiation from the First Cause through many intermediate steps to its objects, which in their turn are, in a limited sense, 'causative.' At no point is Wyclyf concerned with what might be called 'pantheism.' Rather, he constructs two nonidentical realities, and here his concern for a proper understanding of time comes into play. There are, he says, two senses in which time exists.

The first is the eternal form, the universal, which we may call *duratio*, which is extratemporal in the human sense of measuring and counting. God apperceives 'durative time' at the moment of the Creation, and will see it again after the Last Judgement. Wyclyf comes to grips with the problem faced by Augustine and many others, that of the "foreknowledge" of God of all 'futures.' Do we encounter here a statement of determinism? To a degree, yes, but on the other hand, we need to see the 'other' side of reality—that which we may call 'temporal.' Here *tempus* is the particular, successive time, as conceived by men. As susbstance, it is mutable, corruptible. It has no identity with *duratio*, the universal, but it has a correspondence and a relation with it. Man measures hours, days, weeks, months, and years, and thus individuates, thinking simultaneously of the discontinuity as well as of the flow of that quantity which we call time.

We speak and think of the continuous flow of matter, and the continuity of time from one instant to another. But neither, says Wyclyf, takes place. We see (or think about) an infinity of space or an eternity of time, but all are measurable to God, to whom the Universe is fixed and unchangeable. Which sense of time, then, is the "really real"? According to Wyclyf's system, although we speak here on Earth of the 'movement' of things, or of their various parts or qualities, in *duratio* they have neither parts nor qualities, nor can they be truly moved in eternity. What we therefore perceive as time, that is, movement from one instant to another, just as movement from one physical point to another, is really part of a single present, as Idea, *cum omne preteritum vel futurum inplicat esse in eodem instanti.*

None of this reasoning was, however, intended to negate the importance of writing history: rather it always emphasized its moral dimension. The "mighty acts of God" in time thus became at once knowable and revealed at a lower level, and inscrutable and imponderable at another level.

The debate between Wyclyf and his opponents and their successors continued until the first quarter of the fifteenth century, by which time the opposing cause had pretty much swept the day philosophically. The state was actively prosecuting the Lollards who formed the underground of the later English Reformation. The separation between the sciences and the humanities, between philosophy and religion, so familiar to us in our time, had by the advent of another century or so come about.

REFERENCES

1. Bernard L. Manning, "Wyclif," in *The Cambridge Medieval History*, Vol. VII, pp. 486–507.
2. H. B. Workman, *John Wyclif: A Study of the Medieval Church*, 2 vols. (Clarendon Press, Oxford, 1926).
3. *Shirley's Catalogue of the Latin Words of John Wyclif*, revised by Johann Loserth (The Wyclif Society, London, n.d.).
4. O. Odlozilik, "Wycliffe's Influence Upon Central and Eastern Europe," *Slavonic Review* VII, 634–48 (1928-9).
5. S. Harrison Thomson (ed.), *Johannis Wyclif Summa de Ente Libri Primi Tractatus Primus et Secundus* (Clarendon Press, Oxford, 1930).
6. Allen D. Breck, *Johannis Wyclyf De Trinitate* (University of Colorado Press, Boulder, Colorado, 1962).
7. S. Harrison Thomson, *The Order of Writing of Wyclif's Philosophical Works* (Prague, 1929).
8. C. S. Lewis, *The Discarded Image: An Introduction to Medieval and Renaissance Literature* (The University Press, Cambridge, 1964).
9. J. A. Robson, *Wyclif and the Oxford Schools* (The University Press, Cambridge, 1961).

Chapter XV

The English Background to the Cosmology of Wright and Herschel

Michael Hoskin

University of Cambridge

By a remarkable coincidence, the two most notable contributors to cosmology in the middle and late eighteenth century had each learned his astronomy from the popular English textbooks of his day. The achievements of Thomas Wright of Durham and William Herschel were in stark contrast: Wright, an eccentric steeped in religious symbolism, was led to ask speculative questions about the location of Heaven and Hell and about the large-scale distribution of the stars which contemporary astronomers had neglected; Herschel, arguably the greatest observer of all time, constructed monster telescopes and turned them like long-range artillery against the problems of the construction of the heavens.[1] But the two men had in common their limited formal education, their insatiable curiosity, and access to the popular English astronomy books of the time. What, then, was to be learned from such books, and how successfully did they define the frontiers of knowledge of the "fixed" stars and of the universe in the large?

Herschel's sources we know (Hoskin,[6] pp. 20–24): *A Compleat System of Opticks* by Robert Smith, who had succeeded Cotes as Plumian professor of astronomy at Cambridge, and whose two-volume work, published in 1738,[16] included a "history of telescopical discoveries in the heavens"; and the immensely successful *Astronomy Explained upon Sir Isaac Newton's Principles* of James Ferguson[17] (himself an autodidact), the 1757 edition of which contained

---

[1] The principal published and manuscript works of Thomas Wright of Durham on astronomy are *An Original Theory or New Hypothesis of the Universe*,[1] "A Theory of the Universe,"[1] *Clavis Coelestis*,[2] and "Second or Singular Thoughts upon the Theory of the Universe."[3] A survey of his place in the history of astronomy is given by Hoskin.[4] William Herschel's complete papers, all but one from *Philosophical Transactions*, were reprinted in *The Scientific Papers of Sir William Herschel*.[5] The important cosmological papers are reprinted with discussion in Hoskin.[6]

a short chapter "Of the fixed stars." Wright's sources are less clearly established. He flattered himself that his fundamental insight into the structure of the Milky Way owed nothing to "Reading or Study," but it can be shown that he drew on William Whiston's *Astronomical Principles of Religion*[30] when writing his *Clavis Coelestis*,[2] and his Journal relates that in his teens he studied books on astronomy with such passion that "Father, by ill advice, think him mad. Burn all the Books he can get and endevour to prevent Study."[7] These books may well have included Christiaan Huygens' popular *Cosmotheoros*, of which English editions appeared under the title *The Celestial Worlds Discover'd*,[22] David Gregory's substantial *Astronomiae Physicae et Geometricae Elementa*, which was translated into English as *The Elements of Physical and Geometrical Astronomy*,[21] William Whiston's less demanding Cambridge lectures *Praelectiones Astronomicae*, translated as *Astronomical Lectures*,[23] John Keill's Oxford lectures *Introductio ad Veram Astronomiam*, translated as *An Introduction to the True Astronomy*,[31] and William Derham's *Astro-theology*.[24] Though other works could be added to the list, these books formed the basis of the astronomy readily available to the nonexpert. What did they have to say about the sidereal universe, and how competently did they define the frontiers of knowledge?[3]

That the stars are self-luminous like the Sun, and that they are scattered out to distant regions of space, was a commonplace by the beginning of the eighteenth century. The teaching of Descartes to this effect had had a lasting impact,[4] summed up by James Gregory when in 1668 he described our Sun as *stella fixa vicina*—"the local fixed star" (Gregory,[9] p. 147); and all our texts agree that the stars are distant suns. A more controversial question is that of the closeness of the similarity between the Sun and the stars—of the extent to which the stars are uniform. For Whiston, as we shall see, it seemed there was direct evidence of major intrinsic differences among stars. Other writers, undeterred by the known differences in size among the *planets*, were tempted to hypothesize that the differences in brightness which we observe among the stars are (largely or wholly) the result of the unequal distances at which they lie; so that one star looks to us fainter than another, not because it is truly smaller, but simply because it lies at a greater distance. This hypothesis could unlock the door to knowledge of the distances separating the stars from us, and so reveal their distribution in three-dimensional space. For light falls off with the square of the distance, and so if the Sun is measured as being n^2 times brighter than the star Sirius, then the Sun if removed to n times its present distance from us would look exactly as bright as Sirius. Basing ourselves on the hypothesis, we can go on to say that Sirius does in fact lie at n times the distance of the Sun from the Earth (or at n "astronomical units").

[2] See the introduction to the reprint of *Clavis Coelestis*.[2]
[3] It seems inappropriate at a conference to present a narrow piece of completed research. What follows represents "work in progress" and the conclusions are provisional.
[4] On the decisive role of Descartes, see Donahue.[8]

The problem of stellar distances had inspired the following notable British contributions in the closing decades of the seventeenth century:

(1) In 1668, in his *Geometriae pars universalis*, James Gregory set out a method for solving the challenging technical problem of actually determining n^2, the ratio of the apparent brightness of the Sun as compared with that of a bright star such as Sirius. He proposed using a planet as an intermediary between the Sun and Sirius. We are to observe the planet at a time when its brightness exactly equals that of Sirius, so that the problem then reduces to one of comparing the brightness of the Sun with that of the planet. But the planet's brightness depends upon the light it receives from the Sun (and therefore upon the brightness of the Sun), and upon quantities such as the size and reflectivity of the planet and distances within the solar system (quantities which we suppose to be accurately known). A simple calculation then yields the required n^2. Gregory himself obtained $n = 83,190$, but he tells us that with more accurate information on the solar system the figure would be greater still (Gregory,[9] p. 148).

(2) It had long been realized that if the Earth travels around the Sun every year, observers on the moving Earth could expect to see the stars apparently moving with an annual cycle; furthermore, if a star were observed every six months—that is, from opposite ends of a baseline equal to the diameter of the Earth's orbit—then the difference in its observed position (its "parallax") would, by elementary trigonometry, yield its distance. In 1674 Robert Hooke published "An attempt to prove the motion of the Earth" in which he discussed at length the frightening technical difficulties facing an attempt to make these delicate observations of an annual cycle with instruments themselves subject to physical alteration each year from seasonal climatic changes.[10] To avoid variations in atmospherical refraction, Hooke himself had given his attention to the star Gamma Draconis, which passed overhead to an observer in London; but he had made only four observations, which Continental commentators found woefully few as a basis for establishing the distance of the star.[5]

(3) John Flamsteed, the Astronomer Royal, carried out a meticulous series of observations of the Pole Star spread over seven years, from 1689 to 1697, as a result of which he concluded that for this star n is about 9000. He also examined Sirius and Gamma Draconis, and arrived at similar values for n.[13] However, J. D. Cassini quickly pointed out that the movements observed by Flamsteed could not be those sought because, for a given time of year, they occurred in the wrong direction![14]

One might suppose that the efforts of Hooke and Flamsteed to measure n by direct trigonometrical methods would be applauded by early eighteenth-century writers on astronomy as creditable failures, and a search made for the reason for the anomalous observations of Flamsteed; further, that meanwhile the figure for n derived by Gregory would be refined with the aid of more exact data on the

[5] See the comments by Huygens[11] and Wallis.[12]

solar system. And in the long run, this is what happened. Bradley, following up Hooke's attack on Gamma Draconis, had by 1728 uncovered a complication undreamt of by Hooke, but even after allowing for this, he had failed to detect the required annual movement; from his failure he inferred that, for this star, n must be *at least* 400,000. A preliminary announcement of Bradley's conclusion was inserted in the 1728 edition of Whiston's *Astronomical Lectures*,[23] (p. 37) and a full account published in the *Philosophical Transactions* the following year.[15] The complication Bradley had uncovered—that of the aberration of light—affects all measurements of star positions, and his discovery was therefore of the first importance in sidereal astronomy and immediately recognized as such. Smith reports Bradley's estimate of the minimum distance of Gamma Draconis in his *Opticks*[16] (Vol. ii, p. 454) and Ferguson's *Astronomy*[17] (p. 231) makes this the actual distance of the nearest stars, for "it is easy to prove, that the Sun, seen from a distance, would appear no bigger than a Star of the first magnitude"—an evident recognition that Bradley's *minimum* distance for Gamma Draconis derived from direct measurements fit the *actual* values of n (of some hundreds of thousands) which theoreticians were deriving by Gregory's method for the brightest (and, by hypothesis, nearest) stars (Hoskin,[6] p. 33). Herschel and his contemporaries, then, could obtain from Ferguson an excellent idea of the distances separating the Sun from neighbouring stars.

But in the short run, the situation was very different. Newton owned a copy of Gregory's book[18] and gave an excellent exposition of his method in "The System of the World."[19] Newton there concluded that the brightest stars lie at some 100,000 times the distance of Saturn from the Sun; that is, n for these stars would be about one million. The "System" was intended to be part of his *Principia*, but he replaced it with more technical material in which the stars were scarcely mentioned. As a result, this treatment of Gregory's method and the distances Newton thereby derived for the brightest stars were published, not in 1687, but posthumously, in 1728.

In the 1690s Newton embarked on an extensive revision of the *Principia* in which he incorporated an account of Gregory's method and again concluded that for the brightest stars n is about one million.[6] But this revision was never published. However, in May 1694 Newton discussed (Turnbull,[20] Vol. iii, p. 312) the method with David Gregory, a nephew of James, and David sets the method out in full in his *Elementa* of 1702[21] (pp. 278-81)—but without the slightest hint of the *results* to which the method led. In consequence, James Gregory's brilliant proposal of 1668, which so quickly led Newton to a correct understanding of the distances to the nearest stars, was effectively in limbo until the second quarter of the eighteenth century.

In its stead, students of astronomy were introduced to the method of Christiaan Huygens (*Cosmotheros*,[22] pp. 135-8), which was based on the same

---

[6] The manuscripts are now in Cambridge University (Add. MS 3965); an edition by the present writer is in preparation.

assumptions but used a much inferior technique for comparing the brightness of the Sun and a Star. Huygens tried to make a hole in a screen of such size that the fragment of the Sun which could be viewed through the hole would appear as bright as a star; the fraction of the Sun's disk visible through the hole would then be equivalent to $1/n^2$. The technique, however, was inadequate: The hole could not be made small enough and the Sun's light had to be further diffused by use of a lens whose effect could only be guessed. Huygens concluded that for Sirius n = 27,664—a result much too small and a poor estimate by comparison with Newton's figures. But it was Huygens' method that was in practice available in the first quarter of the eighteenth century, in his *Cosmotheoros* and in the books by Whiston (*Praelectiones*,[23] pp. 31-33), Derham (*Astro-theology*,[24] 3rd ed., pp. 22-23), and David Gregory (*Elementa*,[21] p. 281).

The contemporary British attitude to the attempts by Hooke and Flamsteed to measure the annual apparent shifts of stars is also different from what one might expect. Whiston, lecturing in 1701, apparently knows nothing of Cassini's fundamental objection to Flamsteed's claims, which Whiston accordingly accepts. Since Flamsteed has concluded that three stars, despite their differing brightnesses, all lie at about 9000 astronomical units, Whiston infers that the differences between these stars are not merely apparent but real (*Praelectiones*, Lect. IV). He therefore rejects both the numerical result arrived at by Huygens and the assumption on which it is based, which he derides as this "ingenious, and barely probable, Conjecture" (*Praelectiones*, Lect. III; *Astronomical Lectures*, 1728, p. 27).[23] On the contrary, "there is probably a mighty Inequality among the Fixed Stars" (*Praelectiones*, p. 42; *Astronomical Lectures*, 1728, p. 39).[23] For his part, David Gregory rejects the conclusions of Flamsteed, not because of Cassini's objection (which is again apparently unknown to him), but because Flamsteed's method of measurement assumes that there is no annual nutation of the Earth's axis. Gregory thinks that a nutation of a kind not foreseen by Flamsteed may result from differences between the northern and southern hemispheres of the Earth (*Elementa*,[21] pp. 275-6).

Never one to shrink from controversy, Whiston counterattacked Gregory in support of Flamsteed in a lecture dated 29 October 1705; but to the criticism of Cassini some concession must now be made:

Scholium. But it is to be noted, that the Famous Mr. *Flamsteed* hath not ordered his Reasonings altogether rightly in this Place, which the *French* have lately noted; and hath sometimes deduced the Parallax of the Fixed Stars from Phaenomena in no wise proving it. But yet when I looked more narrowly into this Matter, Eleven of Fifteen remarkable Observations, which the *French* allow to be true, and agreeing with their own, do even yet shew the Parallax of the Fixed Stars; and of those Four that seem to disagree with it, there is only one of that Quantity as to give us any Trouble in this Business; which therefore it is reasonable to think to be owing to some Mistake, whether in the observing or in the writing. Especially since the like Parallax seems manifestly to appear from the accurate Observations of Dr. *Hooke*. But these things we leave to the further Diligence and Scrutiny of Astronomers.[25]

As this passage shows, even when Cassini's critique was known, it was not necessarily accepted as conclusive.

In his *Astronomical Lectures* (2nd ed., p. 27),[23] Whiston had inferred from Flamsteed's measurements that "a very great Number of these Stars may be placed about the Sun on every side, at or near the same Distance." This is reminiscent of Kepler's model of the universe: "There is therefore an immense cavity in the midst of the region of the fixed stars, a visible conglomeration of fixed stars around it, in which enclosure we are" (Koyré,[26] p. 81). Kepler had argued that the appearance of the sky contradicts any claim that the stars are spread out in a uniform distribution (Koyré,[26] p. 80):

> If the region of the fixed stars were everywhere similarly set with stars, even in the vicinity of our movable world, so that the region of our world and of our sun had no peculiar outline compared to the other regions, then only a few enormous fixed stars would be seen by us, and not more than twelve (the number of the angles of the icosahedron) could be at the same distance from us and of the same [visible] magnitude; the following ones would be scarcely more numerous, yet they would be twice as distant as the nearest ones; the next higher would be three times as far, and so on, always increasing their distance [in the same manner].
>
> But as the biggest of all appear so small that they can hardly be noted or measured by instruments, those that would be two or three times farther off, if we assume them to be of the same true magnitude, would appear two or three times smaller. Accordingly we should quickly arrive at those which would be completely imperceptible. Thus very few stars would be seen, and they would be very different from each other.
>
> But what is seen by us in fact is quite different. We see, indeed, fixed stars of the same apparent magnitude packed together in a very great number. The Greek astronomers counted a thousand of the biggest, and the Hebrews eleven thousand; nor is the difference of their apparent magnitudes very great. All these stars being equal to the sight, it is not reasonable that they should be at very unequal distances from us.

Kepler's misconceptions need not detain us. What is significant is that, in a widely circulated work, he attempts a comparison between (a) a model of the Universe in which the stars are regularly distributed and (b) the observed distribution of the brightest (first magnitude) stars and of stars of succeeding magnitudes. Newton, when driven by Richard Bentley to consider how the stars can be at rest ("fixed") in a Universe in which each of them is subject to a gravitational pull from every other star,[7] took refuge in a roughly regular distribution of stars so that the various pulls would cancel each other out. His treatment of this is to be found in manuscripts of the revision of the *Principia*, which (apart from a sentence added for the 1713 edition) never saw publication (see footnote 6), but once again he discussed the question in 1694 with David Gregory (Turnbull,[20] Vol. iii, p. 312), who gives a balanced account of the evidence in his *Elementa* of 1702 (*Elementa*, pp. 159-60).[21] Gregory argues, following Newton, that the number of first magnitude stars (about 13) and the number of second magnitude stars (about 4 × 13) are what one would expect if in the neighborhood of

---

[7] The exchange of letters is conveniently available in Vol. iii of the Newton *Correspondence*.[20]

the Sun the stars were equal and equally spaced. But thereafter "the Matter does not go on so well, (which made *Kepler* of another Opinion,) as is evident even from hence, that upon the first cast of our Eyes upon the Heavens, some Tracts of the Firmament appear fill'd with innumerable Fix'd Stars, whereas others are found to be almost empty and void of any" (*Elements*, 1728, p. 290).[21]

Newton in his manuscript revision had gone to great lengths to reconcile the numbers of stars of higher magnitudes with his model of a symmetric universe in which the innumerable pulls of gravitation cancelled each other out; but he ignored the nonuniformity evidenced by star clusters and especially by the Milky Way,[8] which is so obvious to Gregory. However, it seems that both men agreed that God had arranged the distribution of the stars so that the system would survive without rapid changes. Newton had told Gregory in 1694 "that a continued miracle is needed to prevent the Sun and the fixed stars from rushing together through gravity" (Turnbull,[20] Vol. iii, p. 334) and in the 1706 edition of his *Opticks* he asks (in what became Query 28) "What hinders the fix'd stars from falling upon one another?" and answers, God. But the implication of Newton's manuscript revision of the *Principia*, and his insertion in the 1713 edition (Book III, Prop. XIV, Cor. 2, ". . . the fixed stars, everywhere promiscuously dispersed in the heavens, by their contrary attractions destroy their mutual actions"), is that God maintains the structure of the sidereal universe, not by countermanding gravitation, but by arranging the stars so that the obvious consequences do not quickly follow. Gregory seems to hold that because the star system is boundless, there is no danger of gravitational collapse: "The indefinite Number of those Systems, included in no Space, is the Reason why they don't run into one, but, being separated from one another, will for ever stand in the Universe, as Marks of the Power and Wisdom of their Almighty Creator" (*Elementa*, p. 483; *Elements*, 1728, p. 856).

Gregory's book would of course be well known to Halley—indeed, Halley's own copy survives (Cambridge University Library). It is a simple step from Gregory's discussion to Halley's famous papers of 1721 (not 1720)[28],[9] in which he explores the implications of an infinite and regular distribution of stars. Such a system, he believes, would be stable under gravity, for "all the parts of it

---

[8] On other occasions Newton did not ignore this evidence. In a manuscript (Cambridge University Add. MS 4005) setting out briefly the Gregory method and its numerical verification, Newton writes,[26] "Yet this is to be understood with some liberty of recconning. For we are not to account all the fixt starrs exactly equal to one another, nor placed at distances exactly equal nor all regions of the heavens equally replenished with them. For some parts of the heavens are more replenished wth fixt stars then as the constellation of Orion wth greater or nearer stars & the milky way wth smaller or remoter ones. For ye milky way being viewed through a good Telescope appears very full of very small fixt stars & nothing else then ye confused light of these stars. And so ye fixt clouds & cloudy stars are nothing else then heaps of stars so small & close together that without a Telescope they are not seen appart, but appear blended together like a cloud."

[9] The date is established from the Society's Journal Book.

would nearly be *in aequilibrio*, and consequently each fixt Star, being drawn by contrary Powers, would keep its place; or move, till such time, as, from an *aequilibrium*, it found its resting place" (p. 23). Halley then goes on to ask questions, in the manner of Kepler, about the apparent diameters of distant stars. As to the aggregate effect of so many stars, he tells of "Another argument I have heard urged, that if the number of Fixt Stars were more than finite, the whole superficies of their apparent Sphere would be luminous" (p. 23)—the effect doubly misnamed by modern astronomers "Olbers' Paradox."[29] Halley has a (fallacious) answer to the problem, which does not concern us here; but his discussion recapitulates so closely the mathematics of Gregory that it seems that the *Elements* must be assigned a role in the history of the "paradox."

Newton, David Gregory, and Halley had all discussed whether the stars form a regular and ordered system, in our immediate neighborhood or even throughout infinite space. That the sidereal universe is something more than a random scattering of disparate stars was a conviction encouraged by the deeply religious outlook of many authors. "This System of the Universe," Whiston writes, "is God's great House, or Family, or Kingdom"[30] (pp. 131-32). Keill tells us that "we are to consider the whole Universe as a glorious Palace for an infinitely Great and everywhere present GOD; and that all the *Worlds* or Systems of *Worlds*, are so many Theatres, in which he displays his Divine Power, Wisdom and Goodness."[10] For Derham, there is a "great Parity and Congruity observable among all the works of the Creation; which have a manifest harmony, and great agreement with one another" (*Astro-theology*,[24] 3rd ed., p. xvi). Accordingly, "these several Systemes of the Fixt Stars, as they are at a great and sufficient Distance from the Sun and us; so they are imagined to be at as due and regular Distances from one another. By which means it is, that those multitudes of Fixt Stars appear to us of different Magnitudes, the nearest to us large; those father and father less and less" (*Astro-theology*,[24] 3rd ed., p. xxxix). Modern telescopes, he says, have revealed many more stars than were formerly known,

> and all these far more orderly placed throughout the Heavens, and at more and due agreeable distances, and made to serve to much more noble and proper ends.... [The stars] are not set at random, like a Work of Chance, but placed regularly and in due order ... they look to us, who can have no regular prospect of their positions, as if placed without any order: like as we should judge an army of orderly, well disciplined soldiers, at a distance, which would appear to us in a confused manner, until we came near, and had a regular prospect of them, which we should then find to stand well in rank and file (*Astro-theology*,[24] 3rd ed., pp. 40, 52, 57-58).

Keill, who acknowledges elsewhere his debt to David Gregory, believes in an overall uniformity among the stars: "... it is more reasonable to suppose that they are spread every-where thro' the vast indefinite Space of the Universe, and that they are at great Distances from one another; so that there may be as great

---

[10]Keill,[31] *Introductio*, p. 43; transl. from 1st English ed., pp. 40-46. The English text amplifies the Latin.

Distance between any two *Suns* that are next to one another, as there is between our *Sun* and the nearest *fixed star*"[11] (*Introductio*, p. 42; *Introduction*, 1721, pp. 39-40).[31] Whiston, who inclines to believe that the stars are finite in number, says that by analogy with the solar system

> it is very rational to conclude, that some regular Order hath Place also amongst the Fixed Stars. There may be a certain orderly and harmonious Disposition of the Fixed Stars amongst themselves, when they are beheld from some other proper Place, altho' that Order appears not when they are seen from this Earth [just as happens with the planets]: Or this Order may consist in certain beautiful Proportions, fitted to the several Systems, which are wholly unknown to us (*Praelectiones*, Lect. IV; *Astronomical Lectures*, 1728, p. 42).

In assessing the origins of the theological approach to the distribution of the stars which led Thomas Wright to his insight on the Milky Way, it would be hard to exaggerate the importance of the climate of opinion represented by these writings.

As we see, the conviction that the star system is orderly raised questions to which answers could not yet be given. If the system is finite, is there one position from which the system is visibly well-ordered? If the system is infinite, does it appear essentially the same to observers anywhere in the universe? Or are there exceptional regions—the Milky Way, the Magellanic Clouds (which Gregory[21] reported to "present us here and there with little *Nebulae*, and small *Stars*"; *Elementa*, p. 162; *Elements*, 1728, p. 294), or the handful of nebulae, that Whiston[30] (p. 80) declared to be "a Company of very small Fixed Stars, as invisible to us with our ordinary Telescopes, as the known Telescopic Stars in the Milky Way are to our natural Eyes, which give such an irregular Appearance of indistinct Light also," but which Halley considered "nothing else but the Light coming from an extraordinary great Space in the Ether; thro' which a lucid *Medium* is diffused, that shines with its own proper Lustre"?[32]

But as challenging as these questions relating to spatial distribution were questions relating to possible changes with time. If Newton was right in holding that gravitation operated throughout the universe, then what were the effects of gravitation over long periods of time? Do the stars, despite gravitation, remain truly at rest, are they as "fixed" as the observations of two millenia suggested and as Newton believed? Is the system essentially unchanging and static, or does it alter cyclicly like the planets in orbit around the Sun, or is its present structure no more than a transition between a quite different past and a quite different future?

Before 1718, there was no evidence that the stars were in motion, unless one chose to use motion to explain the alterations in the apparent brightnesses of stars. The new stars of 1572, 1600, and 1604 had had a decisive impact on the history of science as tangible proofs that change occurred in the heavens as well as on Earth, and so were well known to everyone concerned with astronomy.

[11] The Latin ed. has "Sirius" instead of "the nearest *fixed star*."

Another new star, which appeared in 1596, had disappeared like the others but had reappeared in 1638, and in 1667 Ismael Bullialdus published a study[33] of it in which he pointed out that its brightness was varying with a period of 333 days. Full details of these and other variable stars were included by Nicholas Mercator in his *Institutionum Astronomicarum Libri Duo* (London, 1676), and each of our textbooks gives some account of the observational data. Gregory offers no attempt at an explanation, and Whiston in his *Praelectiones* (Lect. V; *Astronomical Lectures*, 1728, p. 48)[23] declares the very irregular phenomena "wholly insoluble." Derham takes them to be *"Erraticks* of some kind or other," perhaps *"Wandering Suns,"* though "of what use they should be, is hard to imagine," but more probably "Planets revolving round such Suns, as cast a much fiercer and more vigorous Light than our Sun doth" (*Astro-theology*, 3rd ed., pp. 46, 48), though he admits no explanation he can offer is in any way plausible. But his general conclusion is significant:

> And so for the New Stars, which I have said are so many Signals of Planetary Systemes dispersed here and there all over the universe, they are all of them so many manifestations and demonstrations of an infinite Being that hath imparted motion unto them: and they are a sign also that there are other Globes, besides the Sun and its Planets, which are moving Bodies, even that all the Globes in the Universe are such, and consequently so many Proofs of an *Almighty First Mover* (*Astro-theology*, 3rd ed., pp. 70–71).

Whiston, in his *Astronomical Principles*[30] (p. 79), and Keill follow Bullialdus and explain variable stars as rotating bodies like the Sun with its spots. Keill writes of the 1596 star:

> It is probable that the greatest Part of the Surface of this *Star* is covered with Spots and dark Bodies, some Part thereof remaining lucid and while it turns about its *Axis* does sometimes show its bright Part, sometimes it turns its dark side to us: But the very Spots themselves of this *Star* are liable to Changes, for it does not every Year appear with the same Lustre (*Introductio*, p. 59; *Introduction*, 1721, p. 56).[31]

He regards Tycho's *nova* of 1572 as a very long-term variable, and stars which have vanished as having been overwhelmed by their spots (*Introductio*, pp. 60, 61; *Introduction*, 1721, pp. 57, 58). In this way the problem of variable stars encouraged the view that the analogy between the Sun and the stars extended to the stars' having (at least) rotatory motion.

In 1718, however, Edmond Halley pointed out in a short paper[34] in *Philosophical Transactions* that the three bright stars Aldebaran, Sirius, and Arcturus were no longer in the positions (relative to the other stars) agreed upon for them by observers in Antiquity. He remarked: ". . . these Stars being the most conspicuous in Heaven, are in all probability the nearest to the Earth, and if they have any particular Motion of their own, it is most likely to be perceived in them, which in so long a time as 1800 Years may shew itself by the alteration of their places, though it be utterly imperceptible in the space of a single Century of Years" (Halley,[34] p. 737).

Two questions were involved. First, are the effects observed due to the actual movements of the stars in question, or to the movement of the observer in the solar system, or to a mixture of both? Second, what in dynamical terms do these movements represent?

As to the first question, it is well known that in 1748 the dilemma was expressed by Bradley: ". . . it may be of singular Use, to examine nicely the relative Situations of particular Stars; and especially of those of the greatest Lustre, which it may be presumed lie nearest to us, and may therefore be subject to more sensible Changes, either from their own Motion, or from that of our System."[35] A similar view was voiced by Robert Hooke in a lecture published posthumously in 1705, where he comments that if the stars appear to move, "it will be yet a further dispute, whether this has been caus'd by some slow motions of those Stars one among another; or whether by the alteration of the very system of the Sun in respect of them."[36] Newton briefly considers the motion of the entire solar system as a formal possibility at the beginning of Book III of the *Principia* (Prop. XI, Theorem XI), but for Whiston the possibility is real:

> The common Center of Gravity of the Earth, Sun and all the Planets, either rests, or is mov'd uniformly in a right Line. This is manifest from what hath been demonstrated before: But indeed it appears by no certain Token, whether it rests or is mov'd. This only is to be concluded, That if it be mov'd, and with it the whole Solar System, the Motion must needs be very slow unless it be mov'd uniformly and evenly with the Centers of other Systems. For the Fixed Stars, which encompass us on every Side, neither appear greater nor less to us at this Day, than they did to the Ancient Astronomers 2000 Years ago. Which Phaenomena seems to shew the rest, or at least the very slow Motion of the said Center (Whiston,[25] p. 349).

Movements of the Sun and stars, then, so plausible on dynamical grounds in a Universe with gravitational attraction, were by no means unexpected when Halley made his announcement in 1718. For Halley, presumably, such movements were oscillations about an equilibrium position. Whiston, on the other hand, realized that a finite star system initially at rest would collapse into its center under gravity:

> . . . since, withal, the Sun and Fixed Stars do not revolve about one another, or about any common Center of Gravity . . . it follows that the several Systems, with their several Fixed Stars or Suns, do naturally and constantly, unless a Miraculous Power interposes to hinder it, approach nearer and nearer to the common Center of all their Gravity; and that in a sufficient Number of Years, they will actually meet in the same common Center, to the utter Destruction of the whole Universe (Whiston,[30] p. 88).

The impact of Halley's announcement awaits investigation by historians, but in the meantime it seems that no one saw in these tiny movements the stable orbital rotations of stars about the center of their system, until Wright approached the problem with the appropriate religious preconceptions.

To conclude: Bradley's discoveries apart, the early decades of the eighteenth century were unexciting for British students of the fixed stars, and in their knowledge of stellar distances they lacked information of the best contemporary work. But in their gropings for an understanding of the structure of God's universe of stars and of the effects of gravity on that universe, they provided through their writings a sympathetic milieu for the insights of Wright and Herschel.

REFERENCES

1. Thomas Wright of Durham, *An Original Theory or New Hypothesis of the Universe* [London, 1750; facsimile reprint together with first publication of "A Theory of the Universe" (1734), ed. by M. A. Hoskin, London, 1971].
2. Thomas Wright of Durham, *Clavis Coelestis* (London, 1742; facsimile reprint, ed. by M. A. Hoskin, London, 1967).
3. Thomas Wright of Durham, *Second or Singular Thoughts upon the Theory of the Universe*, ed. by M. A. Hoskin (London, 1968).
4. M. A. Hoskin, The Cosmology of Thomas Wright of Durham, *J. Hist. Astron.* i, 44–52 (1970).
5. William Herschel, *The Scientific Papers of Sir William Herschel*, ed. by J. L. E. Dreyer, 2 vols. (London, 1912).
6. M. A. Hoskin, *William Herschel and the Construction of the Heavens* (London, 1963).
7. Edward Hughes (ed.), The Early Journal of Thomas Wright of Durham, *Ann. Sci.* vii, 1–24 (1951), p. 4.
8. W. H. Donahue, The Dissolution of the Celestial Spheres, 1595–1650, Ph.D. Thesis, Cambridge Univ. (1972), unpublished.
9. James Gregory, *Geometriae Pars Universalis* (Padua, 1668).
10. Robert Hooke, An Attempt to Prove the Motion of the Earth (1674), in *Lectiones Cutlerianae* (London, 1679).
11. Christiaan Huygens, *Phil. Trans.* ix, 90 (1674).
12. John Wallis, *Phil. Trans.* xvii, 845 (1693).
13. John Wallis, *Opera Mathematica* (London, 1699), Vol. iii, pp. 705–8.
14. J. D. Cassini, Refléxions sur une lettre de M. Flamsteed à M. Wallis touchant la parallaxe annuelle de l'étoile polaire, *Mem. Acad. Roy. Sci. pour 1699* (1702), pp. 177–83.
15. James Bradley, An Account of a New Discovered Motion of the Fixed Stars, *Phil. Trans.* xxxv, 637–61 (1727–28).
16. Robert Smith, *A Compleat System of Opticks* (1738).
17. James Ferguson, *Astronomy Explained upon Sir Isaac Newton's Principles*, 2nd ed. (London, 1757).
18. R. de Villamil, *Newton: The Man* (London [1931]), p. 106.
19. I. Newton, *A Treatise of the System of the World* (London, 1728); in *Sir Isaac Newton's Mathematical Principles of Natural Philosophy*, F. Cajori, ed. (Univ. California Press, Berkeley & Los Angeles, 1934), pp. 596–7.
20. H. W. Turnbull (ed.), *The Correspondence of Isaac Newton*, Vol. iii (Cambridge, 1961).
21. David Gregory, *Astronomiae Physicae et Geometricae Elementa* (Oxford, 1702; Geneva, 1726); English translation, *The Elements of Physical and Geometrical Astronomy* (London, 1715, 1728).
22. Christiaan Huygens, *Cosmotheoros* (Hagae-Comitum, 1698); English translation, *The Celestial Worlds Discover'd* (London, 1698, 1722, and subsequently).

23. William Whiston, *Praelectiones Astronomicae* (Cambridge, 1707); English translation, *Astronomical Lectures* (London, 1715, 1728).

24. William Derham, *Astro-theology* (London, 1715; 3rd ed., 1719).

25. William Whiston, *Sir Isaac Newton's Mathematick Philosophy More Easily Demonstrated* (London, 1716), p. 238.

26. Johannes Kepler, *Epitome Astronomiae Copernicanae*, Liber I, Pars II; transl. in Alexandre Koyré, *From the Closed World to the Infinite Universe* (Baltimore, 1957).

27. A. R. Hall and M. B. Hall (eds.), *Unpublished Scientific Papers of Isaac Newton* (Cambridge, 1962), pp. 375-6.

28. E. Halley, Of the Infinity of the Sphere of Fix'd Stars, *Phil. Trans.* **xxxi**, 22–24 (1720-21); Of the Number, Order, and Light of the Fix'd Stars, *Phil. Trans.* **xxxi**, 24–26 (1720-21).

29. S. L. Jaki, *The Paradox of Olbers' Paradox* (New York, 1969); M. A. Hoskin, Dark Skies and Fixed Stars, *J. Br. Astron. Assoc.* **lxxxiii**, 254–62 (1973).

30. William Whiston, *Astronomical Principles of Religion* (London, 1717).

31. John Keill, *Introductio ad Veram Astronomiam* (Oxford, 1718, 1721); English translation, *An Introduction to the True Astronomy* (London, 1721 and several times subsequently).

32. E. Halley, An account of several Nebulae or lucid Spots like Clouds, *Phil. Trans.* **xxix**, 390-2 (1715-16), p. 390.

33. Ismael Bullialdus, *Ad Astronomos Monita Duo* (Paris, 1667).

34. E. Halley, Considerations on the Change of the Latitudes of some of the principal fixt Stars, *Phil. Trans.* **xxx**, 736–8 (1717-19).

35. James Bradley, A letter to the Right honourable George Earl of Macclesfield concerning an apparent Motion observed in some of the fixed Stars, *Phil. Trans.* **xiv**, 1–43 (1748), p. 41.

36. *The Posthumous Works of Robert Hooke*, ed. by Richard Waller (London, 1705), p. 506.

Chapter XVI

The History of Science and the Idea of an Oscillating Universe

Stanley L. Jaki

Seton Hall University

At the Spring 1974 meeting of the American Physical Society in Chicago, Jeremiah P. Ostriker, an astrophysicist from Princeton, read a paper[1] which made news all over the United States. In view of the particular problem which gave rise to the paper, the excitement should have seemed out of proportion. It was an old story that rotating gravitational systems like our system of planets and the system of stars composing our galaxy or other spiral galaxies obey Kepler's laws, the third of which states a very specific ratio for the decrease of orbital speed with distance from the center. It is an equally old story that Kepler's laws can be derived from Newton's investigations of the mutual attraction of moving masses according to the inverse square law. If it had been known around the turn of the century that the observed orbital velocities of stars in our galaxy did not fit the ratio in question and if it had been proposed that more mass ought therefore to exist in the galaxy, no particular excitement would have ensued. Around the turn of the century, the Universe was pictured as containing an infinite amount of mass and the value of the average density of matter could not be considered critical.

Quite different has been the case ever since Einstein ushered in the era of relativistic cosmology. The exact value of the average density of matter is of such importance as to play the role of arbiter among cosmological models. To be specific, the value of that density contains the answer to the dilemma whether the Universe, which is in an expanding state, will stay in that state irreversibly or whether the Universe is merely acting like a gigantic oscillator and will go on

---

[1] The technical background of that paper, still to appear in print, can be gathered from a joint publication of Ostriker and Peebles, also of Princeton University.[(1)]

forever expanding and contracting. It remains to be seen whether certain spots in recent photographs of spiral nebulae truly indicate, as Ostriker suggested, a large amount of matter hitherto undetected. Anyone familiar with efforts of the past fifty years aimed at specifying the value of the density in question knows how greatly revisable the figures proved to be. Instances of rashness were not lacking in scientific publications dealing with this problem of cosmology, although the majority of statements displayed a fair measure of moderation.

Of this moderation little could be felt in the radio and press as the news broke about Ostriker's paper. Listening to various newscasters, I could not help feeling that in speaking of the potential consequences of Ostriker's reasoning and findings they were in no sense casting around.[2] They invariably exuded a curious mixture of excitement and satisfaction over the fact that at long last the missing matter has been found. If that missing matter had really turned up, then we might live in a Universe in which the running away of galaxies from one another would yield in eighty more billion years to the mutual approach of those galaxies. But why should this prospect look so exciting? After all, it may be reasonably doubted that the contraction will be done by the very same galaxies that took part in the expansion. Anyone daring to size up the immensity of eighty billion years might conclude right away, even without knowing much about the dynamics and stability of galaxies, that none of them would last that long. We certainly know that no star will last longer than a fraction of those eighty billion years. Our Sun may still have a few hundred millions of useful years to sustain life on Earth, but the ultimate extinction of all life in our solar system and even the ultimate collapse of our planetary system are inevitable. Whether Ostriker will prove right or not, whether the excitement of newscasters was justified or not, nobody of our race, none of the planets, none of the constellations will be there when the expansion turns into contraction. Nobody will be around to derive satisfaction from the knowledge of the fact that the idea of an oscillating universe was found to match reality eighty billion years[3] earlier. So why get so excited about the whole business of expansion and contraction?

To such a line of thought it may rightfully be objected that anthropomorphic and anthropocentric considerations should not be mixed with appraisals of scientific ideas, theories, conclusions, projections, and findings. The objection is certainly well-founded. But this still does not dispose of all legitimate suspicion about the excitement of those newsmen and broadcasters over the presumed finding of the missing matter and the possible verification of the idea of an oscillating universe. Such a truth is certainly a grandiose truth about the cosmos,

---

[2] Having had to do a good deal of driving on that day, I listened, for want of anything better to do, to a number of radio stations, and found the tone and substance of broadcasts about Ostriker's suggestion invariably excited and enthusiastic.

[3] This is the figure most frequently found in the literature, technical and popularizing. The figure is dependent on the uncertainties of evaluating several parameters to be taken into account for the calculation of the length of the period in question.

and one can rightfully be excited over finding a truth of such portent, or even a much lesser truth. But then why is there no similar excitement over another equally grandiose scientific notion, namely, the idea of a simply expanding Universe? Why is it, to carry our psychoanalysis of scientific excitement a little deeper, that Ostriker's paper was reported with hardly any hint of the great uncertainties of his reputed finding of the missing matter, and of the enormous uncertainties implied in extrapolations stretching to eighty billion years and many times over? It seems indeed that newsmen and broadcasters, who were then followed by an army of science writers and science digesters, all became excited because they got hold of something which they had wanted to hear and to have for some time and because they felt that many of their listeners and readers had a similar desire. Whatever their faults, broadcasters and newsmen are often people with sensitive antennas which at times are even better in detecting than in transmitting.

Newscasters and their general audience cannot, of course, be viewed as the original source of this excitement and satisfaction with the idea, if not the truth, of an oscillating universe. The original source should certainly be sought in the scientific community, or rather among scientists most active in modern theoretical and observational cosmology. It is well known that in 1922, five years after Einstein had completed his work on the cosmology of general relativity, A. Friedmann found that its basic equations permit a solution according to which the overall dynamics in a universe composed of a finite number of stars or galaxies is periodic.[2] The formula derived by Friedmann for that periodic motion closely resembled the one for the period of a pendulum, but Friedmann made no effort to calculate the time of that cosmic period which suggested itself naturally as the alternation of a cosmic expansion with a cosmic contraction. Nor did Friedmann connect that cosmic oscillation with the recessional motion which had for over a decade been observed in an increasingly larger number of galaxies.

Friedmann's work found no immediate favor with Einstein, whose first reaction to it was to point out some algebraic errors.[3] It was only a year later that he added, with a touch of apology, that he found Friedmann's conclusion "valid and enlightening."[4] Einstein himself did not investigate in detail the case of an oscillating universe until about eight years later.[5] The scientific world, long used to worshipping the idea of an infinite universe, had already found it very difficult to accept the idea of a finite mass for the universe as demanded by Einstein's memoir of 1917 on the cosmological consequence of general relativity.[4] That a finite universe might in addition be subject to a very specific overall dynamics only added to a keenly felt difficulty of parting with the hallowed idea of a homogeneous static infinity in space as well as in time. As a

---

[4] The English translation of the German original is readily available in Ref. 6. E. Borel, B. Russell, and M. Schlick were some of the more notable figures who expressed in writing their uneasiness about the finiteness of mass composing the universe. For details, see Ref. 7, pp. 218–25.

result, Friedmann's paper found no notable echo. Five years after it appeared in the leading journal for physics of the times, Eddington and McVittie were still unaware of its existence as they started work on a very similar analysis of Einstein's cosmological equations. The same was true of Abbé George Lemaître in Louvain, whose historic paper, published in 1927,[8] contained not only a theoretical derivation of an expanding motion of the universe of a finite number of galaxies, but also the derivation of the rate of expansion. It agreed closely with the value which Hubble and Humason published shortly before on the redshift of galaxies as a linear function of their distance.

Lemaître's paper might have remained buried in the pages of the *Annales de la Société Scientifique de Bruxelles* had it not come, after two years' delay, to the attention of Eddington, who saw to it than an English translation be published in the *Monthly Notices*.[9] Reaction was immediate. Jeans' *Mysterious Universe*, published in 1930, contained the telling remark that most scientists preferred the oscillating model to the idea of a single, once-and-for-all expansion.[10] The next year witnessed an international symposium on relativistic cosmology sponsored by the British Association. Eddington, Lemaître, de Sitter, Jeans, Millikan, and Bishop Barnes were among the speakers. The first three sided with the idea of an expanding Universe, while Jeans and Millikan argued on behalf of a Universe being in perpetual motion going through the same cycles forever. In his turn, Bishop Barnes voiced some reservation on the soundness of an oscillating Universe, but hailed the finiteness of the Universe as permitting its rigorous and complete exploration.[11]

Somewhat earlier there was a similar conference on cosmology at the California Institute of Technology with Einstein and Millikan among the participants. Of the meeting the papers reported that the Universe was oscillating, expanding, tumbling, and rolling in all conceivable ways, if the ideas of all participants were to be considered.[12] Then there followed during the 1930s a period lacking in outbursts of sensationalism either on the part of scientists or on the part of science reporters. The situation drastically changed in 1948 with the publication of a paper by H. Bondi and T. Gold, who contended that the Universe was in a steady state.[13] A further elaboration of the idea was given shortly afterwards by F. Hoyle.[14] It went largely unnoticed that the steady-state Universe was in a sense an oscillating one. At any given area of the Universe the presence of matter varied between a density approaching zero and a density corresponding to that in a fully developed galaxy. The basic contention of the steady-state theory, the continuous creation of matter out of nothing and without a creator, was, however, noted very strongly. The idea produced a very strong opposition both from the scientific and from the philosophical points of view. From the scientific side, Herbert Dingle, to mention only one example, decided "to call a spade a spade" and urged that the perfect cosmological principle, namely, the claim that the Universe should always appear essentially the same in space and in time, be designated as the "perfect agricultural principle."[15] From the philosophical side, Pope Pius XII, in an address entitled, "On the Proofs of the Existence of God in the Light of Modern Natural Science," singled out the idea of a creation out of nothing but without a creator as something wholly gratuitous.[16] Whether the Pope was justified in seeing in the expansion of the Universe an evidence of a creation out of nothing five billion years ago, is another question. As a historian of science I merely wish to register the fact that from the late 1940s on, the expansion of the Universe has been taken by many a scientist, philosopher, and theologian as a strong pointer to the creation of the Universe at a sufficiently specified moment in the past.[5]

Since action is quickly followed by reaction, it should not be surprising that scientists with a dislike for metaphysics instinctively looked for an alternative. Some found it in the steady-state theory; others, unable to grant the idea of a creation out of nothing and without a creator, were naturally driven to the idea of an oscillating Universe. This is not to suggest that a dislike of metaphysics was the whole explanation of the sudden popularity of the oscillating Universe from the mid-1950s on. There were other and purely scientific reasons as well, such as the availability of the 200-inch Palomar telescope and the rapid development of radioastronomy. Both extended enormously man's reach into the Universe, or to be more specific, both made possible a far more extensive count of galaxies. On this count depended a more accurate determination of the average density of matter in the Universe and therefore the choice among rival cosmologies,[6] to recall an expression made popular in the late 1950s. In the early 1960s the champions of the steady-state theory suffered a heavy blow when a specific radiation postulated by the theory was not detected by artificial satellites. About the same time, the popularity of an oscillating universe gained increasingly both in scientific and popular literature. They carried such telltale remarks as the one which appeared in an article by some collaborators of R. H. Dicke in *Scientific American* on the radiation at $3°K$, a leftover from the Big Bang that started the actually observed expansion of galaxies some eight billion years ago: "It is difficult," the remark reads, "to explain the apparently spontaneous creation of matter that is called for if one associates the beginning of the expansion of the universe with its actual origin. Dicke, therefore, preferred an oscillating model in which the present expansion of the universe is considered to have been preceded by a collapsing phase."[20] Did the expression, "spontaneous creation" of all matter eight billion years ago, suggest that a creation out of nothing of the whole Universe by the Creator was more difficult to imagine than attributing eternity to it in the form of perpetual oscillations? Or did the remark merely

---

[5] A prime mover behind this trend was Whittaker with his Riddell Lectures[17] and Donnellan Lectures.[18]

[6] See the published account of the symposium broadcast in 1959 by the BBC.[19]

mean that a cosmologist should always try to find physical antecedents to any apparent initial conditions?[7]

However that may be, the remark seemed to express the hope that the oscillating Universe offered an escape from some major cosmological difficulties. Such hope was somewhat strange because the count of galaxies and other considerations as well gave a distinct though not definitive support to the case of a single expansion. But a historian should not, even indirectly, try to prejudge the ultimate outcome between contending theories. The historian of cosmology cannot, however, overlook a symptom which made its appearance in cosmological literature in the 1960s and early 1970s. The symptom is the recognition that the idea of an oscillating Universe is the reappearance in a new garb of a very old idea, the idea of eternal recurrence. In speaking of the oscillating Universe, Jean Charon notes in his *Cosmology:* "This reminds one of the Great Year in which the Ancients believed; they thought that at the end of this Great Year everything would begin again, in a new cycle which was identical with the old."[22]

On a cursory look this alleged similarity between the cosmology implied in the notion of the Great Year and the cosmology of an oscillating Universe might seem a mere historical curiosity. But things may look very different as soon as one recalls that the idea of the Great Year was the dominating feature of the world-view in all great ancient cultures. Of these ancient cultures the first to be mentioned are the Chinese and the Hindu, partly because they were the most populous, and partly because they have regained a new political and cultural status during our century and are the subject of extensive contemporary studies. Of the two, the Chinese have a literary record which is not only vaster but also permits a very accurate dating due to the Chinese preoccupation with chronology. But for all the vastness of their chronological recordings,[8] the Chinese developed no historical consciousness. They had only a sense of temporal continuity to the extent of considering each dynasty a reaction to the preceding one. Their chronological datings restarted with each new dynasty, a circum-

---

[7] As this paper goes to press, Walter Sullivan reports[21] from the "neighborhood meeting" of the Center for Astrophysics operated jointly by the Harvard College Observatory and the Smithsonian Astrophysical Observatory that there was on the part of some of the participants a "reluctance to accept the concept of infinite expansion, if only on philosophical grounds." One astronomer, E. R. Harrison, was quoted as saying that to conceive of the Universe as marching inexorably toward a "graveyard of frozen darkness" was a "horrible thought." This, he added, "would make the whole Universe meaningless," and were it true, "I would quit [as an astronomer] and spend my life raising roses." It is an old story that scientists committed to believing in an eternally living Universe readily fall back on "philosophical" considerations whenever the facts obtained by their science fly in the face of this belief of theirs. In this historical perspective let it merely be pointed out that cosmology as a science flourished and advanced only to the extent in which its cultivators surrendered to the elementary precept that the touchstone of cosmological truth has nothing to do either with horror or with roses.

[8] According to the noted Sinologist, W. A. Haas, ancient histories of China would fill, if translated into English, 450 volumes of 500 pages each. See Ref. 23.

stance which suggests that for them the flow of time was not linear, but cyclic. Indeed, all events, political and cultural, represented for the Chinese a periodic pattern, a small replica of the periodic interplay of two basic forces in the cosmos, the Yin and the Yang. These two names are best remembered as expressing the male and female factor, but their signification permeates the entire Universe considered as an organic whole. The individual was considered as an accidental product within an organismic Universe which went through its own life cycles determined by the alternating predominance of the Yin or of the Yang. Since the Yin and the Yang were organic, living forces, they could not be categorized quantitatively. Much less could man hope to fathom their secret and thereby dominate them, however moderately. The only thing man could hope for was a passive identification with those forces. Corresponding to that passivity was the age-old satisfaction with rigid social rules and subservience. Within that outlook, things as well as conditions were never to change radically; success was to alternate with failure in an inevitable sequence, as was progress with decay.

The ensuing resignation of the Chinese into practical mediocrity, though not into despair and despondency, was a matching counterpart of their moderate preoccupation with the exact period of the great cosmic cycle, the Great Year. The morbid preoccupation of the Hindus with assigning specific figures to the length of the Great Year is closely related to expressions of utter despair which turn up with great frequency in their ancient literature. The Hindus of old had fully perceived that the eternal recurrence was a treadmill out of which there was no point trying to escape. This treadmill exercised such a stranglehold on Hindu thought that even the Buddhist program to escape it by achieving Nirvana could not effectively remove its mesmerizing perspective. The cycle of existence codified in the doctrine of Yugas, or world periods, was, to recall a phrase from the *Maitri Upanishad*, like a deep, waterless well at the bottom of which man, like a frog, was imprisoned forever.[24] Actually, man was even less than a frog, when the worlds themselves were likened to bubbles appearing randomly in the pores of Brahma's cosmic body which went through its life cycles through infinite ages. This desperate outlook had to appear even more so in view of the fact that it was agreed among the ancient Hindus that the Kaliyuga, or the worst of the four ages, had got underway only a few thousand years ago. Since it was to last 432,000 years,[9] what else could be generated by a look at the future than a radical despair?

This pessimism sets also the tone of the world-view of the cultures of the pyramids, the Babylonian, the Egyptian, and the Mayan. They were steeped in the belief of a cyclic recurrence of everything and the Maya in particular were engrossed in specifications of the length of the cosmic period and of its sub-

---

[9]The 4 yugas constituted 1 mahayuga, or 4,320,000 years; 1000 mahayugas equalled 1 kalpa; 2 kalpas formed 1 day of Brahma, and 1 life cycle of Brahma was equal to 100 years of Brahma, or 311 trillion (3.11×10^{14}) ordinary years. This was the largest unit of time to which the Hindus were driven by their preoccupation with eternal cycles.

divisions. To illustrate the practical impact of the defeatist outlook generated by the idea of eternal recurrence, a reference to a well-documented detail[25] from the last days of the Maya empire should be instructive. One Maya tribe, the Itza, was, due to its remote location in the Yucatan peninsula, still free of Spanish domination as late as 1618. Then two missionaries decided to convert the tribe, centered on the town of Tayasal on the Western shores of Lake Peten Itza. Canek, the ruler of the Itza, told the two missionaries that the time of their surrender had not arrived yet. It would, however, inevitably come in eighty more years with the onset of the fateful age, Katun 8 Ahau. Having received this priceless information, the Spaniards waited patiently until March 1698, when Captain Martin de Ursua appeared with 100 men on the shores of Lake Peten Itza. With hardly a shot fired, the hundred Spaniards overwhelmed 2000 Itza canoemen and 5000 Itza footsoldiers. They fled in a panic generated not so much by the muskets of the Spanish as by the awareness that the wheels of eternal recurrence had just turned into the phase of total catastrophe.

It was in their effort to determine the length of the various phases of the cosmic cycles or Great Year that the Maya practised most their finest scientific feat, the vigesimal system of counting. Again, the decimal system and positional numbering, as developed by the Hindus, found its most prominent use in giving in ordinary years the length of the various Yugas, some of which were supposed to last millions and billions of years. Obviously, the historian of science stands here in the presence of a pathetic misuse of scientific talent which leads directly to the most portentous aspect of all great ancient cultures: the stillbirth of science in each and every one of them. Portentous though it should seem, this stillbirth of science in all great ancient cultures is the least discussed, yet possibly the most instructive problem of the history of science.[10] Such a negative attitude toward this problem should seem even more curious in view of the firm belief of this scientific age of ours that the rise of science in modern times is the most decisive event of human history.

While most historians of science are unwilling to analyze in depth the reasons of the repeated stillbirths of science, the history of science speaks loudly on this topic to anyone willing to listen. This voice sounds clearest in the case of ancient Greek culture, an ancient culture not mentioned yet. No other ancient culture advanced even remotely as far as the Greeks did along the road of science. A brief look at the almost complete absence of geometrical studies among the

[10] In my recently published book, *Science and Creation: From Eternal Cycles to an Oscillating Universe*,[26] I tried to fill this gap by devoting almost half the book to the fate of science in the great ancient cultures, including the Arabic. The main ideas of the following pages can be found there in detail and with an extensive documentation. Needless to say, many factors—geographical, social, economical, and political—played part in the stillbirth of the scientific enterprise in the various ancient cultures. The only common factor in all cases seems, however, to be the commitment to the cyclic world view. This partly cosmological and partly theological factor deserves a special attention all the more as ultimately all science is cosmology.

ancient Chinese and at the geometry of Euclid should illustrate this point with an almost brute force. It is also important to recall that although in the case of Copernicus, Kepler, and Galileo this very same Euclidean geometry was a key tool in their scientific success, the Greeks, the inventors of the science of geometry, could not turn science into a self-sustaining enterprise. Even in the culture of Thales, Pythagoras, Plato, Eudoxus, Archimedes, and Ptolemy science suffered a stillbirth, perhaps the most memorable and most tantalizing one.

Clearly, something else was needed than geometry, but this additional factor could not be had by the Greeks, whose world-view was steeped in the idea of eternal cycles. This brought about a narrowing of their mental horizon which is best illustrated in their smug complacency with the state of advance of their crafts and learning. Life, Aristotle declared, possessed at his time all the conceivable commodities.[27],[11] Other ancient Greek sages spoke in the same vein. They considered themselves to be on the crest of the wave which was soon to carry everything downward along the path of an endless up and down cycle. This cyclic notion is the very foundation of the three main cosmologies developed by the Greeks, the Aristotelian, the Stoic, and the Epicurean (atomistic). These three schools produced much of what is known today as the Greek scientific *corpus*. The reason why it had no built-in vitality of its own might be better seen from the historical fate of Greek science than from philosophical analysis which, however well founded, might have no convincing force in this age of positivist scepticism.

The best known fact about the historical fate of the Greek scientific *corpus* is that it got into the hands of the Arabs, who with an astonishing zeal tried to acquire any and all Greek manuscripts. Their zeal was matched by their scholarship, which soon produced critical editions and translations from at times widely differing texts. By the early 800s the Arabs had whatever there was available in Greek learning. They hardly made a step beyond. Instead of trying to learn intelligently from the Greeks, they became the worshippers of Greek learning. Averroes and the trend he initiated was not an accidental event but a logical development. The logic in question is very simple: it consisted in the inability of the Arabs to extricate themselves from a cyclic concept of the Universe embodied in the doctrine of the Great Year to which they referred approvingly on numerous occasions.

That many of the Arabs, ardent monotheists, became trapped in Greek pantheism when it came to science and philosophy, should seem surprising. The explanation lies in the *Koran*, in its concept of a capricious God.[12] To save the all-willfulness of Allah, devout Arabs, like the Mutakallimun, chose to deny the notion of physical law, in fear that it might put constraint on the will of Allah.

---

[11] Theophrastus, Aristotle's successor at the Lyceum, voiced the same opinion even more emphatically in his *On Pleasure* which survives only in fragments. See Ref. 28.

[12] The classic passage is Sura XXXV, which was often quoted by the Mutakallimun as discrediting the notion of physical law.

The classical Arab expression of this stance was the *Tahafut al-falasifah* ("Incoherence of Philosophers") written by the foremost and very influential Muslim mystic al-Ghazzali. The Mutazalites, or Muslims ready to trade faith for learning, had in the same act jettisoned the only factor, the belief in the Creator, that could have helped them to steer clear of the pantheistic shallows of the Greek world-view, including the notion of Great Year, that is, the notion of a world existing by itself and going through the same cycles forever in great intervals. Such a world-view invited an *a priori* approach to its understanding and discouraged the role of experimental investigations. The most celebrated representative of this trend was Averroes, author of the *Tahafut al-tahafut* ("Incoherence of the Incoherence"), a lengthy polemic against al-Ghazzali's work.

That a very different course of events was to follow from the confrontation of biblical monetheism with the Greek world-view can be seen from the first major phase of that confrontation, which occurred during the third and fourth centuries of our era. From Origen to Augustine the Fathers of the Church kept decrying the idea of the Great Year as a pernicious doctrine utterly irreconcilable with such cardinal points of the faith as Creation, Incarnation, and final resurrection. With an eye on the Greek world-view, the Church Fathers considered time and again the possibility of Judas betraying Christ not once but time and again in each succeeding world age, nay in an infinite number of times, and denounced it as wholly inconceivable within the Christian outlook for which existence is not cyclic but linear in the spiritual as well as in the physical realm. As to the pagan Greek philosophers, the Church Fathers called their attention to such implication of their belief in the Great Year as Socrates, the noblest of all Greeks, drinking the hemlock not once but time and again, in an infinite number of ages.[13]

The linear concept of existence written in many pages of the Old and New Testaments received its first major theological formulation in Augustine's *De civitate Dei*, a great and almost lonely classic during the early Middle Ages. No wonder that when the Greek *corpus* reached Western Christendom through the Arabs during the twelfth and thirteenth centuries, it was accepted with only some major corrections, a pattern very different from the way in which most learned men among Muslims had accepted Greek science and philosophy. Christians could not, for instance, accept the divinity of the heavens implied in the doctrine of the Great Year and especially in its Aristotelian formulation. As a result, the heavenly and earthly regions were bound to be conceived as being of the same perishable matter and ruled by the same dynamics and mechanics. From the thirteenth century on there appeared more and more frequent references to the world as a clockwork mechanism. Equally decisive was the simultaneous impact of the notion of a linear existence deriving from the concept of creation out of nothing by a rational and provident Creator. It gradually led to

---

[13] Many of these patristic texts are quoted in Chapter viii of my *Science and Creation*.[(26)]

the notion of linear inertia both in isolated motion and in mechanical constructs. A most telling example of this connection occurs in Oresme's famous commentary of 1378 on Aristotle's *On the Heavens*. The context is in Oresme's parting with the Aristotelian notion of intelligences moving the celestial spheres. The situation, Oresme wrote, is "much rather like that of a man making a clock and letting it run and continue its own motion by itself. In this manner did God allow the heavens to be moved continually."[29], [14]

Oresme, it might be noted, explicitly rejected the idea of the Great Year,[15] which experienced a curious comeback a century later following the establishment of Ficino's school of Neoplatonism. This school made its imprint on the Renaissance not only by making Plato available in Latin, but also by advocating the ancient Greek doctrine of the "renaissance" or rebirth of ages, as well as of souls in endless sequence. Interestingly enough, those Renaissance humanists who adopted much of antique paganism were also the least sympathetic to the scientific movement. Ficino's school, in clear evidence of its sympathies with Hellenistic paganism, was also in charge of the translation of a third-century collection of mystical, magical, and cabbalistic writings, mistakenly ascribed to the mythical Egyptian priest, Hermes Trismegistus. The translation was chiefly responsible for the obscurantism, magic, and cabbala that flourished in many quarters during the sixteenth century. The most publicized and most pathetic spokesman of such vagaries was Giordano Bruno, whom late nineteenth-century Italian anticlericalism turned into a champion of reason and science. Of science there was very little in Bruno's works. Frances A. Yates, author of a book of prodigious erudition[31] that finally put Bruno in his own obscurantist light, rightly noted that had Copernicus been alive in 1584, he would have bought up and destroyed all copies of Bruno's freshly printed *Cena de le ceneri* (*The Ash Wednesday Supper*), the first book on Copernicus.[16] It contained, like Bruno's other works, an emphatic endorsement of the idea of the Great Year and of eternal recurrence.

Contrary to the claim often printed in run-of-the-mill histories of science, Bruno exercised no influence on the science which Kepler and Galileo were developing toward maturity.[17] This is equally true of the empiricist and rationalist method of science as articulated at that time by Bacon and Descartes, re-

---

[14] John Philoponus was the first to subject Aristotle's scientific philosophy to a searching criticism in the light of the biblical doctrine of creation. Philoponus did not, however, become sufficiently known in the West until about 1500. Crescas' famed critique of Aristotle was steeped in the same belief in a transcendental Creator.

[15] He did so both in his *De proportionibus proportionum* and *Ad pauca respicientes*.[30]

[16] The first English version of Bruno's *Cena* has just been published.[32]

[17] Bruno's renown as a scientist is largely a fiction foisted upon modern times by some strongly anticlerical figures of the Italian Risorgimento and by uncritical twentieth-century historians of science. See on this the introduction to Ref. 32. A. Koyré, who noted with a measure of disappointment[33] that Bruno was "not a very good philosopher" (Ref. 33, p. 54), strongly doubted that Bruno had influenced his contemporaries at all.

spectively. The balanced fusion of these two trends was achieved by Newton. If he was more successful than Bacon and Descartes in formulating or rather practising scientific method, it was due in part to his deeper perception of the consequences for scientific epistemology of a world-view based on the creative act of God. Such a world was for scientific methodology a rational product of a rational Creator, the supreme craftsman. While this notion might have had its shortcomings as actually formulated at that time, it assured a confidence in a rational investigation of nature and made also an empirical investigation of nature necessary in view of created nature's contingency. Needless to say, the Universe was viewed as a once-and-for-all product, a linear process[18] insofar as speculations were offered on cosmology.

All these are plain facts of the record of the history of science during the seventeenth century, the century of genius. Much of that genius was spent in turning science into a robust enterprise. The chief achievement was Newton's physics, which saw its further elaboration during the eighteenth century. In these two centuries, as if by an uncanny logic, the notion of the Great Year had almost completely disappeared from the printed pages of respectable literature. The return of that notion into the limelight was due to German Idealism. This is not the place to enter into the fantasies offered on science by Schelling and Hegel. Let it merely be noted that bad, very bad science went once more hand in hand with enthusiasm for eternal returns and with a categorical denial of the Christian tenet of creation. The Marxist offshoot of Hegelianism remained true to its roots by proclaiming, in materialist terms of course, the formation and dissolution of worlds in an eternal sequence as can be seen in Engels' *Dialectics of Nature*.[37] It is no coincidence that one of the most memorable statements on eternal returns in the nineteenth century came from the pen of a Marxist, L. Blanqui, who for a while was regarded as a potential rival of Marx for leadership of the proletariat. As all revolutionaries, Blanqui, too, was a visionary whose thought was not to be constrained by the hard facts and truths of science. Much of what he offered on science as a basis of eternal returns for worlds as well as for individuals was centered on his obsession with some obscurantist notions on the constitution of comets.[35], [19]

Nietzsche, the best known nineteenth-century spokesman of eternal recurrence, betrayed a similar amateurism in matters scientific as he tried to shore up his passionate poetizing on his living the same life over and over again an infinite

---

[18] Kubrin's effort[34] to present Newton as a believer in a cyclic world view rests on slender evidence, on a conspicuous myopia concerning the strong evidence to the contrary, and on an oversight of the fact that whatever cyclic elements there might have been in Newton's cosmological speculations, they were never submitted by him as hints opposing the idea of a once-and-for-all creation.

[19] The contents of this book are amply presented and discussed in my *Science and Creation*,[26] pp. 315–19.

number of times.[20] It should be of no surprise that, like Engels, Nietzsche saw in the relatively new doctrine of entropy a threat which both, and especially Engels,[21] tried to neutralize by pouring ridicule on some of the scientists most closely connected with the formulation of the entropy principle. They were William Thomson, the future Lord Kelvin, Clausius, Helmholtz, and Maxwell, who also clearly perceived the cosmological relevance of the dissipation of energy in all physical processes. One point that went largely unrecognized was the contradictoriness of the notion of a Universe composed of an infinite number of stars. In connection with the entropy principle such a Universe posed a curious dilemma if it existed since eternity. Could its infinite energy-content be reconciled with the dissipation of its energy that had already been underway since eternity?

Failure to recognize this problem was all the more curious as the optical paradox implied in an infinite Universe had already been brought to the center of attention by a paper of Olbers published in 1823.[22] But neither Olbers nor others perceived that there was something essentially wrong with the idea of such a Universe. Partly because of this, some outlandish theories were put forward in support of a local regeneration of dissipated energy. Thus, for instance, Rankine, who did work in thermodynamics, imagined the existence of concave ether walls in interstellar space so that the dissipated energy could be refocused and new stars might be born in those focal points.[39] Others tried to stake their hope of circumventing the prospect of a running-down Universe on statistical mechanics. The demon which Maxwell described in his book on heat as sitting on a stopcock connecting two vessels filled with different gases was meant to be a forceful and sarcastic reminder of the futility of reversing the direction in which entropy guided all physical processes.[40] The scientist who knew best Maxwell's thought and still refused to follow him on this point was Boltzmann. The last section of his lectures on molecular dynamics is a cosmology based on statistical mechanics and on an infinite Universe. Boltzmann bravely pictured the statistical recombination of scattered atoms into new worlds, an effort which might have carried conviction as long as the Universe was erroneously believed to be constituted of an infinite amount of matter.[41]

The twentieth-century sequel of cosmology has already been outlined in the

---

[20] At one point in his life Nietzsche wanted to devote several years to studying physics to gather scientific evidence for the eternal recurrence of worlds. He did not get beyond the point of gathering many, largely popularizing books on science in his library. The long monograph by Mittasch[36] is an unwitting evidence of Nietzsche's amateurism in matters scientific.

[21] Not surprisingly, Engels' chief target was Clausius. See Ref. 37.

[22] Before Olbers, Halley in 1720 and Chéseaux in 1743 discussed in print the paradox but without creating any echo. On the historical antecedents of Olbers' paper, its contents, and repercussions throughout the nineteenth century, see Ref. 7 and for some additional details Ref. 38.

opening section of this paper. It now remains to restate, in a qualified form with the help of this historical survey, some earlier remarks. First, the alleged identity of the ancient doctrine of Great Year and the idea of an oscillating Universe. Clearly, it would be a great mistake to see any meaningful parallelism between the fantasies on individual and collective rebirths as implied in that doctrine and the mathematical formalism of the cosmology of an oscillating Universe in its strict scientific sense. The oscillating Universe as a physical theory, let alone a true description of physical reality, is a different matter. As a physical theory, the oscillating Universe is beset with extraordinary difficulties. One is the enormous extrapolation in time which it implies. It takes not only mental boldness but sheer fancy to have oneself convinced about the actual turn of physical processes as they ought to occur in eighty billion years from now. In addition, there is the further difficulty of assuming that the same process will repeat itself in spans of twice eighty billion years in an endless sequence.[23]

Another extraordinary difficulty lies in that superdense state into which all matter of the Universe will have to be concentrated at the end of each contraction. About the physics of that superdense state nothing is known today except the ominous possibility of the whole Universe turning into a gigantic black hole. To eliminate such a possibility, one has to postulate a mechanism which would put an end to contraction some time before all galaxies had approached one another within a critical distance, or more specifically, before all galaxies had formed a four dimensional spacetime continuum with a radius so small that gravitation would gain an upper hand over all other physical forces and would do so irreversibly.

There is in addition a third major difficulty, the question of entropy in an oscillating Universe, which seems to have received little attention so far. Expansion means work to be done and so does contraction. Why is it then to be assumed that the enormous work implied in pushing apart the galaxies and pulling them again together is free of the dissipation of energy? Should it not rather be assumed that due to this dissipation of energy the oscillating Universe represents an oscillator with a vibration of ever-diminishing amplitude? Would not, therefore, an oscillating Universe tend toward a state of complete quiescence in close resemblance to the heat-death of the Universe as proclaimed by Helmholtz and others a century ago?[24]

---

[23] Actually, as was shown by Hawking and Ellis,[42] any statement on more than one cosmic cycle (expansion and contraction) is not so much physics as mere speculation. I am indebted for this information to Misner, one of the participants at the Colloquium.

[24] Those very few, like Novikov and Zel'dovich[43] who looked into the problem of entropy in an oscillating universe, found it very much to be a threat to perpetual oscillations. About the argument that the entropy cannot be meaningfully applied to the Universe as a whole, two remarks seem to be in order. The first is that such an objection is invariably based on an empiricist philosophy for which the Universe as a whole is not a meaningful concept. Copernicus, Galileo, Newton, and Einstein were certainly not empiricists in that sense. On the contrary, their science was steeped in a view of the Universe as a fully

The enormity of such difficulties and the almost complete absence of any clues as to how to cope with them can be minimized only at the price of becoming trapped in a most unscientific attitude. It is with such an attitude that one is confronted when reading or hearing rave accounts about the alleged truth of an oscillating Universe. Such a poor scientific attitude does not originate in purely scientific considerations: It originates in a philosophy which is determined to see the physical world as the ultimate entity, as an absolute self-containing, self-explaining, self-perpetuating being. To look at the world as the ultimate entity is an age-old urge in man. It propels him to seek identification with the blind forces of nature and to find in that identification a relief from certain metaphysical and ethical questions. Such an urge is fully satisfied in the idea of a Great Year and in certain presentations of the idea of an oscillating Universe. In that respect there is a close identity between the two. In both, man is nothing more than a bubble on unfathomable dark waves that rise and fall with no purpose and without a beginning and an end. A bubble of such origin has no obligation to make persistent queries about the reason of its existence here and now. As long as this urge dominated man's thought, science was condemned to repeated stillbirths, and in whatever circles this urge made its reappearance in modern times, it gave rise to a parody of science.

Science was born when an urge of a very different type became the dominating cultural feature. It was the urge to seek the ultimate source of intelligibility and being in a factor beyond the realm of change. This urge asserted itself in a most unique way in the biblical vision, for which existence was rational because it had an absolute beginning and an absolute consummation set by a rational Creator. It was an equally unique development when this vision of rationality became a broadly shared cultural matrix of which medieval Western Christendom was its tangible superstructure. The connection of that cultural matrix with the only viable rise of science in history can best be seen in the manner in which Buridan, Oresme, Copernicus, Kepler, Galileo, Boyle, and Newton anchored scientific methodology in man's ability to perceive from the order of particular, changing beings the ultimate unchangeable source of order and being, the Creator of all.

While history shows forcefully the superiority of this second urge over the first, history also shows that this superiority must be taken in a qualified sense. Anthropomorphism and anthropocentrism nowhere play greater havoc than in a theologically oriented thinking. Indeed, the vision of rationality anchored in the Creator time and again degenerated into an illusion which equated this or that rational world-view with the rationality of the Creator. The result was a painful

---

ordered whole, and they did not postpone the construction of scientific cosmology until the clarification of each and every empirical difficulty. In fact, Galileo found Copernicus' greatness in his courage to commit a rape of his senses! The second remark is a reminder of the crucial role which the recognition of the impossibility of *perpetuum mobile* played in the progress of physical science.

narrowing of the horizons of which the late sixteenth and early seventeenth centuries saw more than enough in that classic clash between an ossified scholasticism and a science full of its newborn strength. Ironically, this strength of science derived from the same source which enabled centuries earlier some great scholastic philosophers to perceive major shortcomings in the Greek philosophical and scientific world view.[25]

The vitality of that source could not be unfolded in full by any single genius, but only by that collective genius which evidences itself through the historical development. This development brought out two main lessons: One was the recognition that the richness of nature, precisely because it is the handiwork of the Creator, cannot be at any time known exhaustively by a created mind, which has, therefore, always a profound reason to look beyond what is already known. The other lesson consisted in the recognition that the processes of nature are running a course which is linear in the sense in which entropy revealed itself as a universally valid direction of time's flow. This direction runs from a maximum level of energy to a maximum level of entropy, a situation that readily suggests an absolute beginning and an absolute end for physical processes. Needless to say those who have the kind of urge that gave rise to science find no difficulty with the idea of an absolute physical beginning and end. They should, however, try to find difficulty with any physical specification of either of those endpoints. Science and theology demand eyes ready to see beyond any physical barrier, however impenetrable it may appear.

Does this mean that one ought, therefore, to fancy that one can indeed perceive beyond the barrier of that cosmic black hole and other similar barriers and that one is therefore entitled to endorse the idea of an oscillating Universe? To do so would amount not only to an oversight of some enormous difficulties posed by scientific evidence but also to a disregard of a major lesson of scientific history. This lesson concerns the ambivalent role which uncritical affirmations of infinity played in science. The principle of plenitude fell far short of its cosmological promises.[26] For all we know, the Universe is not the realization of all possible existences but only of a relatively few forms. Reflection along these lines should play a sobering corrective to visions in which the Universe is the plenitude in time, a kind of cosmic *perpetuum mobile.*

Apart from all these considerations, it still remains true that the idea of an oscillating universe bespeaks of a totality of things which is a most particular being in its content and laws. As such a being it cannot derive its specificity from itself. For even if it was shown that given, say, the hydrogen atom, all other fundamental particles are determined and even all features of the cosmos, the particularity of the hydrogen atom still remains to be accounted for. The explanation of a particular universe can only start with some particularity which is

[25] I plan to discuss this point at some length in my Gifford Lectures to be given in the Springs of 1975 and 1976 at the University of Edinburgh.
[26] As argued and documented with great forcefulness by Lovejoy.[44]

not accounted for by the particularity itself if circularity in argumentation is to be avoided. Furthermore, a specific being like a particular universe is a specific event that happens, and because it does happen, and does so in a radical sense, it cannot exist necessarily. It therefore owes its existence to a being that does not happen, that is, not subject to change which is happening, in any sense. Such a being *is* without any restriction. In a profound philosophical insight the name of that being was given long ago as the ONE WHO IS. The same insight also perceived that the ONE WHO IS is the Lord of all—the entire realm of change—precisely because He IS and therefore transcends any and all change.

Such are some of the reflections that might be entertained in the midst of excited reports that we live in an oscillating universe. Such a universe, or any universe, is a universe of change with all the far-reaching consequences of this elementary fact. Those willing to ponder them will have every right to be excited on finding that the physical possibility and the conceptual history of an oscillating universe are transfused of the very same message which should seem of paramount importance for the understanding of science, its past as well as its future.

REFERENCES

1. J. P. Ostriker and P. J. E. Peebles, A Numerical Study of the Stability of Flattened Galaxies: or, can Cold Galaxies Survive?, *Astrophys. J.* **186**, 467–480 (1973).
2. A. Friedmann, Über die Krümmung des Raumes, *Z. Physik* **10**, 377–86 (1922).
3. A. Einstein, *Z. Physik* **11**, 326 (1922).
4. A. Einstein, *Z. Physik* **16**, 228 (1923).
5. A. Einstein, Zum kosmologischen Problem der allgemeinen Relativitätstheorie, *Sitzungsber. Preuss. Akad. Wiss., Phys.-Math. Kl.* **[XII]**, 235–37 (1931).
6. H. A. Lorentz, A. Einstein, H. Minkowski, and H. Weyl, *The Principle of Relativity: A Collection of Original Memoirs on the Special and General Theory of Relativity*, with notes by A. Sommerfeld; transl. by W. Perrett and G. B. Jeffrey (Dover, New York), pp. 177–88.
7. S. L. Jaki, *The Paradox of Olbers' Paradox* (Herder and Herder, New York, 1969).
8. G. Lemaître, Un univers homogène de masse constante et de rayon croissant rendant compte de la vitesse radiale des nébuleuses extra-galactiques, *Ann. Soc. Sci. Bruxelles* **47A**, 49–59 (1927).
9. G. Lemaître, *Monthly Not. Roy. Astron. Soc.* **91**, 483–90 (1930–31).
10. J. Jeans, *The Mysterious Universe* (Macmillan, New York, 1930), p. 182.
11. Discussion on the Evolution of the Universe, in *British Association for the Advancement of Science: Report of the Centenary Meeting, London, 1931, September 23–30* (Office of the British Association, London, 1932), pp. 560–610.
12. P. Michelmore, *Einstein: Profile of the Man* (Dodd, Mead & Co., New York, 1962), p. 164.
13. H. Bondi and T. Gold, On the Steady-State Theory of the Expanding Universe, *Monthly Not. Roy. Astron. Soc.* **108**, 252–70 (1948).
14. F. Hoyle, A New Model for the Expanding Universe, *Monthly Not. Roy. Astron. Soc.* **108**, 372–82 (1948).
15. H. Dingle, Science and Modern Cosmology, *Monthly Not. Roy. Astron. Soc.* **113**, 406 (1953).

16. Pope Pius XII, On the Proofs of the Existence of God in the Light of Modern Natural Science (English translation, National Catholic Welfare Conference, Washington, D.C., 1951).

17. E. T. Whittaker, *The Beginning and End of the World* (Oxford, Oxford University Press, 1942).

18. E. T. Whittaker, *Space and Spirit* (Henry Regnery, Hinsdale, Illinois, 1948).

19. *Rival Theories of Cosmology* (Oxford University Press, London, 1960).

20. J. E. Peebles and D. T. Wilkinson, The Primeval Fireball, *Sci. Am.* **216,** 36 (1967).

21. W. Sullivan, *The New York Times*, Nov. 2, 1975, p. 56, col. 2.

22. J. Charon, *Cosmology*, transl. by Patrick More (McGraw-Hill, New York, 1970), p. 237.

23. W. A. Haas, *The Destiny of the Mind: East and West* (Macmillan, New York, 1956), pp. 38–39.

24. R. E. Hume (transl.), *The Thirteen Principal Upanishads*, 2d rev. ed. (Oxford University Press, London, 1934), p. 414.

25. Sylvanus G. Morley, *The Ancient Maya* (Stanford Univ. Press, Palo Alto, Calif., 1946), p. 445.

26. S. L. Jaki, *Science and Creation: From Eternal Cycles to an Oscillating Universe* (Scottish Academic Press, Edinburgh; Science History Publications, New York, 1974).

27. *Aristotle's Metaphysics* transl. with commentaries and glossary by H. G. Apostle (Indiana University Press, Bloomington, 1966), pp. 13–14.

28. *Atheneus, The Deipnosophists*, with an English transl. by Charles B. Gulick (W. Heinemann, London, 1955), Vol. V, p. 299.

29. Nicole Oresme, *Le Livre du ciel et du monde*, ed. by Albert D. Menut and Alexander J. Denomy; transl. with an introduction by Albert D. Menut (University of Wisconsin Press, Madison, 1968), p. 289.

30. Nicole Oresme, *De proportionibus proportionum and Ad pauca respicientes*, ed. with introductions, English transl., and critical notes by Edward Grant (University of Wisconsin Press, Madison, 1966), pp. 306 and 382.

31. Frances Yates, *Giordano Bruno and the Hermetic Tradition* (University of Chicago Press, Chicago, 1964).

32. Giordano Bruno, *The Ash Wednesday Supper*, translation, introduction, and notes by S. L. Jaki (Mouton, The Hague, 1975).

33. A. Koyré, *From the Closed World to the Infinite Universe* (The Johns Hopkins Press, Baltimore, 1957).

34. D. Kubrin, Newton and the Cyclical Cosmos: Providence and the Mechanical Philosophy, *J. Hist. Ideas* **28,** 325–46 (1967).

35. L. Blanqui, *L'éternité par les astres: Hypothèse astronomique* (Librairie Germer Baillière, Paris, 1872).

36. A. Mittasch, *Friedrich Nietzsche als Naturphilosoph* (Alfred Kröner Verlag, Stuttgart, 1952).

37. F. Engels, *Dialectics of Nature*, transl. by Clemens Dutt with preface and notes by J. B. S. Haldane (International Publishers, New York, 1940; fourth printing, 1960), pp. 202 and 216.

38. S. L. Jaki, *The Milky Way: An Elusive Road for Science* (Science History Publications, New York, 1972).

39. M. Rankine, On the Reconstruction of the Mechanical Energy of the Universe (1852), *Miscellaneous Scientific Papers*, ed. by W. J. Millar (Charles Griffin & Co., London, 1891), p. 202.

40. James Maxwell, *Theory of Heat*, 3rd ed. (Longmans, Green & Co., London, 1872), p. 308.

41. L. Boltzmann, *Lectures on Gas Theory*, transl. by S. G. Brush (University of California Press, Berkeley, 1964), pp. 447–48.

42. S. W. Hawking and G. F. R. Ellis, *The Large-Scale Structure of Space-time* (Cambridge University Press, 1973).

43. I. D. Novikov and Ya. B. Zel'dovich, Physical Processes near Cosmological Singularities, *Ann. Rev. Astron. Astrophys.* **11**, 387–410 (1973).

44. A. O. Lovejoy, *The Great Chain of Being: A Study of the History of an Idea* (Harvard Univ. Press, Cambridge, Mass., 1936).

Chapter XVII

Heaven and Earth—The Relation of the Nebular Hypothesis to Geology

Philip Lawrence

Harvard University

One of the main concerns of the geologist has always been to interpret the history of the Earth. Prior to the nineteenth century, earth scientists generally believed that such a history would go back to the creation of the planet. By the early nineteenth century, however, most geologists admitted that the nature of their science did not allow for considerations of planetary origin. The topic, as a scientific problem, passed into the realms of physics and astronomy. Consequently, it would seem, the influences which theories of planetary origin might exert on geological thought appears to have also passed from the concern of historians of geology. The result has been some serious misunderstandings with regard to the geology of the nineteenth century. Even though most geologists did not hypothesize on the origins of the Earth, their work was, nevertheless, significantly affected by such considerations; a misunderstanding of this point leads to a distorted view of nineteenth-century geological theory. Though the geologist himself could not treat of the origin of the Earth, he was free to utilize what the physicist and the astronomer believed could be proven, just as he always accepted from the physicist and chemist his knowledge of the immanent properties of matter. In point of fact, the character of much of nineteenth-century geological thought was determined by two factors: (1) the explanation of the Earth's origin accepted by most astronomers, and (2) the geophysical processes most physicists believed had to follow from such an origin. An understanding of nineteenth-century geology is not possible without a comprehension of this relation to physics and astronomy.

My main purpose in the following pages is to trace the development of this relationship from inception to mature form. The relationship originated at the beginning of the second decade of the nineteenth century. When it achieved its mature form in the fourth decade, the relationship had given rise to that conceptual element which bestowed meaning and coherence upon the conclusions the geological community had gleaned from their studies of purely geological evidence. The terrestrial consequences of established physical law and the range of geological observations harmonized together in a unified theory of planetary history.

During the second half of the nineteenth century, this theory, in its turn, would strongly influence developments in biology and natural theology, areas of thought inseparably linked to prevailing views of Earth history. The influences the theory exerted is perhaps the best illustration of its significance. The important role it played has been hitherto almost completely lost to historians of science because of a commonly held misconception that Charles Lyell "established" the science of geology. Thus, while my paper is principally devoted to the analysis of the relationship between geology, physics, and astronomy, I shall turn briefly toward its end to both these issues. In order to fully understand the geology of the nineteenth century and its relation with physics and astronomy, one must also see that Lyell, rather than revolutionizing geology, challenged the supporters of a theory of great scientific scope and refinement. One must see, in addition, that he was actually unsuccessful in his attempt to supplant this theory with his own.[1] The commonly held misconception seems to have arisen from a lack of precision in assessing the relation in which geology stood with astronomy and physics. It is, indeed, the main burden of the following pages to illuminate the nature of the relationship.

The source of this relationship was the nebular hypothesis of Laplace. Initially put forward as a speculative concept in 1796, the work of the observational astronomer, William Herschel, had, at the beginning of the second decade of the nineteenth century, converted this idea into a serious scientific hypothesis, in which the origin of the Earth and of all planets was held to be that of an incandescent mass.[2] By the end of the second decade of the nineteenth century, the nebular hypothesis was for all intents and purposes an accepted tenet of astronomy. With the exception of a period of diminished approbation during the latter 1840s and the 1850s,[2] the nebular hypothesis would continue to be so regarded throughout the rest of the nineteenth century and into the twentieth (see, e.g., Clerke[3]).

---

[1] An important statement of this point of view, upon which a number of historians of science have been working, has been made by Rudwick.[1]

[2] Astronomical observations made with improved instruments in the period (the first by the Earl of Rosse, in 1845) indicated that definitely in many cases, and perhaps in all instances, "nebulae" could be resolved into clusters of remote stars. Grave doubts were consequently thrown on the validity of the nebular hypothesis for a large number of scientists because this challenged Herschel's conclusion that nebulae were uncondensed stellar matter. These doubts, however, were dispelled in the early 1860s when William Huggins established beyond doubt the existence of true gaseous nebula.

The explanation of the Earth's origin contained in the nebular hypothesis possessed implications of potentially great import for geology. The origin of the Earth could be traced back to a nebulous state. Beginning as an incandescent mass, the Earth had cooled and condensed into man's present abode. It had long been known, at least since the seventeenth century, that the temperature in mines increased with depth. This empirical evidence could be interpreted to suggest that the Earth still possessed a central thermal reservoir representing the "residual" heat of its initial incandescent state. These geological implications, moreover, were obvious, at least in broad outline, to Laplace's contemporaries. Buffon had seen to that. In his well-known theory of "origins," published in 1778,[4] Buffon had worked out a comprehensive set of geological implications for an Earth whose initial planetary state was the same as the one Laplace postulated. Inherent in the nebular hypothesis was the key to a theory of geological dynamics.

By the second decade of the nineteenth century, geology had emerged as a rapidly developing and expanding field of natural history. It was also becoming an immensely popular subject and attracted all types of people to its study. Within this vast collage of ability and interest, a body of geologists possessing "professional" status is clearly discernible, both to the historian and contemporary. Geology was the science *par excellence* in the first half of the nineteenth century. Accordingly, in that age of scientific organization and professionalization, its institutional arrangements and accepted values were generally in advance of other specific disciplines. Early in the century, geology had already established types of acceptable data, their proper ordering, and forms of presentation; it had also identified the key phenomena requiring explanation. Within this framework different theories could and did arise. The shared normative framework supplied the criteria for individual competency and the evaluation by competent individuals of proposed theoretical concepts. This implicit but rather well-developed framework allowed distinctions to be drawn between the serious geologist, the dabbler, and occasional instances of the fantastic.

The geologists of "professional" status were united in the belief that geological causes had been and were naturalistic and, unless there was strong evidence to the contrary, actualistic. The advocates of "supernatural" occurrences or ultimately inexplicable events were no longer accorded recognition by serious geologists. Though many geologists in the course of the nineteenth century would have argued that the intensity of geological agencies had varied or that unique combinations of agencies had occurred at specific past times, none believed that the "laws" or the basic patterns of nature were ever violated.[3] Geo-

---

[3] Any predominant influence religion may have had on geology, it should be added, was also passing from the scene. "Biblical" geologists there had been and they would continue. But they would no longer constitute an important influence. While many of the major geologists, especially in England, would be concerned with the relation of their science with *Genesis*, the facts of the Earth always came first.

logical theories, accordingly, had to be couched in terms of analogs with processes observable at present—in terms of the principle of "actualism." Aware of both their role as historical scientists and the relative infancy of their science, moreover, serious geologists in the first half of the nineteenth century did not view all such theoretical explanations as of a piece. During this period they appear to have made a distinction between explanations in terms of "history" and explanations in terms of "dynamics."

In the first half of the nineteenth century, the geologist could theorize on one or both of two interrelated levels. First there was "historical theory."[4] Such considerations began with the basic data of geology and ended with the formulation of geological histories. Historical theory, therefore, dealt with the events that have shaped the earth as well as the geological processes and rates of activity inferred from the evidence of these events. What historical theory did not consider was the actual nature of the dynamic agents and their causes. This was the realm of "dynamic theory," the goal of which was to fathom the workings of the forces which had produced the events of geological history. Needless to say, the two forms of generalization are intimately connected. This is also, obviously, a relationship wherein one mode of theory can overbalance or set the terms for the other. The position of predominance is not logically necessary, but open to the predilection of geologists.

The place of the nebular hypothesis in geology was clearly in the realm of dynamic theory. At the beginning of the nineteenth century, dynamic theory had been the dominant branch. The debates between the Huttonians and Wernerians were first and foremost concerned with rival systems of causal dynamics. Historical material had been utilized in the formulation of these systems, but their basic structure had been determined by other philosophical and scientific factors.[5] The major role of historical material in the rival theories of the Earth had been to elucidate arguments and to fortify the special pleading of disputed points. Though fervent at the beginning of the nineteenth century, by 1810 the debates between the advocates of fire and water had passed their peak and were rapidly waning, with victory denied both sides. Wearied of the confusing and elusive search for a comprehensive theory of geological dynamics, the majority of geologists in the decade 1810–1820 increasingly eschewed such considerations. The geological implications of the nebular hypothesis were no exception.

The nebular hypothesis achieved general acceptance as an explanation of the solar system's origins in the decade 1810–1820. While there was growing agreement on the origins of the Earth, there was corresponding discord over the tenability of the geological implications of the nebular hypothesis. In the period

---

[4]This term and its counterpart, "dynamic theory," express the sense contemporary geologist had of their activities. Similar though not identical terms are used, for example, by William Whewell.[5]

[5]For Hutton see Davies.[6] For Werner and a good brief summary of the debates see Wilson.[7]

1800-1815, the caloric theory of heat reached and passed the peak of its influence. Prominent among the topics argued by adherents of rival caloric systems was the physical possibility of the continued action of a residual heat over time.[6] The doubts engendered by such debates were reinforced by systematic studies of the long-known apparent tendency of heat to augment with depth.[7] Increasing knowledge concerning terrestrial heat patterns also appeared to belie the existence of an adequate "central" source of thermal energy.[8]

Prevailing concepts of the nature of thermal phenomena during the second decade of the nineteenth century, in short, implied that the Earth's incandescent origin as set forth in the nebular hypothesis could not exert any significant influence on geological history. Doctrines of a "residual" heat provided no ready successor to the declining causal systems of Hutton and Werner. Lacking a firm foundation on which to build concepts of comprehensive dynamics, the majority of geologists came to avoid such issues altogether. There was too much uncertainty and speculation involved. The tone was set by the Geological Society of London, founded in the midst of the Huttonian and Wernerian debates. By

[6]For general background on this subject see Cardwell [8a] and Fox.[8b] The best example of the debates in the geological context is the dispute waged in the pages of the *Transactions of the Royal Society of Edinburgh* between John Playfair and John Murray.[9]

[7]See, for example, the papers on this subject by the French geologist D'Aubuisson De Voisins.[10]

[8]Excellent examples of this line of thought are found in three papers dealing directly with geology that Laplace himself published between 1817 and 1819.[11] Laplace did not offer a concise or comprehensive argument. The elements of his position were distributed among the several papers and none of them were totally devoted to geology. Laplace had no intention of turning the nebular hypothesis into a theory of geology. The concept of geological implications he possessed was so like that of Buffon that Laplace's sole purpose seems to have been to clear away possible geological objections to the validity of the nebular hypothesis as an astronomical theory. That Laplace apparently felt that a defense was required in this area is a strong indication of the scientific standing of the objections to theories of central heat.

Like Buffon and in contrast to the Huttonians, Laplace thought of the Earth's geological history as simply that of a cooling mass, its surface features primarily shaped when the crust solidified. The central heat did not lift mountains or shape continents as it did in the Huttonian system. Laplace could thus maintain that the maximum heat loss from the Earth had occurred soon after the planet began to consolidate. For most of geological time, the diminishing remanent of central heat had not possessed the requisite power to perform dynamic processes on the Huttonian scale. The flow of heat during geological time, however, could well have sufficed to influence climatic patterns and here the strata record offered strong corroboration (see below in the body of the paper). The existence of an internal heat gradient, in turn, indicated that however feeble, an internal reservoir still might possibly exist. Laplace could thus, also argue that the flow of central heat would have sufficed, if channeled into and concentrated in a favorable geological structure, to produce a pattern in which igneous activity progressively decreased in extent, though not necessarily intensity.

Laplace, in short, attempted to develop the most tenable connection possible between the nebular hypothesis and geology in the decade 1810–1820 and, in so doing, pose answers to what he considered to be possible sources of objections to his nebular hypothesis.

1811, reaction was apparent and attested to by the famous preface to the first publication of the Society's *Transactions:* "In the present imperfect state of this science, it cannot be supposed that the Society should attempt to decide upon the merits of the different theories of the earth that have been proposed."[12]

The emphasis was on evidence, not old speculation. The new normative principle was that men of opposite theoretical persuasions should, by objective observation, produce identical descriptions of the same phenomenon. Then, working from these accepted descriptions, fruitful debate on the merits of theory could follow. As one member of the Geological Society put it, speaking for almost the whole of his science:

> As I have related the facts I observed, independent of any theory, if they are at all valuable in the geological history of this country, their value will remain undiminished whether the speculations I have entered into are just or fallacious. If the geologist strictly guards himself against the influence of theory in his observations of nature, and faithfully records what he has seen, there is no danger of his checking the progress of science, however much he may indulge in the speculative views of his subject.[13]

Discussions of dynamic theory formed only an undercurrent in the geological literature of the decade 1810-1820. The dominance passed to the historical branch of geological theory.

During the second decade of the nineteenth century, a historical picture of geological history, acceptable to the majority of geologists, emerged. The predominant influence in the formulation of this consensus was generated by the development of the study of stratigraphy. In England, the Geological Society of London learned the principles of stratigraphy from the surveyor William Smith. Concurrently, these ideas were also developed by the French paleontologists George Cuvier and Alexandre Brongniart. These earth scientists had found that the ages of different strata could be determined and correlated by their fossil assemblages. Previously, the succession of strata had only been based on the lithological characteristics and these correlations had proved far less than conclusive in too many cases. The principle that each formation contained certain fossils with such regularity that these "type" fossils could be used to recognize the different strata did prove conclusive. An accurate succession of strata was the obvious key to the geological history of an area. Such histories, in turn, were recognized by geologists as the key to historical theory. The paramount task of geology became the determination of the full stratigraphic succession.

Geologists interpreted the record of the strata with none of the leaps of imagination that would characterize the later methodological philosophy of geology.[9] Disconformities, unconformities, the violently folded strata of moun-

---

[9]The most significant of these interpretations was put forward by George Cuvier. His theory was the true parent of all the nineteenth-century expressions of geological theory termed "catastrophic" by historians of science. The theory first appeared as the preface to Cuvier's great paleontological and stratigraphic work, *Recherches sur les Ossemens Fossiles* (1812).[14] This preface, entitled *Discours sur les Révolutions de la Globe,* was reprinted as a separate work in several editions. For a treatment of the philosophy of geology see Albritton.[15]

tain chains, the progression by stratum of distinct organic populations, and, most important, the remains of what is now recognized to be glaciation, all appear to commonsense to imply violent breaks in the strata record. The evidence appeared to argue for the conclusion that the surface of the Earth had periodically been subjected to the work of forces not now operational or no longer operating at the same intensity. These forces had produced a series of sudden geological "revolutions." Stratigraphic evidence also indicated that the history of the Earth, at least from the time it first became inhabited, had been directional and "progressive." The fossiliferous rocks presented evidence of a succession of environments that had progressed from a generally tropical state to the present configuration. In this succession of environments, appropriately adapted organisms had successively come into existence, flourished, and then passed away as geological "revolutions" set the stage for yet another "world." The succession of living forms had been progressively of a higher order and greater complexity.

Historical geological studies also yielded insights into the types of geological forces that had shaped the apparently episodic, saltatory, and directional course of Earth history. One class of geological agency was in a logical position to be accorded pride of place. The importance of tectonic and volcanic activities was obvious and they had long been actively studied in their own right. It was equally obvious that almost all such activities were products of igneous phenomena. Studies of unstratified rocks during the early nineteenth century tended to corroborate the conclusions geologists were gleaning from stratified deposits. They emphasized that modern igneous activities did not appear to match the extent and magnitude of igneous evidences in earlier epochs. Granitic intrusions were more common and extensive among older strata and became progressively less apparent as one advanced through the succession. Metamorphized rocks seemed to follow the same pattern. There could only be one reasonable conclusion. When Robert Bakewell drew this conclusion in 1815, he spoke for most of his colleagues: "During a former state of our planet, this internal fire [the cause of igneous activity] must have been more intense than since the records of authentic history."[16] Moreover, epochs of intense tectonic activity seemed to alternate with quiescent periods over much of the globe.[16, 17] This, indeed, harmonized well with the evidence stratigraphers were gathering from the sedimentary rocks; geological history was episodic and directional. Igneous forces came to be regarded as the major agents of uplift and subsidence.

In sum, then, work in historical geology during the decade 1810-1820 presented increasingly persuasive evidence for a directional Earth history in which diminishing igneous forces played a crucial role. By the early 1820s—the geological literature reveals—the directional interpretation of earth history constituted the consensus of geological opinion.[10] Quite clearly this interpretation cannot be properly considered as a kind with the comprehensive theoretical systems proposed by Hutton, Lyell, or even Buffon. The directional interpretation of

[10] This theoretical consensus embraced both former Neptunians and Plutonists. See, for example Ami Boué (former Neptunian),[18] and John MacCulloch (former Plutonist).[19]

Earth history was very simply what it was: a coherent and comprehensive "description" of the geological past. Having repudiated the speculative discussions of causality which had marked the Huttonian–Wernerian debates, earth scientists tended to restrict theoretical considerations to areas where they believed sound generalizations could be based on empirical data. The acceptable geological data in this period were "historical." A theory of the Earth emerged from the study of this evidence, but it could only be a theory of Earth history. Though the types of geological processes that had shaped the rocks could be inferred from the historical evidence, as well as the rates of intensity at which they had worked, the rocks said nothing of the origin or causes of these forces.[11] The history contained in the rocks was that of the deposits themselves, nothing more. Growing out of a research tradition (out of a widely shared gestalt of the appearance of the strata record), established in the decade 1810–1820, there had come into being by the early 1820s what can only be described as a "historical theory," accepted by most geologists. This theory would continue to be the dominant element in geological thought during the ten years directly preceding the appearance of the first edition of Charles Lyell's *Principles of Geology*.

The directional interpretation of geological history reached the peak of both its influence and scope as a consensus around the year 1825. The mid-decade also saw a general resurgence of interest among geologists in the causal dynamics of geological history. It became clear that the "description" of the history of the Earth could only be carried to a point, without verging in its own turn on speculation. All forms of "historical" evidence appeared to indicate that Earth history had been punctuated by sudden and violent revolutions, the evidence

---

[11] The majority of geologists avoided such issues altogether. There was too much uncertainty and speculation involved. Geologists of the period (1810–1820) were cautious with "dynamic theory." The subject of igneous forces was no exception. To take theoretical ideas about tectonics and volcanism beyond concepts of historical effect was to deal not only with the immediate but with the ultimate nature of the igneous forces. Some natural philosophers did, nonetheless, attempt to pose at least hypothetical explanations. The most notable and popular of these was the "theory of central fire" offered by Humphrey Davy.[20] Davy suggested that the dense core of the Earth must consist principally of the highly inflammable metals united with oxygen or "pure aire." The action of water, of which the crust of the Earth abounded, on these metallic materials would produce a massive chemical combustion seated at the Earth's center. All igneous phenomena could be traced to this great "central fire." By the mid-1820s the majority of geologists had rejected such ideas. They were conclusively shown to be inadequate with the publication of books on igneous phenomena by Charles Daubeny[21] and George Scrope[22] in 1826 and 1825; respectively. However, it should be noted that Daubeny, while rejecting all other proposed explanations, continued to espouse Davy's concept of central fire. He persisted in this opinion even after Scrope had taken special pains to demonstrate the inadequacy of this cause and Davy himself had rejected it.[23]

of which could not be accounted for by present geological agencies. That such events had occurred was nothing more than an empirical conclusion; the causes of the catastrophes were simply unknown, not unknowable. Special "creative" geological agents or supernatural occurrences were not considered as dynamic forces.

The great unknown of geology was the break in scientific analogy, analogy of scope and magnitude, but not in kind, between existing agents and evidence of past events. Yet, in merely "describing" the nature of geological revolutions, the scope of effect attributed to dynamic forces was often stretched to the point where the difference between past and present manifestations became almost those of kind. [12] This implied that the physical cause of the geological revolutions might well have been forces not inferable by analogy from observations of present forces and hence in violation of the principle of actualism. The most promising means of resolving this difficulty was to circumscribe closely the possible causes of the geological revolutions by adding careful studies of geological dynamics to the inference on the historical interpretation of the geological past. For this reason, geologist in the 1820s began to devote much more effort to the study of dynamic agents than they had in the previous decade. The purpose of these geologists was to demonstrate that the present geological forces were suf-

---

[12]The major and most notable instance of this was the interpretation of the Drift by the British geologist and natural theologian, the Rev. William Buckland. To the geologists of the early nineteenth century, the Drift (the term used in the first half of the nineteenth century to describe those phenomena now attributed to the Pleistocene glaciation) was clearly a product of the most recent of the geological revolutions. Buckland developed a strong line of reasoning based on historical geology, in which this most recent revolution was depicted as, in effect, a gigantic violent tidal wave (the two best compendiums of Buckland's ideas are *Vindiciae Geologicae* and *Reliquiae Diluvianae*[24]). Drift was the most pronounced of the relatively superficial geological changes that this catastrophe had produced. As such, the Drift was also the key datum for Buckland's most important conclussion. Drift was found extensively throughout the Northern Hemisphere and at high altitudes. The last catastrophe, consequently, had totally overwhelmed at least most of the globe and, Buckland argued, it in fact had been a worldwide revolution. Buckland's conclusions on the nature of the formation of the Drift were a plausible and persuasive interpretation of the empirical data. While he was criticized for extending the scope of the catastrophes he described beyond the locales where Drift was encountered, his general rendering of the evidence was widely accepted.

Buckland's interpretation of the Drift, nonetheless, stretched the principle of actualism to its very limits. Buckland's arguments persuaded most of his colleagues that if the last catastrophe had been capable of completely overwhelming at least much of the continental areas, then it was only philosophical to conclude that this had also been the dynamic nature of all the preceding revolutions. If this were the case, then the analogy between the energy of past and present geological forces was stretched to the point where the differences became almost those of kind.

ficient, by means of extrapolations of intensity and possible combinations, to account for the events read from the strata.[13]

Geologists became especially interested in the precise dynamics of tectonic and volcanic agencies, acknowledged as the main constructive agents of geological change. During the eleven years between 1820 and 1830, geologists considerably enlarged their knowledge of the evidences of past tectonic and volcanic activities. Such investigations lent strong support to the reaffirmation of the actualist principle by enhancing the conception of igneous forces as the prime constructive dynamic agents of geological history. The continued studies of the evidence of these forces expanded geologists' perceptions of their past extent; the indications of extensive activity were worldwide. Studies of the fossiliferous strata found in the principal mountain chains showed that they had been uplifted at different points in geological time.[14] These historical evidences seemed to clearly trace the steady directional decline in intensity and extent of the tectonic and volcanic agencies.[15] There were also strong indications that igneous action had brought about the metamorphisms, dislocations, and contortions of the strata—in sum, all the phenomena associated with mountain-building and lesser uplifts. The prime constructive agents of geological change, igneous forces, had, in addition, left a record of their activities that appeared to show no violation of the actualist principle. The energy and scope of these agents had varied with time, but there was no evidence of a break in analogy between their past and present manifestations.[16]

It was, moreover, in the area of igneous forces that the most important developments in concepts of geological causality were to arise. These conceptual developments would culminate in a system of causal dynamics which both confirmed and supplied an explanatory pattern for the directional interpretation of Earth history. The foundation of this system of dynamics was the nebular hypothesis of Laplace. The dominant influence which converted its inherent

---

[13] This outlook was summed up in 1825 by one of the leaders of the movement, George Scope (*Considerations on Volcanoes*,[22] pp. iv–v): "As the idea imported by the term Cataclysm, Catastrophe, or Revolution, is extremely vague, and may comprehend any thing you choose to imagine, it answers for the time very well as an explanation; that is, it stops further inquiry. But it has also the disadvantage of effectually stopping the advance of science by involving it in obscurity and confusion.

If, however, in lieu of forming guesses as to what may have been the possible causes and nature of these changes, we pursue that which I conceive to be the only legitimate path of geological inquiry, and begin by examining the laws of nature which are actually in force, we cannot but perceive that numerous physical phenomena are going on at this moment on the surface of the globe, by which various changes are produced in its constitution and external characters; changes extremely analogous to those of earlier data, whose nature is the main object of geological inquiry."

[14] The first of these studies was that of Alexandre Brongniart.[25]

[15] Examples are Charles Daubeny,[26a] John MacCulloch,[19] Robert Bakewell.[26b]

[16] Prominent examples are William Fitton and Adam Sedgwick.[27]

apparent relation to geology into a definite relation in fact belonged to Joseph Fourier.

Fourier was the most influential of a comparatively young group of French scientists who around 1815-1825 successfully wrought a major reappraisal of the nature of heat.[17] This reappraisal attained its culmination and best expression in Fourier's *Analytical Theory of Heat* (1822).[28] Fourier's program was to develop the invariable mathematical properties of heat with no concern for the first causes of heat phenomena. Rival theories based on different axioms would no longer be possible, only a progressively unified body of mathematically expressed knowledge about the behavior of heat.

In 1816 Fourier published a full account of his theory in extract.[29] He then proceeded to elaborate major aspects of the theory in papers that appeared in 1817 and 1818.[30] In 1822 the complete exposition of the theory was issued. Though he explained his most important conclusions in plain words, Fourier's treatment of thermal phenomena was highly mathematical and therefore beyond most geologists. They had to take the accuracy of Fourier's derivations on faith, faith engendered by the reactions of natural philosophers. As the parts of the theory were worked out in detail, it gained rapid acceptance. With acceptance came the popularization of the theory for the intelligent and informed reader. Because it represented a most elegant and important application of the new theory, one aspect singled out for special attention in this area, both by Fourier and others (see, e.g., Ref. 31), was the portent for the nebular hypothesis and the concept of central heat.

Fourier explicated the terrestrial application of his heat theory in two papers published, respectively, in 1820 and 1824. The content of the two papers was basically the same. In the first paper,[32] however, Fourier developed the nature of the principal mathematical formulas and their variables that were necessary for the analysis of the motions of the heat involved in the Earth's various thermal phenomena. This paper was written primarily for the physicist. As a result, much of the content was beyond the limited mathematical knowledge of most geologists. The second paper,[33] covering the identical ground, was purely descriptive. This second paper was probably intended to be, and was in fact, the main instrument for carrying Fourier's ideas into the science of geology.[18]

The primary thermal phenomena of the Earth were three in number. The first was the radiant heat received from the Sun. This heat nourished life and by its unequal distribution caused the diversity of climate. The second heat motion consisted of the miscellaneous radiant heat received from the stars. Third and

---

[17]The major figures in this group were Joseph Fourier, Dominique Arago, Pierre Dulong, Augustin Fresnel, and Alexis Petit.

[18]It should be noted that Fourier originally intended that the terrestrial application of his heat theory should comprise a section of the *Analytical Theory of Heat*, but did not in fact, include such an explication in the book.

finally, its surface manifestations masked by the two other heat flows, was the Earth's own primitive heat, still preserved in the interior. Fourier analyzed each of these heat phenomena in detail. Each of the three discussions was of importance to physics. The discussion of the third was also of major import to geology.

The most important aspect of Fourier's treatment of the Earth's internal heat for geology was the starting point of his treatment, his interpretation of this phenomenon as "residual heat." He accepted the validity of the nebular hypothesis as axiomatic: "The earth preserves in its interior a part of that primitive heat which it had at the time of the first formation of the planets" (Fourier,[33] p. 137). This was an extremely logical decision, for two connected reasons. The first was simply that by the 1820s the nebular hypothesis was an accepted tenet of astronomy. Second, the nebular hypothesis fitted closely with the terrestrial application of Fourier's heat theory. Starting with the planetary origin predicated by the nebular hypothesis, the heat theory could readily account for the observations of the heat gradient of the Earth:

> ... the latest progress of mathematical science enables us to reach, is to ascertain exactly by what laws a solid sphere heated by a long immersion in a medium would discharge that primitive heat if it were transported into a space of a constant temperature inferior to that of the first medium (Fourier,[33] p. 156).

The existence of an internal thermal gradient proved that the Earth could be considered as such a cooling sphere.

The great conceptual innovation that Fourier made in his theory of heat was in the determination of how heat distribution took place within thermal systems. The general consequences of his mathematical analysis of the cooling Earth, thus, were those already abstractly outlined for similar cases in his papers on heat. The cooling of the Earth was a picture of refrigeration along exponential curves. Fourier's laws of thermal diminution demonstrated that progressively less and less of the remaining transferable heat had been passed along the Earth's thermal gradient and into space. At first, the Earth's initial heat had diminished rapidly, but in the course of geological time the remaining heat had moved with mounting slowness. As the flow of heat slowed, the curve tracing the temperature pattern in the thermal gradient would have become sharply steeper, beyond the easy curve found at shallow depths. Such a configuration would have started to occur with the commencement of the process of temperature diminution of the gradient itself. This diminution would progress from the periphery to the center and parallel the progressive decreases in the rate of heat transfer.

Fourier's new analysis of the Earth's internal heat greatly modified the specific geological consequences derivable from the planetary origin predicted by the nebular hypothesis. His mathematical analysis proved that the planet still possessed an extensive igneous reservoir:

> ... the same theory shows that the excess of temperature, which is almost nothing at the surface, can become enormous at a distance of some thousands of meters, so that

the heat of the intermediate strata would exceed by far that of substance heated to incandescence (Fourier,[33] p. 164).

The present existence of an extensive igneous reservoir, in turn, demonstrated that the Earth had always possessed a significant store of central heat. Fourier had thus established the groundwork for the resurgence of theories of central heat in geological dynamics. The rapid acceptance of his theory of heat had done much to dispel the long-standing objections to concepts of central heat. Fourier's own scientific prestige and that of his theory of heat guaranteed that the terrestrial application of his theory would also exert a strong influence on collateral subjects. It is only in the light of this exposition on the existence of a residual heat that subsequent developments in dynamic geological theory can be understood.

In the years between the appearance of Fourier's two articles, his treatment of central heat received strong independent support in the work of the Royal Geological Society of Cornwall. The miners of Cornwall were driving their complex workings deeper into the earth than men had ever before penetrated. Accompanying the miners were the members of the region's geological society. Prominent among the geological phenomena they observed in their descents was the temperature gradient. The fathoming of this phenomenon was taken up by the Cornish geologists as their special contribution to their science; though they were ignorant of the physics that readily made sense of findings yet enigmatic for them,[19] the collective research and debates of the Cornishmen produced a refined picture of the gradient. The command of technique and the copious observations acquired by the Cornish geologists caused their work to supersede all previous investigations.[20] Their empirical studies showed that the tendency of heat to increase with depth was broadly both constant and universal, that is, that there was a "geothermal gradient."

---

[19] The seven major and numerous minor studies of the geothermal gradient produced by the Society during the years 1820–1824 elicit but one reference to Fourier and that one in passing, with no comprehension of the significance.[34] This inability to appreciate the significance of Fourier's work should be compared with the immediate realization of its importance expressed by other, more learned British and French geologists. Some of the most astute Cornish observers still talked vaguely of strange "aqueous vapours" and the like.

[20] The great contribution of the Cornishmen was technical. The findings of their predecessors had been based on temperature readings of the air or water found in mines. The Cornishmen determined that such readings were influenced by extraneous causes such as the body heat of the miners or their animals. The truly important temperature in mines, the Cornishmen found, from the geological standpoint, was not the temperature of the "mine" but of the earth, the rocks themselves. The way to measure the geothermal gradient was not to carry a thermometer about and take readings, but to lodge the instrument as deeply as possible in the walls of the shafts, free from all outside influences. Such measurements carried out on a large scale revealed that not only was there the long-known tendency of temperature to increase with depth, but the temperature increase seemed to express itself as a ratio of the degree Fahrenheit for every sixty or so feet of depth (this ratio is quite close to the present determination of 30°C per kilometer).

To determine that a measurable geothermal gradient increased constantly in temperature with depth was an important contribution to geology in its own right. Assuming that this regular increase continued without limit at progressively greater depths, a new basis was discovered for all studies of "interior" phenomena. Viewed in conjuction with Fourier's papers, which chronologically bounded the main efforts of the Royal Geological Society of Cornwall, the Cornishmen's work on the geothermal gradient took on an even higher importance. Fourier had based his work on the earlier studies of the gradient. He knew only that there was a tendency of temperature to increase with depth and had structured his analysis in terms of an empirically hypothetical, though mathematically predicted, constant temperature increment per unit of depth near the surface. The Cornish studies indicated that there was indeed such an increment. This ratio of temperature increase, worked out by means of Fourier's differential equations and along the exponential curves he had derived, lent persuasive verification to Fourier's conclusion that there existed a vastly hot igneous reservoir seated at the core of the Earth.

The great import of this blending of empirical geology with abstract physics was immediately grasped by natural philosophers and geologists alike.[35] The Earth had to be viewed as a progressively cooling, hence contracting, body whose temperature had decreased exponentially through time until the present, when the surface manifestations of this process were insensible compared with the effects of solar radiation. The extrapolation of the observed geothermal gradient by means of Fourier's equations equally demonstrated, nevertheless, that the Earth still possessed a vast igneous reservoir. In fact, the appearance of the geothermal gradient indicated that the crust of the Earth was actually quite thin and incapable of self-support unless buttressed by contact with the incandescent core.

Fourier's theory of heat established the groundwork for a comprehensive theory of geological dynamics predicated on the existence of a "central heat." If such an explanatory dynamics could be successfully arrived at and then connected, in turn, with the prevailing interpretation of the stratigraphic record, geologists would thus possess a complete actualist synthesis of the course and nature of Earth history. During the latter half of the 1820's, a number of attempts were made to arrive at such a synthesis.[22, 36, 37] Each of these attempts extended geological knowledge of the precise workings of the igneous forces. It was only at the end of the decade, nevertheless, that a viable theoretical synthesis actually did emerge with the work of the French geologist, Leonce Élie de Beaumont. Two types of difficulties had beset the efforts which preceded his theory. In the first place, these attempts at a theoretical synthesis had each involved errors in the application of physics to geology.[21,22] As a result, the

---

[21] Scrope based his theory of central heat on the interaction of the expansive force of the thermal reservoir with the structure of the overlying crust. His theory proper, thus, only commenced after the originally totally incandescent Earth had cooled sufficiently for complete crustal consolidation. The incandescent core, contained by the crust, built up vast expansive force and extensive lava pockets at, compared with the present, vastly ac-

celerated rates. At the same time, the extent and diffusion of these processes enhanced the ability of the overlying rocks to resist the expansive pressure. When and where the strata did yield, a geological revolution would occur. As the cooling of the core proceeded, the extent and magnitude of these revolutions decreased until the present levels were attained. This same cooling process also accounted for the directional development of global climate.

The principles of physical science, however, could not justify Scrope's application of an igneous dynamics, constituting the interactions of the expansive force of central heat with geological structure, to the whole of geological history after crustal consolidation. Neither the incandescent core nor the crust would have behaved the way Scrope postualted they would. The tendency of a cooling body, as Fourier's heat theory clearly showed, was to achieve equal heat distribution. In the case of the Earth, this tendency would express itself as a heat gradient extending from the planet's center to the void. At those periods of planetary history when the heat flow was pronounced enough to manifest its dissipation by means of tectonic activity on a continental scale and shape global climate, then local structural features would have proved no deterrent to heat transfer. Structure would only begin to play a part in tectonic activity when the heat flow had decreased to relatively current levels. The basic tendency of an incandescent mass of the Earth's magnitude, for most of the cource of cooling would be simple contraction, not expansive force.

22 Cordier, for his part, based his theory on the principal conclusion which had emerged from his own exhaustive studies of the geothermal gradient[36]: The crust of the Earth was actually quite thin and incapable of self-support unless buttressed by the incandescent core. As the Earth cooled, it necessarily contracted. Cordier argued that because the process of solidification proceeded from the periphery progressively downward, the rate of cooling had followed this same pattern. Contraction was, therefore, greatest at the surface, with the rate progressively dropping off as the curve of the temperature gradient mounted with depth. Crustal contraction was the agent of mountain-building and all other major plications of the solidified rocks. They were all the product of the wrinkling of the crust. This same process was also the origin of the other primary agent of geological dynamics, volcanism. Cordier had asserted that contraction of the solidified crust proceeded much faster than contraction of the incandescent core. The result of this difference in rate would logically be that the crust continuously exerted great pressure on the core. When critical pressure levels were reached at weak points in the crust, the upper portions of the incandescent core would break through in the form of lava, producing volcanic activity. The clearly discernible belts of past and present volcanic activity represented the crustal weak points of less thickness and resistance. Determined by the progressive diminution of central heat and the rate of dissipation, the course of Cordier's geological dynamics was therefore directional.

Cordier's dynamic system was, in fact, no stronger than that of Scrope and in one respect markedly weaker. In contrast to Scrope's theory Cordier's actualist dynamics could not fully explain the appearance of the strata. This was the most obvious and glaring defect of his system. While the strata seemed to chronicle an episodic, saltatory geological history, Cordier's theory, if anything, seemed to imply that this history had been a smooth, directionally linear progression. The energy of his major dynamic agents would have apparently decreased at uniform rates incapable of producing a dynamic pattern of long epochs of relative quiescence puntuated by brief episodes of intense volcanic and tectonic activity. There was also a less easily discernible, but no less serious flaw. Cordier, as had Scrope before him, made an error in the application of physics to geology. A geologist's ignorance of physics made a proposed comprehensive actualist dynamics untenable. By the terms of Fourier's physics of heat, the crust of the Earth attained, not approached, the limits of refrigeration when it finally reached the state where it was completely solidified about the core. Once so formed, the crust them simply transferred heat. A system of geological dynamics predicated on the concept of the thermal contraction of the crust was physically impossible.

dynamic theories offered were at least in part untenable. More significant, however, was the fact that these attempts at a theoretical synthesis were developed
strictly within the context of the study of dynamic processes and their evidences. The connection with the geological history revealed by stratigraphy was
only the claim that the proposed dynamics offered causes adequate to account
for appearances. There was no really solid evidence that these proposed forces
had been, in fact, the actual causes. In short, these systems had been respectively
more or less able to invoke a claim of sufficiency, but never one of necessity.
Élie de Beaumont, on the other hand, was able directly to correlate the actions
of a soundly formulated dynamics with the geological history gleaned from
stratigraphy.[23]

Élie de Beaumont presented persuasive stratigraphic and structural evidence
which strongly suggested that Earth history had been punctuated by 12 world-
wide orogenies which had produced 12 global systems of mountains. The orogenies, he was further able to show, appeared to coincide exactly with the major
discontinuities of fossil species. The episodes of elevation were, therefore,
directly linked to the episodic appearance of the fossil record. Élie de Beaumont
asserted that his researches proved that the former were the cause of the latter.
His findings seemed to indicate that the parallelisms of the constituent elements
of a given mountain system were so exact and the strata so contorted, that the
beds could have only been thrown up at the same time by a single tectonic
event—a massive ridging of the crust. Such an occurrence of crustal elevation
would necessarily produce violent tidal waves and other secondary effects on a
massive scale, true "revolutions" of the globe. Saltatory events of this nature
offered a completely satisfactory explanation of the episodic appearance of the
strata record.

Élie de Beaumont had succeeded in adding an unprecedented degree of ex-
actness to the common view of geological history. The stratigraphic evidence

---

[23]Élie de Beaumont read his paper to the Académie in 1828 and then published it a year
later.[37] The fact that Élie de Beaumont first read his paper is significant because this
oral presentation differed from the published account. When expounding his ideas, Élie de
Beaumont presented both his evidence and his dynamic system; in sum, his complete
theory. Charles Lyell, visiting France at that time, was one of those who learned of the
new system from listening to Élie de Beaumont (Wilson,[7] p. 259). Those geologists who
had direct contact with Élie de Beaumont communicated his theory to yet others. In this
manner the geological community at large learned of his ideas. Élie de Beaumont did not
include his dynamics in the published paper. He chose instead to concentrate totally on
his stratigraphic evidence: "Le cause des phénomènes passagers que je viens de rapeller
n'est entrés pour rein dans l'object de mon travail actuel: les questions que je me suis pro-
posé de résoudre n'étainet que des questions d'époques et de ćonincidences de dates (Ref.
37b, p. 225). The attendant dynamics did not appear in print until 1831, when Élie de
Beaumont sent an English abstract of his complete theory to his friend, Henry De la
Beche, who published it in the *Philosophical Magazine.*[37c] Because geologists were cogni-
zant of Élie de Beaumont's complete theory from 1828, the following discussion makes
no distinction between the two papers.

from which he induced his results also strongly supported the dynamics he proposed to explain this course of events. Only a worldwide cause could produce a worldwide effect. There was but one adequate source of energy for global geological dynamics: the residual heat of the Earth's incandescent origin. Élie de Beaumont was the first of the geologists who attempted to formulate "theories of central heat" to correctly apply Fourier's heat theory to the question of the cooling Earth. This question resolved itself into the relationship that existed between the solid crust and the still incandescent core.

Élie de Beaumont correctly realized that the solid crust, once formed, had for all purposes attained the limits of refrigeration and would simply transfer heat. Consequently, the temperature and volume of the core would progressively decrease while the crust remained thermally stable. The thin crust possessed only a limited capacity for self-support. When periodically this capacity was overcome, the crust would collapse back around the supporting core. Merely the force of gravity would produce this effect. While with the decrease in the rate of heat flow, the time between orogenies would lengthen, the conditions for collapse were a constant. These events were, therefore, always uniformly quick. The mountain systems were the chief structural products of the crustal collapses.

The orogenies were the main events of geological history. These orogenies were also very much "special" events produced by conditions that only occasionally prevailed. Still, they represented to Élie de Beaumont no violation of the actualist principle. Just as he had carefully inferred his version of an episodic geological history from present evidence, he had striven to produce a scientifically sound dynamic system based on the application of accepted physics to geology.[24] Along with other serious geologists, Élie de Beaumont understood that actualism was the only legitimate guide to geological inquiry. Thus, because there was no evidence to the contrary, Élie de Beaumont believed that in the long, tranquil intervals between the orogenies, geological conditions had been at an order of magnitude closely analogous to those now prevailing:

> Everything tends to lead us more and more to separate the facts offered for our observation by the sedimentary strata into two distinct classes: one comprising the facts relating to the tranquil and progressive manner in which the accumulation of each of the sedimentary strata has occurred and the other including those facts connected with those sudden interruptions which have established lines of demarcation between different consecutive deposits. After having subtracted, so to speak, the role of the violent and transitory phenomena, one would perceive more easily the analogy which those phenomena, which have occurred repeatedly on the surface of the globe during the different periods of tranquility which have ensued there, appear to have presented to phenomena of the present time (Élie de Beaumont,[37b] pp. 224–25).

The astronomical fact of the Earth's incandescent origin dictated, of course, that the events of this level of activity had also been channeled along a directional

---

[24] See, for example, Élie de Beaumont,[37c] p. 263.

pattern of their own.[25] The slow and gradual exponential fall of the central heat flow would have produced both a progression of global climate from generally hypertropical conditions to those of the present and a slowing down of the rate and intensity of the "basic" geological agents that have always been at work. Through a succession of directional events both saltatory and gradual, the Earth had passed through a number of distinct episodes and arrived at its present state.

Élie de Beaumont first presented his theoretical synthesis in a paper read before the Académie in 1828. He first began to publish his ideas the following year. Charles Lyell had been developing an independent view of the nature of geological change during the period that saw the reemergence of theories of central heat. In 1830, he brought out the first of the three volumes of his major geological work, *Principles of Geology*. The publication of Lyell's theory followed so closely upon that of Élie de Beaumont, that, in fact, the geological community responded to the two theories at almost the same time. A well-articulated theoretical system of a scope sufficiently comprehensive to rival the one Lyell would thus offer did not emerge until just prior to the publication of the first volume of the *Principles*. This late appearance of a clearly discernible "rival" theory unquestionably has been a major factor in fostering the misunderstanding with regard to geology of the nineteenth century I mentioned at the outset. It has lent persuasive support to Lyell's own description of his predecessors as somehow not quite "scientific" geologists.[38] The absence of explicit descriptions of a comprehensive theoretical system in the geological literature of the 1820s supports the conclusion that one did not exist. In the same vein, the fact that references to Élie de Beaumont's theory only begin to appear after 1830 has led to the argument that his theoretical system arose not prior to, but as a "reactionary" response to that of Lyell.[26] Consequently, Lyell is generally credited with "establishing" geology as a true "science." His theory is most often portrayed as receiving an enthusiastic reception and rapidly attracting a large number of supporters (most likely because Charles Darwin largely adopted his view of geology). As we have seen, such views simply cannot be substantiated by a study of the geological literature.

Quite clearly, a theoretical system with a scope and explanatory power comparable to that of Lyell did not appear until the end of the 1820s. Yet, it is erroneous to argue that such a theoretical system did not exist prior to the publication of Lyell's *Principles of Geology*. A close examination of the geological literature seems to indicate, at least to me, that during the course of the 1820s there *arose* a theoretical system of equal scientific stature to the one Lyell offered. Élie de Beaumont's work completed a process of conceptual develop-

---

[25] Élie de Beaumont did not discuss those "basic" igneous agencies that had always been at work. An explanation of such activity, however, was readily derivable from a concept of central heat. See, for example, the treatment of present igneous phenomena in Scrope's *Considerations*.[22]

[26] For example, see Gillispie.[39]

ment that can be traced through the course of the entire decade. The theoretical system he presented, nevertheless, was not the culmination of a unilateral research tradition. His work represented, instead, the synthesis of two traditions whose histories, while strongly related, were nonetheless distinct. The dominant form of geological endeavor during the first half of the 1820s was "historical" studies. The directional interpretation of Earth history that emerged from these studies found support in all forms of geological evidence. The directional theory which expressed the consensus of geological opinion was both a self-consciously limited analysis of the empirical evidence and a general historical model capable of accommodating new data. Considerations of geological dynamics in this period played a distinctly secondary role. Once the possibility of a unified geological dynamics cordial to historical theory was again opened up at mid-decade by the general acceptance of Fourier's heat theory, however, there was a rapid resurgence of interest in dynamic explanations. The research tradition which culminated in Élie de Beaumont's version of a theory of central heat, though less venerable than the one which produced the historical theory, was nonetheless coherent and continuous. The foundations of the causal dynamics which conjoined with historical theory to form a comprehensive explanatory pattern was the nebular hypothesis. Physical science independently confirmed the historical conclusions of geology. The explanation of the Earth's origin contained in the nebular hypothesis became the conceptual element which gave explanatory coherence to the directional patterns geologists believed they had discerned within the whole span of Earth history. The concept of "central heat" replaced historical theory as the general model shaping geological thought. A specific dynamics of central heat was merged with historical theory to become a single coherent theoretical synthesis owing to the work of Élie de Beaumont. This synthesis was obtained at almost the same time that Lyell began to promulgate his own theoretical system.

The basic nature of Lyell's system of geology is well enough known so as not to require elaborate summary.[27] On the other hand, it is important to note that while for present readers Lyell's arguments seem to form an integrated whole, this was not how his colleagues regarded his system. In order to understand the influence the *Principles of Geology* exerted on geologists, one must recognize, as Martin Rudwick has pointed out,[28] that to Lyell's contemporaries there were three distinct and separable elements involved in his system. Though distinct and separable, these three elements were logically connected in a sequential dependence. Two of these elements were derived from Lyell's reinterpretation of the actualist principle. The first of these was his methodological approach to the study of past geological phenomena. Lyell believed that past geological occurrences could only be explained by literal, not analogical, reference to causes now

---

[27] For a detailed analysis see Rudwick[40] and Wilson.[7]

[28] Rudwick.[40] I am indebted to Rudwick's book for the following analysis of the conceptual elements contained in the *Principles*.

in operation. Any attempts to infer geological causes of a kind at all different from those now readily observable were deemed "unscientific." The second element of Lyell's theoretical system was deduced directly from this first. Such extreme adherence to the actualist principle necessarily stipulated the corollary that the past intensity of the observable set of geological forces could vary only within very narrow limits from their present levels. Therefore, all geological phenomena had to be conceived in terms of the "gradual" processes chronicled in historical records.

Actualism was a methodological first principle accepted by all serious geologists. If Lyell's first two conceptual elements were an extreme rendering of the concept, they were also not a real departure from contemporary thought. The third element of the theory, however, was very much of a revolutionary departure. Lyell advocated that the always constant set of geological forces necessarily pointed to the conclusion that throughout geological history the dynamic condition of the Earth had been that of steady state. Lyell's marshalled evidences of the great power and efficiency of present agencies, when operating over long spans of geological time, were arrayed to show also dynamic balance and override the evidence supporting a directional interpretation of geological history. Uplift equaled subsidence, erosion equaled deposition, the surface of the Earth was constantly changing, but these changes formed together a self-regulating dynamic equilibrium. If Lyell's first two conceptual elements extended a basic premise to its extreme limits, this third element openly challenged all of his contemporaries' most firmly established conclusions.[29]

Clearly, there was no place in Lyell's geology for concepts of a progressively cooling Earth. Not illogically, then, this concept became one of the principal *foci* of his colleagues' response to his theory. Lyell's colleagues, almost without exception, were not persuaded that his reinterpretation of the actualist principle was scientifically justified. There was, of course, no question that geology could be a true science only if the methodological policy of its practitioners was always to infer the nature of past events by analogy with processes observable at present. Actualism in this form had long been one of the philosophical first principles of the serious geologist. The point of debate, as they saw it, was the wisdom of extending the principle to the point where one attained Lyell's steady-state dynamics. The invariable laws of nature, physical regularities of the same order as those now observable, could very well have acted with different effects under different conditions and in so doing modified these conditions.

The overwhelming majority of Lyell's colleagues believed that all the forms of geological evidence supported this conclusion.[41] The results of current research in the 1830s, quite simply, were decidedly in favor of a directional interpretation of geological history. Far more significant, however, was the fact that this same interpretation seemed to follow necessarily from the application of accepted physics and astronomy to terrestrial phenomena. Lyell could pose a

[29] For a good example of the strength of the ideas Lyell challenged see Sedwick.[27]

challenge to the directional interpretation of geological evidence by offering an alternative view. Though he would try in the course of the 1830s, he could offer no substantial points of refutation or viable alternatives to the theory of central heat.[42] Élie de Beaumont's theory, on the other hand, was rapidly accepted as the most impressive dynamic synthesis yet produced.

Lyell's attempt to contrast his theory favorably with the directional synthesis forged by Élie de Beaumont made little impression on his colleagues. They judged his arguments in the light of the contemporary state of geological knowledge and found them wanting in almost every particular. Élie de Beaumont's historical interpretation of geological history was superior to the one Lyell offered and his dynamics of central heat was a superior dynamic explanation. The consensus of opinion was, in the words of Ami Boué,

> that the creating of modifying causes have formerly shown an energy superior to that with which they have been acting since the establishment of actual societies, and that there have been periods of comparative tranquility ... The elevation of mountain chains cannot be ascribed to the continued operation of Plutonic action, but we must rather ... seek for the cause in the "refroidissement séculaire," that is, the gradual diffusion of the primordial heat to which our planet owes its spheroidal form. [43]

Charles Lyell failed in his attempt to win acceptance for his system of geology. Even as the geological community, during the mid-1830s, was rejecting his theory in favor of that of Élie de Beaumont, nonetheless, Lyell's ideas were already in the process of exerting a powerful influence on the science of geology. Though the system of geology Lyell had devised appeared untenable to his colleagues, the arguments and evidence he had raised to support two of the conceptual elements comprising this theory quickly came to appeal to them. The context of this appeal, however, was not the quest for a theoretical synthesis. Rather, Lyell's elements of "actualism" and "gradualism" were viewed in the perspective of the defense of the actualist principle initiated in the middle of the 1820s. If Élie de Beaumont had produced the most successful theoretical synthesis to date, Lyell, in turn, was seen as the author of the most successful application of the actualist method to the natural world. Lyell had persuasively demonstrated that much of the evidence of geological history, hitherto referred to more powerful agencies, could be reasonably interpreted in terms of very close analogies with present forces. In so doing, he also established the methodological desirability of carefully proving that such close analogies did not obtain before attempting to infer broader ones. Henry De La Beche was representative of the firm believers in the reality of saltatory geological events when he acknowledged this influence:

> It is anything but desirable to have constant recourse to comparatively great power in explaining geological phenomena, when the exertion of a small force, or of an accumulation of small forces, will afford sufficient explanation of the facts observed ... We should be careful not to suffer any pre-conceived opinion to bias our views either way.[44]

On the issues of "actualism" and "gradualism," then, the only modification Lyell's fellow geologists applied to his ideas was the qualification that not all geological phenomena could be explained in terms of precisely the present forces.[30] The concept of dynamic steady state, on the other hand, was the real heart of Lyell's theory and the one element which was totally rejected by his colleagues. Instead of establishing a new theoretical synthesis, the *Principles of Geology* had come to be seen as a redoubtable contribution on a much broader philosophical plane, that of method. The *Principles of Geology* would influence how specific geological phenomena were interpreted. The conceptual framework into which such interpretations were fitted would be the theory of central heat. As Roderick Murchison would write in 1839, "the theory of central heat as propounded by the mathematician, finds its best supporter in the geologist."[46]

During the course of the early twentieth century, geologists came to realize that the discovery of radioactivity and of the overthrusts of the Alps negated, respectively, the concepts of residual heat as a dynamic agent and crustal collapse as the source of tectonic activity.[31] Until that time, the planetary origin predicated by the nebular hypothesis would be looked upon as the "source" of constructive geological dynamics. The geological consequences of the hypothesis as set forth by physical science had been confirmed by empirical geological research. Embodied in the theory of central heat, these consequences were the foundation of the dynamics that unified the diverse evidence of earth science. For the geological community, however, this theory as constituted in the middle 1830s, though satisfactory, was by no means complete. The purpose of Élie de Beaumont, the primary architect of the dynamics of the theoretical synthesis, had been to set forth only a tectonic history of the globe, not a complete dynamics such as Lyell had attempted. If a complete synthesis of dynamic and historical theory appeared to be derivable from Élie de Beaumont's work, much remained to be done. The process of internal modification which the directional synthesis would incessantly undergo was one of the main themes in considerations of geological theory after the mid-1830s.

Further extensions of geological research continued to validate and strengthen the conclusion that the directional synthesis was the correct interpretation of geological history. Consequently, the essential character of the synthesis did not undergo significant change after the fourth decade of the nineteenth century.

Increasingly, on the other hand, geologists read in their evidence indications that the tectonic history of the Earth, at least since the current stratified rocks had begun to be laid down and hence for all of organic creation, was less dis-

---

[30]Prominent examples are Henry De La Beche and Roderick Murchison.[45]

[31]The discovery of radioactive heating negated all arguments based on the presence of a dissipating thermal reservoir or "central heat." The computations of the heat being generated by radioactive substances, in fact, indicated that the Earth could well be heating up rather than cooling down. The crustal shortening involved in the overthrusts of the Alps was far beyond the compression the calculated rates of contraction were capable of generating.

jointed than had been previously believed. Certainly this history had been both directional and progressive, with long periods of comparative repose alternating with relatively brief orogenies. The perspective on the nature of tectonic activity during the saltatory episodes and the duration of time encompassed by them was what changed. Orogenies came to be seen as "brief" only in comparison with the epochs of repose; in the actual time spanned they were of long duration. The intensity pattern of tectonic forces at work during orogenies was still held to have progressively decreased, while the indications grew that cases of tectonic activity formerly held to be single occurrences actually involved complex series of lesser events. Rather than a single "revolutionary" occurrence, the combined effects over time of a series of distinct events, each with an intensity perhaps much greater than present levels but of a close order of magnitude, came to be seen as the explanation of tectonic evidence. The historical interpretation of tectonic episodes in terms of gradualist processes replaced the "catastrophic" interpretation Élie de Beaumont had originally advocated.

Historical conclusions of this nature were complemented by the development of geophysical analysis of plication.[32] Analyses were carried out in order to determine more precisely the properties of crust and core. This work showed that in order to maintain its shape against all the manifold influences exerted, the Earth would have to possess a crust much thicker than had previously been thought. Though the precise properties of the crust which prevented the residual heat from reducing its lower segments to incandescence remained mysterious, the analyses which led to this conclusion were persuasive. Such a conclusion, however, did not invalidate the basic form of the tectonic dynamics Élie de Beaumont had formulated; only the concept of tectonic intensity was affected. The crustal collapses that had occurred throughout geological history would have proceeded more slowly and in a more complex, piecemeal manner than had initially been envisioned. The results of historical and dynamic geological theory thus admirably combined in a single trend of thought. This melding of directionalism and "gradualism," a trend established by the end of the 1830s, would continue as one of the predominant influences in considerations of geological theory for the remainder of the century.

As geology's dynamic "paradigm," the directional synthesis, in addition, often played a crucial part in determining the reception other attempts to pose theories concerning natural history would receive. The most notable instance where the directional synthesis would play this role in the second half of the nineteenth century was in the quest for a natural causal mechanism for the origin of species. Of logical and obvious necessity, the science of geology supplies the biologist with his concepts both of "time" and of the nature of the inorganic environments in which life exists.

The best known and most important of the theories of organic development proposed in the second half of the nineteenth century, was, of course, Charles

---

[32] For a general treatment of this trend into the 1850s see Cannon.[(47)]

Darwin's theory of evolution by natural selection. Viewed objectively, with no "Darwinian" leaps of imagination, the historical aspects of the directional synthesis rendered a basically neutral verdict on the adequacy of Darwin's theory. When Darwin introduced his theory in 1859, geological history, interpreted from a gradualist perspective, provided a backdrop upon which evolution by natural selection could have occurred. The course of geological events, on the other hand, naturally yielded no indications as to whether such a process had or had not transpired. The fossil record, in turn, was indeed broadly progressive but also lacking in substantive positive or negative evidence. Along with their growing belief in the gradual nature of geological change, geologists in the second half of the nineteenth century had largely come to believe that the sharp faunal breaks between formations were more apparent than real. Their continuing research indicated that the breaks reflected simply the fragmentary nature of the fossil record. Whether the record was inherently incomplete, or merely man's present knowledge of it, remained a moot point. Toward the end of the century, fossil evidence would provide examples of phylogentic development, linking genera and families, cordial to Darwin's theory. However, evidence supporting evolution by natural selection on the interspecific level or the levels of class and phylum remained elusive and inconclusive. The state of knowledge of the fossil record supported the position that evolution had occurred while providing little indication of the biological processes involved.

The dynamic aspects of the directional synthesis, on the other hand, told quite a different story. In order to accentuate the fragmentary nature of the fossil record and hence give adequate scope to the extremely slow and gradual operations of natural selection, Darwin and his followers employed chronological estimates of vast amounts of geological time. Geologists promptly challenged these estimates as unduly excessive based on present rates of erosion and deposition (see, e.g., Phillips[48]). A by far more serious criticism came couched in terms of the theory of central heat. The author of this attack was William Thomson (later Lord Kelvin), the most eminent physical scientist of the day. His papers rebutting Darwin's notions of geological time constitute the most famous chapter in the relation of the nebular hypothesis to geology.

During the 1860s, Thomson turned his attention to applying the recently formulated laws of thermodynamics to the problem presented by the cooling Earth, and, in concert, extending the precision of the terrestrial application of Fourier's heat theory. One of the major conclusions of Thomson's first papers[33] on this subject was that between twenty and four hundred million years had passed since the formation of the Earth's crust. Most probably the span of geological time was one hundred million years. By way of comparison, it should be noted, Darwin had estimated the formation of the Wealden denundation to

[33]Thomson's most important papers on geochronology are collected in his *Popular Lectures and Addresses.*[49]

require alone more than three hundred million years.[34] While Thomson's calculations were based on some rather dubious assumptions, his personal prestige coupled with that of physics was sufficient to carry his point among the geological community. Sufficient unto itself, then, Thomson's estimates were subsequently confirmed by those of absolute geological time which emerged from the development of geomorphology.[35] The truncated time scale could no longer accommodate the vast vistas of geological time which Darwin's evolutionary mechanism required.

The revision of the time scale was the most radical aspect of Thomson's critique. From this revision another important consequence necessarily followed. The restriction of geological time set definite limits on the degree to which geological history could be interpreted in gradualist terms. By 1860, the geological community had come to conceive of the rate of geological change in almost Lyellian terms. The chronological bounds Thomson's work imposed caused them to once more increase their estimates in order to account for observed effects within available time. Though these estimates did not return to the levels inferred in the first half of the century, the rates of geological change imposed by the restricted time scale required concurrent rates of organic evolution beyond the potentials of natural selection. Developments in the theory of central heat thus delivered a crushing blow to Darwin's explanation of the evolutionary process. They had shown that neither the geological time nor the environmental conditions had existed upon which the viability of the theory depended. Scientists remained convinced that evolution had occurred. Until Thomson's geochronology was rejected at the end of the century, they also remained convinced that the mechanism of change could not be natural selection.

The role the directional synthesis played, both in other areas of geology and in collateral areas, while of unquestioned importance, was not, however, the most significant effect the synthesis would have on nineteenth century thought. The dynamic system of the synthesis, the theory of central heat, contributed greatly to a strong philosophical and theological influence which the nebular hypothesis itself exerted.

The conceptual framework of the nebular hypothesis had philosophical implications which were inherent in any theory which embraced the same topic. Buffon had presented the most prominent development of this theme prior to the nineteenth century. The form of Buffon's development had made this

---

[34] Darwin's precise figure was 306,662,400 years. The formation of the Weald encompasses, roughly, upper Jurassic and lower Cretaceous time. The estimate of over three hundred million years was included in the first edition of the *Origin of Species* (London, 1859).[50] Darwin promptly dropped the estimate from the third edition (London, 1861), which appeared at the beginning of the time-scale controversy.

[35] See Davies.[6] Later in the century, Thomson would further reduce his estimates below limits the geological community felt were acceptable. This aspect of the debates over geochronology in the late nineteenth century, however, is beyond the scope of this paper.

general theme of his work unacceptable to the majority of scientists. In the mid-nineteenth century the theme, now identified with accepted science rather than speculative cosmogony, reemerged shorne of all objectionable features. The integral factor in this revival was the working out of the geological implications of the nebular hypothesis. Louis Cordier, intimately connected with this process, was among the first to attest the revival when he wrote at the conclusion of his essay of 1827[36] (pp. 554-55),

> We arrive at a result of the highest importance, since it seems applicable to all the celestial bodies; and thus we obtain stronger proof of the existence of a great principle of *universal instability* . . . a principle, superior to those grand rules, which we have been accustomed to regard as constituting exclusively the laws of nature, from the security which we see in it, above the longest and apparently perfect revolutions of the solar system:–a principle, which appears to rule the universe, even in its smallest parts; which incessantly modifies all things, which changes, or displaces them, and without return; which carries them along through the immensity of ages to new ends. . .

The nebular hypothesis which allowed the geologist to fathom the causes of tectonic history also linked geology, by means of physical science, with astronomy and the broader field of cosmic history. The theoretical synthesis toward which geologists were working in 1827 was at once the explanation of Earth history and the "case" history for all planets. There was a basic pattern discernable in the Universe, a pattern of birth, development, and dissolution.

This theme would be repeated with increasing frequency as the century wore on. More often the promulgators would be British, for in this pattern "design," so important to so many nineteenth-century Britons, was clear and apparent. There was no chance or randomness in the progressive and directional processes encompassed by the nebular hypothesis. Everything that occurred was the result of the inexorable workings of the secondary laws of nature. Here was the product of a definite plan revealing the designing hand of God.

The importance this concept had attained in theological thought by the middle of the 1830s was attested to by the increasing mention it received in popular works on natural theology. During the first half of the decade, for example, there were testimonials in no less than two of the most influential Bridgewater Treatises, those of Whewell and Buckland,[51] to the power of the nebular hypothesis to exhalt convincement in the existence of a presiding Intelligence. Nor did attempts to turn the theme against this conviction seriously detract from its appeal to theologians throughout the nineteenth century. Though Herbert Spencer would use it as the cornerstone of a blatantly materialistic world-view,[52] James Ward in his attempt to totally rebut this world-view made the theme one of few exceptions.[53] Writing near the end of the nineteenth century, Ward still found the appeal of the nebular hypothesis' theological implications strong enough to here merely accuse Spencer of conceptual misuse rather than utter scientific unsoundness. For Ward and the many other theological thinkers who expounded on this theme and embellished it during the course of the nineteenth century, here without doubt was a pattern of purposeful design which no amount of association with the doctrines of materialsim could tarnish.

This was well for theologians because associations with openly materialistic doctrines or concepts which appeared to imply or support them was not infrequent.[36] Spencer was an extreme but far from isolated case. On at least the philosophical level the "historical theme" of the nebular hypothesis caught the imaginations of almost all thinking men and strongly influenced the forms of scientific and historical explanations they were willing to entertain. The same pattern of development was detected not only in the inorganic world but in the course of human history, in the growth of societies, and the fruits of their cultures. The fossil record clearly revealed the history of living forms to have been both directional and progressive. Why in this sphere should there be an exception to the pattern of the Universe? Why was the development of life not also a result of the workings of the secondary laws of nature? The philosophical import of the nebular hypothesis helped foster the very theories of organic development which the directional synthesis played a major role in evaluating.

The dynamic basis of the directional synthesis, the nebular hypothesis and the theory of central heat which derived from it, exerted strong influences in other areas of thought during the second half of the nineteenth century. As I stated at the outset, the influences these concepts exerted are perhaps the best illustrations of their significance. The important role these ideas clearly played throughout the nineteenth century has been hitherto almost completely lost to historians of science, because of a common misconception of the history of geology after 1830. This misconception remains quite common among historians of science. Geology before "Lyell" is still most often portrayed as imbued with religious preoccupations and not yet a "true" science. Such claims simply are not substantiated by a study of the geological literature. The *Principles of Geology* can in no way be construed as marking a watershed in geological thought. Lyell did nothing more than challenge the adherents of an established scientific theory. His colleagues quite objectively judged this theory superior to his own. While the geological community at large greatly admired Lyell's demonstration of the value of the actualist principle, the directional synthesis expressed the consensus of scientific geological opinion. Unquestionably, then, an appreciation of this product of the nebular hypothesis' relation to geology is crucial in order to understand earth science and collateral subjects in the nineteenth century.

REFERENCES

1. Martin Rudwick, Uniformity and Progression in *Perspectives in the History of Science and Technology*, ed. by D. Roller (University of Oklahoma Press, 1971); *The Meaning of Fossils* (London, 1972).
2. William Herschel, Astronomical Observations Relating to the Construction of the Heavens, *Phil. Trans. Roy. Soc. London* **1811**, 269–336; P. Laplace, *Exposition du Système du Monde*, 5th ed. (Paris, 1824).

[36]Perhaps the most controversial example was the anonymous (Robert Chambers) *The Vestiges of the Natural History of Creation* (1844).[54]

3. Agnes Clerke, *A Popular History of Astronomy* (London, 1902), p. 309.
4. George Louis Leclerc, Comte de Buffon, *Supplements, Histoire Naturelle*, Vol. 5 (Paris, 1778).
5. William Whewell, *History of the Inductive Sciences* (London, 1837).
6. G. L. Davis, *The Earth in Decay* (New York, 1969), Chapter 6.
7. Leonard Wilson, *Charles Lyell, The Years to 1841* (New Haven, 1972), Chapter 4.
8. (a) D. Cardwell, *From Watt to Clausius* (Ithaca, 1971); (b) Robert Fox, *The Caloric Theory of Gases* (Oxford, 1971).
9. John Murray, On the diffusion of heat at the surface of the Earth, *Trans. Roy. Soc. Edinburgh* 7, 411–434 (1815); John Playfair, On the progress of heat, *Trans. Roy. Soc. Edinburgh* 6, 353–370 (1812).
10. J. F. D'Aubuisson De Voisins, Observations sur la chaleur souterraine, *J. des Mines* 21, 119–130 (1807); Quelques observations thermometriques faits à la mine de Beschet-Gluck, *J. des Mines* 11, 517–520 (1801–1802); Sur la température dans les mines de Freyberg, *J. des Mines* 13, 113–122 (1802–1803); Notices sur la température de la terre, *J. de Physique* 62, 443–461 (1806).
11. P. Laplace, Addition au mémoire sur la figure de la terre, *Mém. Inst. France* 3, 489–502 (1818); On the Figure of the Earth, *Phil. Mag.* 54, 371–376 (1819); Sur la figure de la terre et la loi de la pesanteur à sa surface, *Ann. Chim. Phys.* 8, 312–318 (1818).
12. Anon, Preface, *Trans. Geol. Soc. London* 1, viii–ix (1811).
13. Leonard Horner, On the Mineralogy of the Malvern Hills, *Trans. Geol. Soc. London* 1, 321 (1811).
14. George Cuvier, *Recherches sur les Ossemens Fossiles* (1812).
15. Claude Albritton (ed.), *The Fabric of Geology* (London, 1963).
16. Robert Bakewell, Queries and Observations Relating to the formation of the Superficial Part of the Globe, *Phil. Mag.* 45, 453 (1815).
17. Robert Bakewell, *Introduction to Geology*, 2nd ed. (London, 1815); *Travels*, 2 vols. (London, 1823).
18. Ami Boué, Mémoire géologiques sur les terraines anciens et secondaires du sud-ouest de l'Allemagne, *Ann. Sci. Naturelles, first series* 2, 173–203 (1824).
19. John MacCulloch, On Certain Elevations of Land, *Quart. J. Sci.* 14, 262–295 (1823).
20. Humphrey Davy, Lectures on Geology, *Phil. Mag.* 37, 392–398, 465–470 (1811).
21. Charles Daubeny, *A Description of Active and Extinct Volcanoes* (London, 1826).
22. George Scrope, *Considerations on Volcanoes* (London, 1825).
23. Humphrey Davy, On the phaenomenon of Volcanoes, *Phil. Mag. new series* 4, 94 (1828) (Taken from *Phil. Trans.* for 1828).
24. William Buckland, *Vindiciae Geologicae* (Oxford, 1820); *Reliquiae Diluvianae* (London, 1823).
25. Alexandre Brongniart, Sur les charactères zoologiques des Formations, *Annales des Mines* 6, 537–572 (1821).
26. (a) Charles Daubeny, On the Volcanoes of Auvergne, *Edinburgh Phil. J.* 3, 359–367 (1820); (b) Robert Bakewell, *An Introduction to Geology*, 3rd London ed. (1828).
27. William Fitton, Presidential Address [for 1828], *Proc. Geol. Soc. London* 1, 133–134 1829; Adam Sedgwick, Presidential Address [for 1829], *Proc. Geol. Soc. London* 1, 205 (1830).
28. Joseph Fourier, *Théorie Analytique de la Chaleur* (Paris, 1822) (English translation by Alexander Freeman, 1878).
29. Joseph Fourier, Théorie de la Chaleur, *Ann. Chim. Phys.* 3, 350–376 (1816).
30. Joseph Fourier, Note sur la chaleur rayonnante, *Ann. Chim. Phys.* 4, 259–303 (1817); Sur la température des habitations et sur le mouvement varié de la chaleur dans les prismes rectangulaires, *Bull. Soc. Philomathique* 1818, 1–11.
31. Editors, Sur la Température de l'interieur du globe, *Ann. Chim. Phys.* 13, 183–199 (1820).

32. Joseph Fourier, Extract d'un memoire sur le Refroidissment seculaire du Globe terrestre, *Ann. Chim. Phys.* **13**, 418–438 (1820).

33. Joseph Fourier, Remarques génerales sur les Température du Globe Terrestre et des Espace Planétaires, *Ann. Chim. Phys.* **27**, 136–167, (1824).

34. John Forbes, On the temperature of mines [read 1820], *Trans. Roy. Geol. Soc. Cornwall* **2**, 211 (1822).

35. Editors, Nouvelles observations sur la température de la terre à différentes profondeurs, *Ann. Chim. Phys.* **16**, 78–85 (1821); William Conybeare and William Phillips, *Outlines of the Geology of England and Wales* (London, 1822), pp. vii–xix.

36. Louis Cordier, Essai sur la température de l'intérieur de la terre, *Mém. Inst. France* **7**, 473–556 (1827).

37. Leonce, Elie de Beaumont, Recherches sur quelquesunes des Révolutions de la surface du Globe, *Ann. Sci. Naturelles* **18**, 5–25, 284–415 (1829) (b) **19**, 5–99, 177–240 (1830); (c) Researches on some of the Revolutions which have taken place on the Surface of the Globe, *Phil. Mag., new series* **10**, 241–264 (1831).

38. Charles Lyell, *Principles of Geology* (facsimile reprint of original first edition, London, 3 vol. 1830–1833) (London, 1969), Vol. I, originally issued 1830, Chapters I–V.

39 C. C. Gillispie, *Genesis and Geology* (New York, 1959).

40. Martin Rudwick, *The Meaning of Fossils* (London, 1972), Chapter IV.

41. William Conybeare, Letter on Mr. Lyell's Principles of Geology, *Phil. Mag., new series* **8**, 215–219 (1830); An examination of those phenomena of geology which seem to bear most directly on theoretical speculation, *Phil. Mag., new series* **8**, 359–362, 402–406 (1830); On the phenomena of geology, *Phil. Mag., new series* **9**, 19–23, 111–116, 188–197, 258–270 (1831); William Whewell, *Principles of Geology* . . . by Charles Lyell, review of Volume I, *British Critic* **9**, 188–206 (1831); *Principles of Geology* . . . by Charles Lyell, review of Volume II, *Quart. Rev.* **47**, 103–132 (1832); George Scrope, *Principles of Geology* . . . by Charles Lyell, review of Volume I, *Quart. Rev.* **43**, 411–469 (1830); Adam Sedgwick, Presidential Address [for 1830], *Proc. Geol. Soc. London* **1**, 281–316 (1813); George Scrope, *Principles of Geology* . . . by Charles Lyell, review of the third ed., *Quart. Rev.* **53**, 406–448 (1835).

42. Charles Lyell, *Principles of Geology*, 1st ed. Vol. III (London, 1833), Chapt. XXIV; *Principles of Geology*, 3rd ed. (London, 1934), 4th ed. (London, 1835), *passim*.

43. Ami Boué, On the theory of the Elevation of mountain chains as advocated by M. Élie de Beaumont, *Edinburgh New Phil. J.* **17**, 149 (1834).

44. Henry De La Beche, *Researches in Theoretical Geology* (New York, 1837; American ed. of the first British ed. London, 1834), pp. 109–110.

45. Henry De La Beche, *Geological Manual* (London, 1831); Roderick Murchison, Presidential Address [for 1832], *Proc. Geol. Soc. London* **1**, 438–464 (1833).

46. Roderick Murchison, *Silurian System* (London, 1839), pp. 575–576.

47. Walter Cannon, The Uniformitarian and Catastrophist Debates, *Isis* **51**, 38–55 (1960).

48. John Phillips, *Life on Earth* (London, 1860).

49. W. Thomson, *Popular Lectures and Addresses*, 3 vols. (London, 1894).

50. Charles Darwin, *Origin of Species*, 1st ed. (London, 1859).

51. William Buckland, *Geology and Mineralogy considered with reference to Natural Theology*, 2 vols. (London, 1936); William Whewell, *Astronomy and General Physics considered with reference to Natural Theology* (London, 1833).

52. Herbert Spencer, *First Principles* (London, 1862).

53. James Ward, *Naturalism and Agnosticism*, 2 vols. (London, 1899).

54. Anonymous (Robert Chambers), *The Vestiges of the Natural History of Creation* (London, 1844).

Chapter XVIII

Laplace As a Cosmologist

Jacques Merleau-Ponty

University of Paris, Nanterre

Was Laplace actually a cosmologist? There is no easy answer to this question. Laplace was certainly not a cosmologist in the sense in which we can say that Aristotle or Lambert were; nor in the sense of Hubble or Bondi—for the sole reason that he never proposed any definite theory or model, or even any conjecture, concerning the Universe as a whole. Yet, in a broader but still legitimate sense, he was a cosmologist: He believed in a philosophy of nature of which one of the main tenets is that the laws of mechanics and the law of gravitation are absolutely and rigorously true throughout the Universe. He thought he had proved beyond any doubt that all the observed motions in the solar system can be explained by these laws—and thus he may be defined a cosmologist.

Of course, it is a mere question of words, but it is a significant one, because it makes us aware that the very concepts of the Universe and of the aim and scope of cosmological science were, and probably are and will be, understood differently, depending on many circumstances, e.g., the means of astronomical observation, the prevailing mathematical techniques, the general trend of ideas about nature, the relationship between science and philosophy or religion, and so on.

Now, Laplace is especially interesting in this respect: Although an outstanding physicist, his main concepts are those which prevailed in the philosophy of nature at the end of the Enlightment. Moreover, his fame was so great that his scientific work as well as his philosophical ideas may have, at least partly, determined a very curious feature in the history of cosmology, namely the disappearance of cosmological science as such in the nineteenth century, that is, the investigation of the properties of the Universe considered in its totality—until its surprising revival in the twentieth century.

Using Kuhn's terms and categories, which can be useful here, one may say that Laplace certainly did not achieve any "scientific revolution," but that he did more than "normal science" not only because of the size of his contributions under the Newtonian "paradigm," but because he transformed the "paradigm" itself, bringing it to a sort of philosophical perfection. What he did was certainly not cosmology proper; the sort of technical work he achieved and his philosophical ideas might be considered, in his time, as an equivalent—and, indeed, the only possible equivalent—of cosmology, within the field of positive science. The astronomers and physicists who followed him and a number of philosophers (among them Comte) understood him in that way.

William Herschel, whose long scientific career was parallel to that of Laplace, was undoubtedly a cosmologist, being interested in the "construction of the Heavens," and not only in the motions of the solar system. But, as a cosmologist, Herschel is a rather isolated figure; his work owes little to the speculative cosmologists of the eighteenth century, and his followers in the nineteenth century indeed praised and continued his work as a telescope builder and an astronomer, but almost forgot the cosmological aim that Herschel had bestowed upon it; in this respect, Herschel is closer to Hubble than to his own son John.

In contrast, Laplace can be placed, so to speak, on a scientific trajectory which had started long before his birth and continued after his death down to the end of the century, without any interruption, not only in celestial mechanics, but also in physics, insofar as analytical methods were progressively extended to larger fields of physical research.

To see that in some detail, we will have to examine three points:

1. The situation which prevailed regarding cosmological science about 1770 among scientists working in theoretical astronomy—the point of departure from which Laplace started. Regarding this point, d'Alembert's concepts are the most useful reference, because of the central position he occupied in French science and philosophy at that time.

2. The building up of Laplace's cosmological conceptions during the most creative period of his life, the last thirty years of the century.

3. Laplace's philosophy of nature, as we find it in the great synthetical treatises he published in the second part of his scientific life: the *Exposition du système du monde*, the *Mécanique céleste*, and the *Théorie analytique des probabilités*; those works were enormously influential in the nineteenth century.

1. (a) The first point worth mentioning about the status of scientific cosmology at the beginning of Laplace's career is the then exceptional and radically new characteristics of the Newtonian theory (meaning the deductive theory built upon Newton's laws of motion and law of gravitation). These characteristics were clearly appreciated, at least among working scientists: Not only was the success of that theory—in the description and forecast of astronomical phenomena—already dramatically evident, but its epistemological character was understood to be quite unusual in existing science, namely its ability to combine "l'observation et le calcul," observation and calculation, the two basic elements of natural science—as Laplace himself repeatedly put it.

In some places, d'Alembert goes even so far as to apply a kind of "Popperian" criterion of falsifiability: Comparing the Cartesian theory of vortices with the Newtonian theory, he maintains that the former is indeed able to explain, at least partly, the motions of the planets. But, if these motions were different—d'Alembert guesses—the Cartesian theory would probably be able to explain the latter as well, whereas predictions furnished by the Newtonian theory are so definite and precise that if observation denied them, it would hardly survive.

But the difference between the scientists of that century and those today in that respect is that the former could not imagine, as we do, that a theory may be excellent within a definite, but limited, experimental field, and wrong outside of it; so that for them the Newtonian theory—if it could match the available observations—was to be considered as "le vrai système du monde," the true system of the world, as Laplace often states. Then, any question related to the validity of the Newtonian theory acquired in effect a cosmological significance, even if it did not directly pertain to the structure of the Universe as a whole.

(b) The second important element of the situation we are trying to describe was the law of gravitation; d'Alembert, as indeed Newton himself, remained Cartesian in that he thought that the only type of physical interaction really intelligible was impulsion, that is, the process by which a particle of matter transmits its motion to another by contact. Now, gravitation was not of this type and, as yet, could not be reduced to it by any hypothesis. Newton had cautiously not made any, and those who had tried, like Le Sage, had not been very successful. Thus, one had to rely on observation, both to assert the fact of gravitation and to formulate its mathematical law; and because the law played such a basic role in the system, it was felt as something obscure and unsatisfactory by d'Alembert and many others.

Many questions could be raised and were indeed raised regarding this law: Was it strictly correct, or only approximate? Was it only valid between big and roughly spherical bodies, like planets, or could it be applied to any pair of particles, of any size, shape, and location? Now, these questions could not be solved *a priori*, the *lumen naturale* providing no answer, and they could not be solved *a posteriori* either—the enormous mathematical difficulties of the three-body problem forbidding the rigorous solution of many astronomical puzzles. It was, in fact, impossible to dovetail exactly, on every detail, the results of calculation and the data of observation.

(c) The third element refers to cosmology *stricto sensu* and the question is more complex. In the Newtonian family, little was said about the Universe as a whole—not only for the specific reasons which Hoskin mentioned in his brilliant presentation of the Newton-Bentley correspondence, but also for broader, philosophical reasons: The result of the scientific and philosophical revolution of the seventeenth century had been a sort of reversal in the vistas of rational thinking about the Universe.

Whereas in the pre-Copernican, or even in the pre-Galilean concepts of nature, one of the main aims of natural philosophy was to build up a consistent representation of the Universe—and the understanding of natural phenomena largely depended on that representation—in the new course of natural philosophy, the emphasis was put on the *laws*, that is, the universal and rigorous pattern of connection of natural events in space and time. Now, in this new concept of nature, the structure of the whole was understood as contingent upon circumstances, causes, and conditions to be found, if at all, in some supernatural agency.

This can be seen, for example, in the article "Cosmologie" written by d'Alembert for the "Encyclopédie": The first thing d'Alembert has to say about cosmology, is that it treats of the *laws of nature* and the *chain of beings* (which, as stated by d'Alembert, means the sequence of mechanical events, rather than the hierarchy of living beings). The concept of a *cosmos* simply does not exist in this first definition of cosmology. Of course, it is referred to later in the article and is then associated with theological considerations: The main interest of cosmology, we are told, is to show that the world is put in order by God.

The sincerity of that statement is indeed very doubtful, but this does not matter so much; sincere or not, this sort of remark seems to prove that, to d'Alembert, the cosmic order does not belong to the properly scientific and perhaps rational outlook of nature. This was also Newton's view, at least according to a well-known passage of the *scholium generale* added to the *Principia*. In the Newtonian system, the essential knowledge is that of the laws, which are differential and local; if we are to build up a finite and extended system or process, we must integrate the equations and therefore know the initial conditions, which are quite contingent and escape any attempt to deal with them by logical means.

To come back to d'Alembert, his article "Cosmologie" ends in a long discussion of Maupertuis' principle, and d'Alembert's conclusion is on this point very definite: The principle is a mathematical one which does not tell us more about the world than the laws of motion. It states the same thing "in other words," as is often the custom in mathematics; the integral viewpoint, the concept of totality, does not add anything to the knowledge of nature.

2. (a) Now, how did Laplace build up his own concept of the Universe? His first paper on cosmological topics, written in 1773 (he was 24 years old), is very long and important, especially because it shows how Laplace considered the tasks he had to perform. Among many problems discussed and partly solved in that paper, we find that some deal with the law of gravitation. Laplace feels that some questions about the law remain very obscure. We are unable, he says, to find out the cause of gravitation (meaning unable to reduce it to pattern of impulsion). Yet he tries—in order to bestow, so to speak, a better epistemological standing to the law—to prove it by geometrical reasoning. His argument

sounds very modern, because it is analogous to well-known attempts of cosmologists in this century.

This is the formulation Laplace gives to his problem: What must be the algebraical formula of the law, if it must be independent of the size of the Universe, that is, if the forces it induces are to remain the same, if all the lengths are multiplied or divided by the same number? Now, the inverse square law readily results from those conditions. The proof probably did not satisfy Laplace, since, as far as I know, he does not use it further. Still, he repeatedly mentions that remarkable trait of the Newtonian law in his later works. Anyway, the attempt is significant of the philosophical requisites of Laplace's scientific work.

Although the cause of gravitation is not directly investigated in this first paper, nor in any other, the question arises indirectly, when Laplace discusses the following problem: Is the gravitational interaction the same, irrespective of the relative motion of the interacting bodies? He states that, if the gravitational action is due to the impulsion of very minute particles, this will be the case. He then proves that the irregularities of the observed motion of the Moon could be explained in that way, provided the velocity of the hypothetical particles were at least seven million times that of light.

However, Laplace later found a better explanation of the lunar motions, and it seems as though he realized that this fantastic speed removed any hypothesis concerning the cause of gravitation from the field of reasonable surmise. Eventually he maintained in the *Exposition du système du monde* that our ignorance of the intimate properties of matter deprives us of "any hope" of answering the question of the cause of gravitation. Moreover, the nature of 'force' in general is, according to Laplace, "forever unknown." Now, when a scientist declares that there is no hope of solving a problem, one may guess that he does not, in fact, consider it a problem at all.

(b) Of course, that question of the law of gravitation was linked to a number of technical problems of celestial mechanics, on which Laplace worked for years, more or less concurrently with Lagrange. One had, he thought, to make certain that the doubts of Clairaut, d'Alembert, and others about the scope and value of the Newtonian theory were not justified. Also, one had to be sure that the theory could explain away any alleged anomaly or irregularity in the orbits of planets and satellites, and further that it was able to give a correct account of gravitational phenomena on the surface of the Earth.

In that respect, the main problems were: the observed acceleration of the mean motion of the Moon, the observed inequalities in the motions of Jupiter and Saturn, and the shape of the Earth and the tides. Laplace thoroughly investigated those problems. It took him many years, but the beautiful results he eventually obtained convinced him, and scientists in general, that the Newtonian theory was in agreement with all the available celestial and terrestrial observations, and that the so-called inequalities did not contradict the theory if all the gravitational interactions were taken into consideration. In the general context of the Enlightment, the Newtonian theory then appeared as "le vrai système du monde."

(c) Another result of Laplace's researches was significant to cosmology: Without having, perhaps, deliberately looked for it, he had discovered that all the inequalities were periodic, so that they do not alter the general existing structure of the system, which can be considered steady, despite the very complex motions that take place in it. For example, the acceleration of the mean motion of the Moon does not imply that the Moon will eventually fall on the Earth. This acceleration depends, Laplace proved, upon a slow variation of the eccentricity of the Earth's orbit. This variation is itself periodic, so that the acceleration of the Moon will be followed by a slowing down and so on. The same sort of conclusion was reached by Laplace about terrestrial phenomena. For instance, he proved that the oceans are stable; we do not have to fear that oceanic waves will grow indefinitely until they eventually overflow the mainland. The discovery of this general feature of steadiness was most remarkable, since it was roughly contemporary with the French revolution, which created the general feeling that kingdoms, governments, and social structures do not remain steady at all.

From the cosmological viewpoint the importance of those proofs appears when we go back to the crucial tenet of Newton, which we found again in d'Alembert's writings, namely that natural science is mainly concerned with the laws, the connections of events, and not with the structure and order of the whole Universe, so that the existing order is contingent upon something else. When he proved the stability of the Moon's orbit, Laplace did not explain, by natural theory, why the Sun, the Earth, and the Moon were arranged in this specific way. He made nevertheless a great step forward in this direction by showing at least why the 'arrangement' goes on, and by demonstrating that the laws of mechanics are consistent with a lasting orderly system.

Perhaps we are not in the best position to appreciate the significance that such arguments possessed in the eighteenth century. Many of us are no longer prepared to find something marvellous in the geometrical and kinematical arrangements of the solar system. Besides, here we encounter a type of reasoning which is familiar to the modern mind but which Laplace's contemporaries, except perhaps very few of them, were probably not inclined to accept easily: If we observe the present structure of the solar system, or of any other cosmic system, that observation by itself makes it probable that the arrangement is permanent. Were this not the case, then—as human observation is practically instantaneous—we would have only a negligible chance of observing it.

But, as the use of probability in astronomical science was just beginning (in the researches of such scientists as Michell, Herschel, and, of course, Laplace), this type of argument was virtually unknown, and Laplace's 'proofs' of stability must have seemed very striking. More than fifty years later, Arago, reporting to the 'Chambre des députés' about the projected national edition of Laplace, still dramatizes this point, saying that it had been a great relief to Laplace's contemporaries to learn from him that they lived in a stable world.

(d) In the progress of Laplace's cosmological ideas, those proofs of stability led to the ultimate step, namely the building up of a cosmogony—in the limiting sense of the formation of the solar system—completely deduced from observation and the Newtonian theory; that cosmogony appeared at the end of the *Exposition du système du monde*. Although representing merely a very small part of Laplace's works, it became the most popular, and for obvious reasons: Laplace, still remaining within the limits of natural theory, proved that he could go beyond the Newtonian dichotomy between laws on one side and cosmic order on the other, thus opening the way to a nontheological cosmology. In this respect, a detailed comparison with Kant's theory should be made; it would show that theological reasoning has still a great place in Kant's work, whereas it has none in Laplace's.

Yet the naturalist-philosophers of the nineteenth century misunderstood Laplace's cosmogony on an important point: They construed, and I think misconstrued, it in a transformist outlook; but Laplace's view is not transformist; he does not imagine the Universe to be engaged in an endless process of more or less creative evolution. He only intends to convey that the formation of the solar system has been a process of equilibration through which an unstable structure was transformed, by the spontaneous play of physical interactions, into a more stable one, which seems bound to persist indefinitely, so that his world is more static than transformist.

3. We will now summarize Laplace's philosophy of nature and cosmological concepts as they are presented in the synthetical treatises he published in the second part of his scientific life, when he thought that the main problems of celestial mechanics had been solved.

Like many of the philosophers and scientists of the Enlightment, he maintains a mechanical view of nature. The basic structure is the interaction by contact of massive points in accordance with the laws of mechanics. That was one of the reasons why he prefers the classical, Newtonian theory of light to Huygens' wave theory, revived in his time by Young and Fresnel.

But Laplace is not a reductionist: He does not think that gravitation nor any other type of attraction or repulsion can be reduced to the basic pattern of mechanical interaction. Now the impossibility of that reduction does not deprive the Newtonian law of its fundamental role in natural philosophy. Laplace thinks that, in this respect, he has completed (with Lagrange) the work of Newton, Clairaut, d'Alembert, and Euler, by proving that the law is universally, rigorously, absolutely true, without being reducible to a mechanical model.

One important consequence is that the mathematical scheme prevails over the mechanical model: The basic structure of nature cannot be adequately and

completely described as a set of inert atoms in motion (interacting by contact)—
it is adequately described by the mathematical device of the physicist, namely
differential equations plus initial conditions. That leads to one of the
best-known tenets of Laplace's philosophy, i.e., determinism, in the strict
sense of theoretical physics. This is historically significant: The deterministic
scheme, so basic in the theoretical physics of the nineteenth century, survived
the ontological model on which it had been built.

From the strictly cosmological viewpoint, the most original contribution
of Laplace is, of course, his attempt to place the problem of cosmic structures
and genesis within the field covered by natural law. And yet, that did not
completely reverse the basic Newtonian order of epistemological values: The
emphasis in Laplace's philosophy is still put on the pattern of connection of
events in space and time. The ancient concept of a relationship between element
and totality, that is, the concept of a *cosmos*, did not recover its previous importance; the differential viewpoint still prevailed in cosmology over the integral
viewpoint. Laplace maintains even more strongly than d'Alembert the subor-
dinate character of the least action principle, at least with respect to the laws
of motion.

Concurrently, in Laplace's cosmology the solar system keeps its prevailing
role which is analogous to that of a laboratory in experimental physics: As far
as we treat of physical laws and their verification, a single system, if it is well
controlled by observation and calculus, is more important than the rest of
the world.

Laplace, though, did not consider only the solar system as a natural laboratory; he also saw it as a paradigm (here, of course, in the sense of Plato, not
of Kuhn). What we observe in that system, the inferences we are able to make
about its formation, can be generalized to other cosmic systems which we may
supposed to be formed by condensation of diffuse matter into discrete bodies
gravitating to one another practically in a vacuum. This is the main point where
Laplace meets Herschel. He does not often quote his contemporaries, and when
he does, it is only about the discovery of special astronomical facts. Now, in
the *Exposition du système du monde*, he quotes Herschel, not only as the
discoverer of Uranus, but also as the astronomer who surmised that some of
the observed nebulae are systems undergoing a process of gravitational condensation (this quotation appears for the first time in the fourth edition, 1814).
There are several indications that Laplace considered Herschel's hypothesis
as very far-reaching; e.g., John Herschel reports that, on meeting Laplace, the
latter especially mentioned that question.

As far as Laplace—despite the mainly analytical character of his cosmology—indulged himself in building up a sketch of the whole Universe, he
probably imagined it as the indefinite repetition, in endless space, of gravitating
systems more or less analogous to the solar system, either in the process of
condensation, or having already reached an ultimate and steady state. Nonethe-

less, he certainly did not venture to develop, as Herschel desperately tried to do, the construction of a very large-scale cosmic system. He probably thought it impossible; perhaps also he did not consider it a very fundamental problem. Laplace's surprising silence—which Jaki noticed—on the structure of the Milky Way (a question on which Herschel's early hypotheses were less conjectural than those on the planetary nebulae) may find thus a plausible explanation. Laplace's concept of cosmology was by far more cautious than ours. And this may account, at least partly, for the definitely noncosmological views of astronomers in the last century.

BIBLIOGRAPHY

1. J. d'Alembert, *Oeuvres* (Paris, Berlin, 1821), Tome I; articles "Cosmologie" and "Cosmogonie" in the *Encyclopédie*.
2. H. Andoyer, *L'oeuvre scientifique de Laplace* (Paris, Payot, 1922).
3. F. Arago, *Oeuvres* (Paris, Gide, 1859), Tome III.
4. R. Hahn, *Laplace as a Newtonian Scientist* (University of California, 1967).
5. M. A. Hoskin, *William Herschel and the Construction of the Heavens* (London, Oldbourne, 1963).
6. P. S. Laplace, *Oeuvres complètes* (Paris, Gauthier-Villars, 1878–1912).
7. E. Whittaker, Laplace, in *Am. Math. Monthly* **56**, 369–72 (1949).

Chapter XIX

Cosmology in the Wake of Tycho Brahe's Astronomy

Kristian P. Moesgaard

University of Aarhus

LONGOMONTANUS AND ASLAKSEN

From 1562 to 1565 Tycho Brahe (1546–1601), still in his teens, studied in Leipzig and made his first astronomical observations there, with crude aids, yet good enough to form the basis for his critical evaluation of the knowledge of astronomy he acquired from literary sources. Within the same period, in remote corners of the Danish-Norwegian double monarchy, two men of modest descent were born, Christian Sørensen of Lomborg (Latin: Christianus Severini Longomontanus) (1562–1647), and Kort Aslaksen (Latin: Cunradus Aslachi) of Bergen (1564–1624), whom three decades later we find among the more prominent assistants at Hveen, where Tyco had in the meantime erected his unique center for astronomical research. With Tycho, practical skill and actual intellect had greater weight than noble birth or academic degrees.

Thus Longomontanus, in addition to the basic knowledge he had acquired at the grammar school of Viborg, had behind him no more than one year's study at the University of Copenhagen when, in 1589, he entered his apprenticeship with Tycho. Eight years later he had grown into a professional astronomer not particularly impeded by the traditionalism which insisted, in most cases, that one must go through a full University curriculum. After his master's death, he inherited the task of founding a complete theoretical doctrine of astronomy upon the exceptional new Hveenian accumulation of empirical data. He spent the second half of his lifetime as a professor at Copenhagen, for a great part busy with that job. But he also inherited the limitation of Tycho's intellectual horizon that marked the 16th century. By and large, he lacked the originality to go beyond Tycho's own plans for the completion of the work.

We can therefore take Longomontanus' writings as reflecting rather closely

Tycho's world of thought. In particular, his seven pages on cosmology, prepara-
tory to his "Tychonian Almagest," the *Astronomia Dancia*[12],1 (pp. 35-41),
provide a primary source for the natural philosophy concomitant to Tycho's
astronomy. Already in 1611 he had published a slightly different version of this
small tract, drawn up as 40 theses for disputation.[11] He also dealt with the topic
in his appendix to the *Astronomia Danica*[13] and in his *Introduction to the
Astronomical Theatre*.[14],2

Aslaksen, always aiming at a theological career, arrived at Hveen in 1590
after twelve years of education equally divided between the school at Malmø
and the University of Copenhagen. In 1593 he set out for six years of post-
graduate studies at numerous universities between the Alps and Scotland. On
his return in 1600 he became attached to the University of Copenhagen, and in
1607 he took the chair of theology. Influenced by Pierre Ramée's philosophy
and showing, moreover, Calvinistic sympathies, he had to face opposition from
the leading Danish clergy. He was a prolific writer on theological matters, and
he lived to see one of his works on the Catholic index of prohibited books.

However, what interests us here is that in 1597 Aslaksen published a treatise,
On the Nature of the Triple Heavens[1] (cf. Refs. 2 and 3), dedicated to Tycho
Brahe. His three heavens are the atmosphere, the space containing the heavenly
bodies, and the eternal heaven, the abode of the blessed. Formally the book is
composed along scholastic lines, but in its contents Aslaksen aims at reconciling,
on one side the Holy Scripture and what appeared worthy of preserving of the
traditional Aristotelian cosmology, and on the other the new scientific outlook
he had acquired at Hveen. As a widely traveled theologian he was of course
inspired from several quarters other than Tycho's world of thought. Nevertheless, his central purpose of unifying theology and empirical science, with give
and take in both directions, was quite in Tycho's spirit. His book therefore is
another important source for our knowledge of cosmology in the wake of
Tycho's work.

CHRONOLOGICAL STAGE SETTING

Important epochs in the history of a scientific discipline are marked by the
innovators who first created, for the different branches of that field, the concepts, routines, or tools still basic for work in it. For the technique of astronomical observations, Tycho Brahe, and for the geometrical representation of
the planetary motions, Johannes Kepler, were such epoch-making figures. Together these two men founded our mathematical kinematics of the solar system.
As for the dynamics of this system, that is, the physical or cosmological causes

---

[1] I have used the 1640 edition. Pages 314-320 contain an appendix, "Concerning the spots
on the Moon and their use."
[2] Folios F4v–G3v contain an appendix, "Concerning the refractions of the stars."

behind the kinematics, nobody gave an explanation still current today until Isaac Newton formulated his laws of motion and gravitation. Thus arrayed, like a string of beads, Tycho's observations, Kepler's ellipses, and Newton's laws make a nice progression toward still deeper insights into the mode of operation of nature. The logical continuation would have been to deduce, from Newton's laws, the concept of a terrestrial globe in motion.

Real history, however, deals with the achievements of human beings and its chronology may cut across the laws of logic. Thus the idea that the Earth moved dated back to Antiquity, from time to time it was subjected to academic discussions among natural philosophers, and two generations before Tycho it became a constituent of Nicolaus Copernicus' architectural arrangement of the solar system still accepted today. Copernicus interwove the motion of the Earth with his doctrine of positional astronomy which was so elaborate as to admit of comparison with Ptolemy's. Astronomers, therefore, for the first time in history now had to take the suggestion seriously. At the beginning of his book Copernicus argued against the futility of the traditional cosmology and doctrine of motion, and he formulated and advocated as more probable bits of an alternative view—rather *ad hoc* in character. But he could not claim, neither did he claim, necessity for his own basis. At any rate, he could advance nothing but kinematics in his defense for this basically dynamical feature of the solar system. Acccordingly, the idea could not appear absolutely compelling.

Such are the reasons why astronomers, natural philosophers, and theologians spent two centuries discussing whether Copernicus was right or wrong, and why 20th-century historians still debate if he ought to have been right or not.

As has been pointed out, astronomers after Copernicus, who were not satisfied by "saving phenomena" only, could not confine themselves to improving upon *the* astronomical doctrine. They had the choice between two essentially different doctrines, whence the division "for or against the Copernican system" characterizes adequately an important part of the scientific activity between Copernicus and Newton.

First the opponents had the game all to themselves. On the traditional premises of physics and cosmology, cemented for centuries by the allied forces of Christian theology and Aristotelian natural philosophy, Copernicus' suggestion was contradictory and effaced itself. The supporters, on the other hand, faced the problem of replacing Aristotle's cosmology and doctrine of motion with a foundation for the new world system which went beyond bare kinematics. Galileo's telescopic discoveries demolished stone by stone the credibility of the Aristotelian construction. Kepler tried to make the five regular polyhedra govern the new architecture of the solar system, to make loadstones out of its constituent bodies, and to rule their motions by musical consonant intervals. The first overall successful alternative was provided by Descartes. Tied to his hierarchy of vortices, the new world system conquered learned Europe. Moreover, this alliance for decades survived Newton's introduction of his solution, mainly because the latter entailed the concept of action-at-a-distance which appeared miraculous and hard to grasp.

Beforehand there is no obvious answer to the question of Tycho Brahe's role within this drama of the history of cosmology. After he published in 1588, in broad outline, his own world system, he remained a stubborn opponent to the Copernican system. But by his works on the New Star of 1572 and on the Comet of 1577 he had himself demolished cornerstones of the traditional cosmology; hence he, too, was against Aristotle. Outside polemic contexts, therefore, he was rather reticent about the physics of the universe. In his correspondence with Christopher Rothmann he entered elaborate discussions on selected aspects of the field. But for systematic and comprehensive accounts we must turn to the works of his two pupils presented above.

LONGOMONTANUS' COSMOLOGY

The title and the opening theses of Longomontanus' dissertation[11] (Theses 1-3) classify its theme as a doctrine prerequisite for establishing suitable models within positional astronomy, and also for astrology. Strictly, the field belongs to the discipline of physics. Nevertheless, the astronomer needs it, and as he is the real judge of its value, he should have the right to discuss it. Two main topics are to be dealt with, namely, the material of the heavens and the essential nature of the major world bodies, with special reference to their activity of motion. *The heavens* are defined as the space embracing in its bosom all *the major world bodies and their motions*. These bodies, in a single class, comprise all the heavenly bodies and the globe of the Earth.

Longomontanus, therefore, aimed at a general doctrine of natural philosophy valid everywhere in the Universe. He went even further than Copernicus by putting under the same label not only the Earth and the planets, but also the stars. At any rate, all notions of an Aristotelian boundary between sublunar and celestial physics have evaporated.

The Heavenly Space

Concerning *the heavens* Longomontanus begins with rejecting three opinions. First the view, held by the Peripatetics, that the celestial region should be stuffed with material spheres, hard and impassable in nature. Probably the supporters of this opinion misinterpreted the Pythagorean doctrine of the harmony of the heavenly spheres. Second, the view of the Stoics, followed by Paracelsus, that the heavens should be made out of fire. To make sense, this suggestion must regard the stars themselves and their origin of primeval light[13] (p. 8) rather than the space containing them. Third, the view that the celestial region should consist of dense air, or of water. Now the most recent comprehension of astronomy, in particular regarding refraction and the new celestial phenomena, i.e., the New Star and comets, has made unnecessary any elaborate discussion of the numerous doubtful opinions about this topic. Thus Longomontanus considers it to be sufficient to propose and explain briefly the conception most accordant with Holy Writ, nature, reason, and experience (Theses 4-5).

The underlying material of the visible and finite Universe, according to its nature, is something everywhere expanded, to the utmost thin and rare, like the very image of the incorporeal and imperceptible, and it is homogeneous all through; in its essence it is stable, but nevertheless passable; it has the quality of cold; it was created by God to serve as the medium, or vehicle—as it were— for the light of the stars, and to separate the waters above from those below the vault of heaven (Thesis 6).

First we learn that Longomontanus' visible Universe is finite, although he later admits his ignorance of the shape of its boundary, and of what is beyond it (Thesis 39). Here Copernicus was more radical, subjecting to doubt the very question of a limited versus a boundless region for the fixed stars[7] (Book I, Chapter 8). In case the latter are thought to cause their own risings and settings by whirling through a full circle every 24 hours, they need be at finite distances from the Earth. Now Longomontanus accepts the daily rotating Earth, as Copernicus had done, so for that matter his visible Universe could as well have been unlimited. Nonetheless, he takes it to be bounded, probably because he knew, from the Bible, about its enclosure, God's eternal heaven.

Within this finite part of the Universe, intelligible by the mediation of vision (Thesis 18), we find dispersed the major world bodies among which some at least are in motion. Furthermore, these bodies emit light, original or reflected (Thesis 28), which travels instantaneously[14] (fol. G1r) and along straight lines through the space between the bodies. Now the demand for a vehicle of the light which does not slow down the planetary motions justifies most characteristics of the *expansum*, as Longomontanus terms the material of the heavens.

Thus the *expansum* should be more than bare geometrical space, or a vacuum (Thesis 7), but less dense than anything, however thin, of an aerial nature (Thesis 19; cf. Brahe,[4] Vol. VI, pp. 135ff.). On the whole, the air cannot be the luminiferous medium, because a ray of light is altogether unaffected by possible motions or vibrations of the air through which it passes (Theses 9-10; cf. Zabarella[20],3). The *expansum*, therefore, must also be present inside all translucent bodies, and to cover with a safe margin also cases like red-hot metals, Longomontanus simply maintains that it is spread out everywhere, including the interior of the Earth (Theses 7, 11). This ubiquity of the *expansum* at the same time does away with all conceivable friction related to the passage of the major world bodies through it (cf. Thesis 19). There is simply no differ-

---

[3] Aslaksen quotes from *Liber de natura cæli*, which is probably a part of this work. Cf. Thorndike,[19] Vol. VI, 369.

ence left between the state of 'being in' and that of 'moving through' the *expansum*; accordingly no explanation is necessary concerning how to cleave it. So in relation to tangible matter, the *expansum* is passable as the very image of the incorporeal.

To support, on the other hand, the rectilinear transmission of light it must nevertheless be stable and homogeneous. In particular, there is no dividing surface between sublunar and celestial types or densities of the *expansum*. Longomontanus argues the nonexistence of such a boundary from the alleged fact that we do not perceive its traces in the form of refractive bendings of all light rays (Theses 8, 11). A little later he directs a similar argument, about the absence of refraction near the zenith, against the view—held by Johannes Pena, Christopher Rothmann, and others—that the material of the heavens be nothing but a kind of highly rarefactive air (Thesis 19). Now one must remember that Longomontanus'—like Tycho's—view of refraction included the erroneous notion of an upper altitude of 45° above which no refraction would occur. But apart from that, both of the above arguments are formally sound on the assumption that he took the effect to depend strongly upon the length of travel of light through the refractive medium. For the length of travel through a comparatively thin layer of air decreases rapidly with increasing height above the horizon, whereas that through regions stretching to the distance of the Moon, or even further, is almost independent of the altitude.

Thus far the *expansum*, filling up space, has been presented as the conglomerate of a luminiferous something and a frictionless nothing. Probably in order to meet the possible charge of having deviated too drastically from time-honored doctrines, Longomontanus laconically states that his *expansum* may in a sense be taken as resembling Aristotle's fifth element (Thesis 23).

Then he passes on to its property of being cold. Among the two constituents of the interbodial space, light is warm, and so, according to nature, the *expansum* ought to be cold (Thesis 24). Its actual coldness is confirmed by experience, because the air turns cold to the point of frosty stiffness when the intensity of light decreases, as is the case during winter or in arctic regions. One also feels the coldness of the underlying *expansum* when the air is blown away (Thesis 25). Finally, Longomontanus adduces an analogy with the microcosmos of an animal whose continuous life depends on the proper balance between warm and cold. He suggests that the pairs of opposites, body/soul, warm/cold, light/*expansum*, are parallel. In consequence, the interbodial space should enjoy the very most temperate state, and, moreover, perhaps be able to absorb surplus heat from the luminary world bodies and thus create endurable conditions of life also for the latter (Theses 26–27).

The Atmosphere

As inhabitants of one of the major world bodies, we are bound to take its particular atmosphere as the neighboring part of our heavens. The constituent material of this atmosphere, *the more clean air*, is composed of the luminiferous *expansum* and the light-refracting air. It reaches 12 or 13 German miles above the surface of the Earth, which Longomontanus concludes from the duration of twilight. As fish need water, the creatures on dry land can only breathe, see, and hear when surrounded by this more clean air. Further salubrious conditions for continuous life are brought about near the surface of the Earth, where the heat of the solar and sidereal light and of the Earth together lifts up vapors from the sea to the clouds, whence they return by precipitation. This lower layer of the atmosphere is called *the more feculent air*. From observations of the clouds it is seen to stretch $1\frac{1}{2}$ German miles above the Earth in the summer, and $\frac{3}{4}$ of a mile during the winter. Probably another similar but much weaker cycle elevates small amounts of water from the upper clouds to the outermost limit of the atmosphere. This would explain certain fiery phenomena in the air and the occurrence of whitened corpuscles which in their turn produce the twilight. On the whole, the atmospheric manifestation of the Earth's life finds again its counterpart in the circulation of humors within the microcosmos of the human body, and it is artificially imitated in chemical experiments of distillation (Theses 13-17; Ref. 14, fol. G1v-G2r; Ref. 12, pp. 125-131).

The Major World Bodies

Longomontanus does not hesitate to suggest as probable that the Moon and each of the other heavenly bodies, in analogy with the Earth, is embosomed in its own production of atmospheric phenomena, adapted to the particular nature of the body in question (Thesis 21). Then he advances his view concerning the generation of new celestial phenomena, namely, that the heavenly bodies, at times determined by divine providence, eject hidden seeds, whence conception may happen in the *expansum* and give rise to the sudden, almost miraculous appearance of a new star or a comet. Tentatively he compares the process with the formation of bubbles, and he rejects as absurd the alternative opinion— held by Plinius, Plutarchus, and others—that certain celestial bodies, ordinarily hidden, from time to time pour out fire (Thesis 22; cf. Ref 13, pp. 2-21).

Longomontanus touches upon a number of problems only to admit his ignorance about their solutions. One concerns the material of the heavenly bodies, where the only evidence available, namely, that provided by Galileo, suggests that the Moon is alike "another Earth." Another question is which bodies are luminous like the Sun, and which borrow their light like the Moon? Further there is the question of distribution in space of the very numerous fixed stars, and that of possible inhabitants of other world bodies. Finally, for what use does Jupiter have its four satellites (Theses 28-29; cf. Ref. 12, pp. 314-320)?

Coming to the more important question of the essential nature of the major world bodies, Longomontanus introduces the distinction interior/exterior, which parallels active/passive or "motive principle"/"bounding shape" (Theses 30-31).

As for their interior essence, he adopts Plato's concept of animate world bodies within an animate Universe. Of course, one can only metaphorically compare their life with that of animals, yet the wonderful harmony of the planetary courses demands, so to speak, that each body has been grafted with its own faculty for performing its particular motion. Thus the planets need not trace oval orbits, as some people think; neither are they bound to follow abundant epicycles (Thesis 32; Ref. 12, p. 39 omits the remark on ovals and epicycles). Since no motion continues by itself, the almighty Creator has to keep them all going (Thesis 33).

Furthermore, the interior nature of any world body includes a gravitational tendency, directed toward its center, and affecting all its parts so as to ensure the individual integrity of the body in question (Thesis 34). This echoes Copernicus' concept of "universal" gravity[7] (Book I, Chapters 8-9), which is seen to be a local occurrence without immediate relationship to the mutual motions of the world bodies. It entails, however, that the exterior essence, or the geometrical shape, of the world bodies is globular, which, in its turn, is confirmed by experience as far as the Sun, the Moon, and the Earth are concerned (Thesis 35). Further arguments in support of their spherical shapes are found in Copernicus' opening chapter (Thesis 39; cf. Copernicus,[7] Book I, Chapter 1).

Being globes, the major world bodies are well suited for free suspension in the *expansum* at definite places, and for being kept in position without any aid from material spheres or axes (Thesis 36). Obviously their material must be subject to an equal distribution, leaving no room for any interior vacuum (Thesis 37). The interbodial space, however, filled with light and *expansum* only, can be taken as a vacuum insofar as this means a place devoid of tangible matter (Thesis 38).

Last, but not least, their spherical shapes make the major world bodies most apt for motion, whether it be rotation at the same place or proper change of position. Thence we have paved the way for dealing later on with their courses in accordance with suitable assumptions (Thesis 40).

"In the beginning God created Heaven and Earth"

The foregoing rather elaborate reproduction of Longomontanus' dissertation appears to the point because, in my opinion, it contains a fairly faithful summary of a complex of Pythagorean-Platonic, religious, and astrological convictions which Tycho himself intertwined with his own scientific experience and intended to put in the place of the no longer trustworthy peripatetic natural philosophy (Moesgaard,[15] pp. 53ff). It should, therefore, reflect the atmosphere not only of a 1611 academic discussion at the University of Copenhagen, but also of debates at Hveen twenty years earlier between the master and his assistants. Certainly it must have been an odd experience for the assistants to witness, in August 1590, on the occasion of Rothmann's visit to the island, how the master defeated his guest, at the time a convinced Copernican, with arrows shot from Aristotelian strongholds.

Obviously the main purpose of Longomontanus' treatise is to build something resembling a cosmological basis for what is really an *a priori* axiom of his positional astronomy, namely that any celestial motion is made up of uniform circular components[12](pp. 163-169). He achieved his end, but on the way he demolished so radically the Aristotelian boundary between the sublunar and the celestial worlds that in addition the path was clear for the full Copernican system. He had so far happily passed by in silence the very opening verse of *Genesis*. Thence, in due time, he could reinstate the Earth in its privileged position asserting that what was created first ought to take up the most prominent places, viz., the circumference and the center of the Universe (Ref. 12, p. 158; cf. Ref. 16, p. 131). Whatever we think of this argument, together with Longomontanus' opposition, inherited from Tycho, against a huge space for no use between Saturn and the fixed stars, it provided his excuse for favoring the Tychonian architecture of the solar system.

ASLAKSEN'S COSMOLOGY

Aslaksen's book *On the Nature of the Triple Heavens* covers more than 200 pages, and its collection of arguments cannot be summarized in a few words. A brief survey, however, of its main conclusions will suffice for our purpose of contrasting it with Longomontanus' treatise.

Many of the pages reflect sacred texts and reveal the author's principal aim of laying a philosophical foundation for theological use, not too much stained by lacks of consistency with reason and experience. In one place he parallels the aerial, the sidereal, and the eternal heavens with the *Court*, the *Sanctum*, and the *Holy of Holies*, respectively, of Solomon's temple (Ref. 1, p. 9). Apart from the Bible and Tycho Brahe, Aslaksen repeatedly quotes Otto Casmann,[5,6] Lambert Daneau (1530-1595),[8-10] Jean Pena (died 1558),[18],4 Giacomo Zabarella (1532-1589),[20] and Hieronymus Zanchius (1515-1590).[21] This handful of his favorite authorities again classifies the work under the heading "Christian cosmology placed in scientific scenery" (cf. Thorndike,[19] Vol. VII, p. 414).

The bulk of Aslaksen's arguments concern the sidereal heaven, which is dealt with in the 24 chapters of his second book (Ref. 1, pp. 40-180), whereas he devoted only four chapters to the atmosphere (Ref. 1, pp. 11-39), and six to God's eternal heaven (Ref. 1, pp. 181-214).

Knowledge about the latter can be acquired only through Holy Writ. Therefore we shall not discuss it here further, apart from mentioning that, according

---

[4] Aslaksen makes several quotations from Pena's preface.[18]

to Aslaksen, it is created—together with the sidereal heaven—on the first day of creation, and from then forward is everlasting and incorruptible. It is corporeal, although the utmost in thinness, and it is located above the two other heavens, where it takes up a finite, but incomprehensibly immense space (Ref. 1, pp. 190-200). Human language was always rich in words for describing the unknown.

The Atmosphere

As opposed to the eternal heaven, both the atmosphere and the region containing the heavenly bodies are temporary. The former stretches from the surface of the Earth to the lunar orbit, and is the domicile for the so-called meteors, e.g., thunderbolts and the "celestial waters" (Ref. 1, p. 9). The actual occurrence of meteors, however, is limited to the lower layers of the atmosphere, not farther from the Earth than 12 or 13 German miles. Within this region the air is feculent from terrestrial exhalations, which also causes the refraction of light rays from the heavenly bodies. Above the said dividing line a more thin and clean air is expanded (p. 14).

For the rest, Aslaksen spends one chapter in explaining away the existence of Aristotle's fourth element, fire (Ref. 1, pp. 16-23), and two in moving the "celestial waters" from the region above the stars to the clouds of the terrestrial atmosphere (pp. 23-39). One of his arguments against the existence of a sphere of fire is bound up with the absence of the refractions to which such a sphere ought to give rise, and it is a corollary of placing the "celestial waters" in the atmosphere that the aerial heaven—or at least the inferior part of it—was created on the second day of creation.

The Sidereal Heaven

The opening chapter of Aslaksen's second book classifies the sources and aids by which knowledge about the heavenly bodies and their surrounding space can be obtained, including sacred texts as well as instruments for astronomical observation. The second chapter defines the sidereal heaven as an expanse of the very most slender nature, pervious for the particular motions of the heavenly bodies, and accordingly destitute of real material orbits, as well as immovable and invisible. This heaven is said to be ethereal, and it stands out from the air by its greater purity and rarefication (Ref. 1, p. 47).

Then Aslaksen devotes no fewer than eight chapters to discussing the matter underlying the sidereal heaven (Ref. 1, pp. 49-90). The reader is hauled through a complete Aristotelian scheme of rejecting impossible suggestions, viz., that the material in question is fire, air, water, or a mixture of these elements. Thus the solution must be a certain fifth essence, unaffected by all elementary change and filth. All this is really a gallant show, because Aslaksen has already done away with the fire, and later his heavenly material proves to be hardly distinguishable in essence from the air. Accordingly, his argument against Jean Pena's opinion that the heavens consist of rarefied air appears to be particularly forced (pp. 67-69). Having obviously attended too many University lectures, he could not, like Longomontanus, confine himself to asserting that in certain respects the heavenly stuff resembles Aristotle's fifth element.

After one chapter concerning the essential nature (*forma*) of the sidereal heaven (Ref. 1, pp. 90-105), there follow eight chapters on its essential property (pp. 105-142). Being corporeal, heaven must have a nature. Like terrestrial creatures it is animate, but in the sense of having been endowed with a certain faculty for action. The essential property of heaven is its utmost tenuity or thinness. Aslaksen, in due order, argues his case from evidence concerning refraction (pp. 105-110), the generation of comets in the heavenly space (pp. 111-118), and the motions of the planets to and fro at certain distances from the Earth (pp. 118-121).

Here refraction in particular plays a decisive role. For the nonexistence of any refraction apart from that caused by the impurities of the lower atmosphere entails that, in the space between the outermost visible stars and the inferior atmosphere, there is no distinct dividing surface between different kinds or densities of matter. This does away with the notion of real material spheres for governing the planetary motions, but in addition the boundary between the outer atmosphere and the sidereal heaven grows more and more imaginary. Accordingly, Aslaksen admits that the air and the heavenly matter are equally translucent and almost of the same thinness (Ref. 1, p. 110). Later he completes the identification, saying that the aerial and the sidereal heavens only have different names, like a pair of adjacent oceans, while the very subject remains everywhere indivisible (p. 148). Nevertheless, he continues to maintain that the heavenly matter is more pure and rare than the air (e.g., p. 164). So we end up with a "universal atmosphere," the density of which gradually decreases with increasing distance from the surface of the earth (p. 7).

Among the objections Aslaksen counters (Ref. 1, pp. 123-142), one is worth noticing, viz., that which maintains that the heaven is solid in order to keep together each of the solid bodies it contains. He meets this argument with a reference to Tycho's acceptance of Copernicus' concept of universal gravity (Ref. 1, pp. 138-139; cf. Ref. 4, Vol. VII, p. 235).

The last five chapters of Aslaksen's second book deal with three inferences from the admitted matter, nature, and essential property of the sidereal heaven, namely, that this heaven is destitute of material spheres (Ref. 1, pp. 142-166), that it is immobile (pp. 166-173; cf. Ref. 16, p. 122), and invisible (pp. 173-180).

The discussion concerning the first topic includes the argument that accepting such real spheres would demand the creation of a new set of spheres to serve any New Star. The Aristotelian axiom that one body can perform only one simple motion is rejected, and Aslaksen accepts the view that the planets move in accordance with their particular capacity for motion granted by divine providence. From the extreme thinness of the heavenly material he argues both its own immobility and its lack of capacity for causing the motions of the heavenly bodies. The latter in turn do not prove that the heaven itself moves; it could as well be maintained that the sea had to move because fish are swimming in it.

CONTINUATION OF THE CHRONOLOGICAL STAGE SETTING

Several differences between Aslaksen's and Longomontanus' cosmological doctrines, and also a certain lack of consistency within each of the doctrines, are easily discernible. These features can be understood and to some extent excused in terms of their different personal backgrounds and the purposes of their works. However, in order to place the two works within a broader historical context, it appears more promising to underline their similarity, and to consider them as different approaches to a solution of virtually the same basic problem.

With Copernicus' work the question concerning the place and the state of motion of the Earth in the Universe was detached from its traditional context of natural philosophy and tied up with positional astronomy instead. Philosophers, therefore, saw themselves debarred from judging immediately the problem, except for the trivial assertion that Copernicus' idea was incompatible with traditional cosmology. Thus the next step had to be taken by the astronomers who, in their turn, had to realize that Copernicus' astronomy was a subtle mixture of reformulating Ptolemy and taking into account a number of long-term variations which, by and large, earlier astronomy had ignored.[17] This explains why the first two generations after Copernicus remained rather reticent about his cosmology, but repeatedly quoted and used his mathematics, observations, and tables.

Returning to the philosophers, it seems natural that Copernicus' search for universality in the description of the visible world appealed to them. For the truly philosophical mind it must have been an attractive possibility to get rid of Aristotle's barrier between sublunar and celestial physics. Possible theological objections ought to become silent with regard to the condition that beyond the outermost stars another dividing line was established between the human world and God's abode.

Obviously two different ways of erasing the Aristotelian boundary had to be considered, viz., making the Earth "a major world body" on an equal footing with the heavenly bodies, or making the region containing the latter a part of a "universal atmosphere" around the Earth. Copernicus himself set his foot on the former track, and in 1557 Jean Pena drew attention to the latter. Longomontanus and Aslaksen confronted these two possibilities of establishing a unified doctrine of physics applicable to whatever human eyes could see with the work done at Hveen.

REFERENCES

1. K. Aslaksen, *De natura cæli triplicis libelli tres, quorum I. de cælo aëro, II. de cælo sidereo, III. de cælo perpetuo, e sacrarum litterarum et præstantium philosophorum thesauris concinnati* (Sigenae, 1597) [English transl. of Book III, R. Jennings, *The description of heaven; or a divine and comfortable discourse of the nature of the eternal heaven* (London, 1623)].

2. K. Aslaksen, *De mundo disputatio prima, secunda, et tertia* (Havniæ, 1605, 1606, and 1607).

3. K. Aslaksen, *Physica et ethica mosaica, ut antiquissima, ita vere Christiana* . . . (Hanoviæ, 1613).

4. T. Brahe, *Opera omnia*, ed. by J. L. E. Dreyer (Vols. I-XV, Hauniae, 1913-1929).

5. O. Casmann, *Psychologia anthropologica, sive animæ humanæ doctrina* . . . (Hanoviæ, 1594).

6. O. Casmann, *Marinarum quæstionum tractatio philosophica bipartita* . . . (Francofurti, 1596).

7. N. Copernicus, *De revolutionibus orbium cælestium, libri VI* (Norimbergæ, 1543); quoted by book and chapter.

8. L. Daneau, *Physice Christiana, sive, Christiana de rerum creatorum origine, et usu disputatio*, 3rd enlarged ed. (Genevæ, 1580).

9. L. Daneau, *Physices Christianæ pars altera, sive* . . . , 3rd ed. (Genevæ, 1589).

10. L. Daneau, *Vetustissimarum primi mundi antiquitatum sectiones, seu libri IV, tum ex sacris, tum aliis autoribus* . . . (Genevæ, 1596).

11. C. S. Longomontanus, *Disputatio prima astronomica, de praecognitis, in qua definitio materiæ cæli, adeoque loci cuncta corpora mundana majora, suo gremio complectentis discutietur; una cum natura et forma ipsorum corporum, imprimis, qua motibus suis apta sunt* (Hafniæ, 1611) (quoted by thesis numbers).

12. C. S. Longomontanus, *Astronomia Danica* . . . *in duas partes tributa, quarum prior doctrinam de diurna apparente siderum revolutione* . . . *explicat; posterior theorias de motibus planetarum* . . . *complectitur* (Amsterdami, 1622, 1640, 1663) (quotations are from the 1640 edition).

13. C. S. Longomontanus, *De asscititiis coeli phænomenis, nempe stellis novis et cometis* (Appendix to Ref. 12 separately paginated).

14. C. S. Longomontanus, *Introductio in Theatrum Astronomicum* . . . *quod Havniæ Metropolis Daniæ modo instauratur* . . . (Havniæ, 1639).

15. K. P. Moesgaard, Copernican Influence on Tycho Brahe, *Studia Copernicana V* (= *Colloquia Copernicana* I), pp. 31-55.

16. K. P. Moesgaard, How Copernicanism took root in Denmark and Norway, *Studia Copernica V*, pp. 117-151.

17. K. P. Moesgaard, Success and Failure in Copernicus' Planetary Theories, *Arch. Int. Hist. Sci.* XXIV, 73-111, 243-318 (1974).

18. J. Pena, *Euclidis Optica et catoptrica* . . . *latine reditta per I. Penam* . . . (Paris, 1557).

19. L. Thorndike, *A History of Magic and Experimental Science* (New York, 1941), Vols. V-VIII.

20. J. Zabarella, *De rebus naturalibus libri XXX* . . . (Patavii, 1589).

21. H. Zanchius, *De operibus Dei intra spatium sex dierum creatis opus* . . . , 2nd corrected ed. (Hanoviae, 1597).

Chapter XX

Chronology and the Age of the World

J. D. North

University of Oxford

> The *Europeans* had no Chronology before the times of the *Persian* Empire:
> and whatsoever Chronology they have now of ancienter times, hath been
> framed since, by reasoning and conjecture.[1]

Isaac Newton's two categories of chronological argument, by reasoning and by
conjecture, are the two poles around which belief about the age of the world and
the chronology of its history have always turned. But one man's conjecture is
another man's reason, and Newton's contemporaries often seem to have had as
much difficulty in deciding whether a particular chronological or chiliastical argument was conjectural, literal, cabalistical, revelational or rational, as they had
in first formulating it. Borrowing from these categories, however, I might say
that I am to consider arguments about the age of the world which were 'rational'
in the sense that Newton's chronology was thought to be so.[1] In other words, I
shall try to bring together a group of specifically astronomical arguments developed out of premises which were for the most part conjectural, if not positively nonsensical.

  Writers today who wish to pour scorn on chronology in this 'rational' tradition tend to select as their target James Ussher (1581-1656), Archbishop of
Armagh–a man, I might add, of an erudition seldom matched by that of his
critics. Ussher's supposed folly is to have set a date to the creation of the world.
This event he placed at the beginning of the night preceding 23 October 4004

---

[1] Newton was more concerned about the end of the world than its beginning, and in his published works does not specify its age; but it was a simple matter for his followers to extrapolate to the Creation, using the *Short Chronicle* which prefaces his *Chronology*.[1] He
usually follows the time intervals of the Jesuit Petavius (and not, as one might have expected, Ussher's), but his dates are four years higher than Petavius', suggesting a Creation
date of 3988 B.C.

B.C.[2] But this date, assigned to the Creation by 'acotholicorum doctissimus' (as he was called by his worthy Jesuit opponent Henry Fitzsimon), whether or not it may be properly seen as a symptom of aberrant intellectualism, like millenarianism among Anglicans, was neither new with Ussher nor was it even the product of characteristically seventeenth-century thought.[3] Chronology had been the key to universal history from the time of antiquity, and Old Testament genealogies stretching from the time of Adam himself, and reinforced by other Hebrew and patristic sources, could always be cited in refutation of Epicurus, Lucretius, and all who considered the world eternal, or even excessively old. The roots of Christianity, however, were to be clearly established as older than the philosophy of the Greeks; and yet it was thought essential that the Christian historian harmonize with Holy Writ the chronicles of the different nations—the Persians, 'Chaldeans,' Egyptians, Greeks, Romans, and the rest[4]—in order to strengthen Faith in a literal reading of scripture. That this judgement was right may be seen from the fact that atheist and theist alike, especially in the eighteenth and nineteenth centuries, chose chronology as a point at which to attack the authority of the Bible.[5] These were the tactics of Thomas Paine in *The Age of Reason* (1795-6): 'that witling Paine' was a description offered of him by a chronologist, William Hales.

The fundamental method of determining the epoch of the Creation from the Bible was, of course, to reckon by generations and reigns. Even where the ages of named individuals are given, there usually remains the problem of assigning an interval between the birth of members of successive generations, and early historians tend to accept three generations to the century, as an average. Reckoning by reigns gives remarkably consistent results over different cultures and historical periods, taking an interval of about 22 years as the mean.[6] A more serious obstacle to accuracy is the large number of discrepancies which exist among the different versions of scripture—the Septuagint, the Samaritan version, and the Massoretic text—not to mention such historians as Philo Judaeus, Josephus, and

---

[2]The year 710 of the Julian Period, for which see below. My edition is that of 1722.[(2)] The preface is rather pedantically dated 1650 in vulgar reckoning, and 1654 from the true nativity (Ussher's reckoning).

[3]Ussher's chronology is widely known by virtue of the fact that it was added (by unknown authority) to the margins of the King James Bible, which first appeared in 1611. It is still therefore readily accessible.

[4]The Jews did not establish an era in the precise sense of the word. The chief of the ancient eras were the era of Nabonassar, 747 B.C.; the era beginning with Coroebos' victory in the Olympic Games, 776 B.C.; and the founding of Rome, for which Varro gave the date 753 B.C. The Olympiads were possibly the last system of the three to be actually used, and the era of Nabonassar was probably used from his own time. It was adopted by Ptolemy for use in the *Almagest*.

[5]For this later history of the subject see the excellent *The Age of the World: Moses to Darwin* by Haber.[(3)]

[6]Newton was certainly conjecturing when he took 19 years.

Theophilus, bishop of Antioch.[7] The literature is truly vast, but the method adopted from the first in all comparative chronologies was to seek one acceptable datum (or more) which enters into the chronologies of two or more nations. Thus the birth of Cyrus, which allows of a correlation with Greek chronology, offers a way into the chronologies of the Persians, Medes, and Assyrians, and in the last of these the destruction of Solomon's Temple by Nebuchadnezzar links with Hebrew sources. Thence the chronology may be extended to the foundation of the Temple, to Exodus, Abraham's birth, and—for the intrepid believer—to the reign of Nimrod, the Deluge, and the Creation of the world.

The synthetic procedure thus outlined was not of a sort likely to recommend itself to men of a rationalist temperament, men to whom the symmetries of the world are more appealing than its historical oddities. The simplest way of introducing *a priori* symmetries into a temporal analysis of world history is of course to make use of temporal cycles, and there are no more notorious cycles connected with world history than those which go under the generic name of 'Great Year.'

The expression 'Great Year' has meant many things to many people. Especially associated with Plato's name,[8] the idea of recurrences in history came in for trenchant criticism by St Augustine, who saw that it would not allow any historical event a unique importance.[9] He saw that the uniqueness of Christ's redemptive power thus ran the risk of being undermined by neoplatonic thought, which was in any case incompatible with the very first verse of the Old Testament. The idea of a Great Year is nevertheless closely related to others which were in due course seen as guides to the age of the created world, and its ancestry is important. It has a Pythagorean look about it, for since the Great Year is the interval between periodic recurrences of grand conjunctions of the Sun, Moon, and all the planets, it must be the least common multiple of the periodicities of those bodies; and the Pythagoreans—so the common treatment of the subject runs—had a weakness for numbers, wherefore . . . and so on. At all events, the third-century (A.D.) grammarian Censorinus tells us that the Pythagorean Philolaus had made a Great Year consist of 59 years, including 21 intercalary months, and an ordinary year of $364\frac{1}{2}$ days.[10] Censorinus gives other figures for

[7]Speaking generally, the Septuagint tends to assign longer lives to those between the Flood and Abraham, and adds other individuals to the genealogies. The discrepancy between the derived total span from the Creation to Abraham is that between 1948 years and 3334 years, or thereabouts. From the Creation to Christ's birth those relying on the Septuagint arrive at about 5500 years, while the majority give an interval of the order of 4000 years. For further data see p. 318 below.

[8]The principal source is *Timaeus*, 39D. Cf. Cicero, *De natura deorum*, II.51-2.

[9]See especially *De civitate Dei*, XII.18, but also XII.14, where Augustine has a little trouble with Origen (*De principiis*, 3-5, 3) and a passage from *Ecclesiastes* (1, 9-11): "The thing that hath been, is it which shall be. . . ."

[10]*De die natali*, 18,8. I owe this reference, and the remainder of the paragraph, to Dicks.(4)

the period, quoting various authorities, but all possibly taking the period as an integral multiple of solar and lunar periods only.[11] Censorinus also mentions that Oinopides of Chios, a younger contemporary of Anaxagoras, made the length of a year $365\frac{22}{59}$, while Aelian and Aëtius ascribe to Oinopides a Great Year of 59 years. These two items of information fit together rather well: If the common figure of $29\frac{1}{2}$ days is taken as the length of a lunation, the number of days in a Great Year (of the Sun and Moon only) must be a multiple of 59, and it must also be a multiple of a number close to 365, the approximate length of a year. A period of 59 years totalling 730 months is an obvious one to take; but how Oinopides knew that this contained 21,557 days is not at all clear.[12]

The connection of these ideas with calendrical problems, and especially with that of intercalation, is never far to seek. Recall the intercalation cycles associated with the names of the fifth-century (B.C.) Athenian astronomers Meton and Euctemon.[13] To correlate the lunar months and the solar year, 235 months (= 228 + 7 intercalary)[14] were equated with 19 years and 6940 days. Better agreement with the truth was attained by Callippus, a century later, when four Metonic cycles of years, that is, 76 years, were made equal to one day fewer than the Metonic total (27,759 as opposed to 27,760). Another advantage of the Callippic cycle is that the number of leap years in it is constant, and this is obviously not true of the Metonic cycle, to its detriment.

The Metonic cycle was used in early Jewish and Greek calendars—but sporadically, not regularly even in Athens—as well as in the ecclesiastical calendars of Christian Europe. Long after its invention the cycle was taken to have been begun at 13 July 432 B.C., which meant, for example, that on 13 July 1 B.C. the 14th year of a cycle began, and that the first new cycle in our era began on 13 July A.D. 6.[15] Tedious and unmemorable though these facts may be, I must point out that other source-dates were accepted for the 19-year cycle throughout the greater part of the Christian era—and these as the result of writings ascribed to Bishop Cyril of Alexandria and Dionysius Exiguus. The cycle assigned to Cyril was supposedly issued in A.D. 437, expiring in 532.[16] It is now

---

[11] Dicks,[4] p. 76, points out that Schiaparelli, dividing the Great Year of Philolaus ($21,505\frac{1}{2}$ days) by suitably chosen integers, derives data for the planetary periods in very close correspondence with modern values. This would suggest—and Thomas Heath follows Schiaparelli—that his 'Great Year' was such in the Platonic sense. Dicks disagrees, but it is difficult to see how one should adjudicate in the debate, since it is at least arguable that Censorinus' information is likely to have been incomplete on a matter involving lengthy numerical data. In 59 years, Saturn completes approximately 2 revolutions, Jupiter 5, Mars 31, the Sun, Mercury, and Venus 59, and the Moon 729.

[12] Dicks,[4] p. 89, leaves this matter very much in the air.

[13] The cycle was probably borrowed from Babylon, although in extant sources it first appears there c. 383 B.C. See Parker and Dubberstein.[5]

[14] Of these, 110 were 'hollow' (29 days) and 125 were 'full' (30 days).[6]

[15] The first Callippic Cycle ran from 330 B.C.

[16] In short, it covered 5 × 19 = 95 years.

thought to be a seventh-century Spanish fabrication.[17] Some later writers took A.D. 532, the year with which Dionysius started his 19-year cycle, as a basis for the continuation of a 532-year cycle. For a cycle beginning with A.D. 532, the year 1 B.C., the nominal year of Christ's birth, is very fittingly the first year of a 19-year cycle, A.D. 1 being the second year, and so on.

The Metonic cycle had by this time lost its prime calendrical purpose, and was no longer associated with rules of intercalation. It had become merely a device for ordering the years with a view to determining the date of Easter, and into the tangled web of Easter computation I certainly have no wish to enter. There is involved in it the cyclical period of 532, however, which I should not pass by without further comment. The most famous of the rules for Easter were the Roman and the Alexandrian, and the practice of the early Church was at first fairly evenly divided between them. The Roman rule took an 84-year cycle (of 3 × 28 years), and the Alexandrian rule the 19 years of the Metonic. I will first explain the 28 year cycle, and then say how the cycles thus far mentioned were to be compounded into yet longer cycles.

The 28-year cycle is to modern eyes a simple consequence of the adoption of certain calendrical conventions—those of the Julian calendar, combined with the selection of a period of seven days for our week. Even the Christian who believes that the week was divinely ordained is bound to admit that the Julian calendar had more than a trace of paganism in its institution. There is therefore nothing either empirical or sacred about a rule ascertaining periodicity of recurrence of the phenomenon of a particular day of the month coinciding with a particular day of the week. The cycle is one of 28 years.[18] I have emphasized its conventional character, because even in the nineteenth century we find chronologists marveling at the fact that a creation of the world in 4004 B.C. implies that in the year 1 A.D. the days of the week repeated their order at the Creation. In the words of Edward Greswell, whose life at Oxford (according to the *Dictionary of National Biography*) was spent "in the systematic prosecution of his studies," and who was writing only a few years before Darwin's *On the Origin of Species:*

> this is another of the singular and hitherto overlooked coincidences, which a true scheme of chronology brings to light; and one among other proofs, thereby supplied, that in the Divine mind, and in the providential constitution and adjustment of time and of its several relations from the first, the Julian reckoning of time itself. . . must have been contemplated from the first. It is no accidental coincidence that B.C. 4004 was thus the first year of the cycle of 28. . . [8]

A sceptic might be excused for wondering whether 4004, a figure at least as early as Basil (A.D. 330–379), was first chosen (in preference to numbers close

---

[17]The exposure is due to Bruno Krusch. See Jones[7] for further details.

[18]A common year exceeds 52 weeks by one day. Month-day and week-day would coincide every 7 years, were it not for the leap-year's extra day. The concept of the 'concurrent' arises in this connection. The concurrent is the excess of the annual Julian cycle over the hebdomadal in the 28-year period. Its numerical values (with multiples of 7 removed), in successive years, are 0,1,2,3,5,6,7,1,3, ...

to it, which could have been fitted to the Bible equally well) for the very reason that it was a multiple of the 'solar cycle' of 28 years.

Combining the 28-year and 19-year cycles gives the cycle of 532 years, a cycle which is often—but mistakenly—said to have been introduced into the West by Victorius of Aquitaine.[19] One of its advantages was that it went some way toward reconciling the computus of the Alexandrian church with that of the early Roman church. According to Greswell, Georgius Syncellus used the 532-year period for the measurement of time from the Creation downward.[20]

There were many other calendar-cycles than those I have mentioned. Few were of greater potential historical use than the Julian Period—as it was named by Joseph Justus Scaliger, who recovered it from an earlier work by Roger of Hereford.[21] This period of 7980 years was a product of the solar cycle (28 years), the Metonic cycle (19 years), and the Roman Indiction cycle (15 years).[22] Every year of the Julian cycle is characterized by a different trio of numbers belonging to the three cycles, beginning (perhaps only theoretically) in 4714 B.C. and ending—or recurring, if the world survives—in A.D. 3266. The Julian Period has never been regarded as a Great Year in the Platonic sense of a period after which were reproduced the original positions of two, or more, and preferably all, of the planets, and perhaps their nodes and apogees as well. The calendar cycles to which I have referred were less comprehensive, being merely lunae-solar, but they did permit the Christian chronologist in principle to rectify the dating of the first Easter week, that is, to make it compatible with the chosen date of Creation.[23] I shall now consider very briefly some of the principal early references to Great Years in the Platonic sense, including eastern examples, before turning to medieval Europe, and ways in which Christian thought became reconciled to one or other *numerical value* for the periodicity, without accepting the underlying rationale of a universe with an ever-repeating history.

Plato gave no value for the length of the Great Year, although his editors have fostered on him such figures as 36,000 years, or 12,960,000 daily rotations.[9] There is no justification for any such ascription. The references to Philolaus and Oinopides by Censorinus, as already explained, yield a figure of 59

[19] Jones,[7] p. 64. The old version is that Victorius (Victorinus, according to Scaliger) derived the cycle from Arianus of Alexandria, a contemporary of Theophilus, and introduced it in A.D. 457. It is clearly unknown, however, either to Victorius or to Dionysius.

[20] Greswell,[8] p. 191, note. Syncellus, who was alive in 810, compiled his general account of Creation from *Jubilees*. He will be referred to on a number of occasions below.

[21] See Ussher's preface (*Lectori*), the third page, in the *Annales*.[2] Ussher has 'Robertus' rather than the correct 'Rogerus.'

[22] The last of these was a taxation period, introduced in A.D. 312 by the Emperor Constantine the Great.

[23] The precise date (within the year) of the world's first week must for this purpose be known. Opinions on this matter are discussed at pp. 328–329 below. The doctrine of the relation of the date of Creation and that of the Easter-moon first appeared in the writings of the African authors pseudo-Cyprian and Hilarianus. See Jones,[7] p. 39, and cf. p. 64 on Victorius.

years, and a reference to Democritus by the same author yields a rather puzzling 82 years.[24] Censorinus again is the source for a figure of 2484 years, drawn from Aristarchus; but Tannery has argued that this is a mistake for 2434 years, 45 times as great as a triple ἐξελιγμός, which is the period defined by Geminus as the shortest time containing a whole number of days, a whole number of lunations, and a whole number of anomalistic months.[25] Whatever the truth of Tannery's suggested correction, there seems to be no ground for thinking that Aristarchus' cycle is anything but lunae-solar.

This rather disappointing beginning to the career of so grand a conception was changed with the infusion of a rather ludicrous historical tradition which began, it seems, with the historian Berosus (fl. *c.* 290 B.C.). Berosus, priest of Bel, was one of those principally responsible for transmitting Babylonian history and astronomy to the Greek world, and his work began with the origins of the world and the Flood. He spoke of Babylonian astronomical observations carried out over a period of 490,000 years.[26] Epigenes of Byzantium increased this figure to 720,000 years, while Simplicius made it 1,440,000 years. All were treated with proper scepticism by Cicero, Diodorus, and later by Georgius Syncellus; but with the intrusion of Stoic ideas, what had originally been perhaps no more than an excessively long historical interval became transmuted into a value for the Great Year of astronomy. The Babylonians, moreover, were now imagined to have made astronomical observations since the commencement of a Great Year, and Palchos, in the fifth century, believed that they had observed in every climate and almost every day during that time! (Bouché-Leclerq,[11] p. 575).

Writers mentioning the concept 'Great Year' are almost invariably reluctant to show how an actual value for its length is arrived at, and this for obvious reasons. By the time of late antiquity, very precise parameters for the planetary motions were known, but this only made the discovery of a least common multiple of the periodicities more difficult. Macrobius, for example, in his commentary on Cicero's *Somnium Scipionis*, would obviously have liked to set forth the arguments, but in the end he leaves it that 'the philosophers' reckon the Great

---

[24]Paul Tannery suggested that this was a mistake for 77 years. See Heath,[9] p. 129.

[25]See Heath,[9] pp. 314–6. On the error in calling the basic period by the name of 'Saros,' see Neugebauer.[10] It has long been widely—but wrongly—assumed that period of 223 synodic months (= 242 anomalistic or draconitic months) was the basis of the prediction of eclipses by the Babylonians and early Greeks. (It is, of course, a genuine eclipse period.)

[26]Variant quotations from different sources give in addition figures of 470,000, 473,000, 468,000, and 432,000. The last two are from Syncellus, and the last is discussed again at p. 315 below. See, for further information, Bouché-Leclercq,[11] pp. 38–9, where figures of 300,000 years (Firmicius Maternus, 4th century A.D.) and 17,503,200 (Nicephorus of Constantinople, *c.* A.D. 750–829) are mentioned. Nicephorus wrote a work on chronology, translated into Latin by Anastasius the Librarian, and included in his *Chronologia tripartita*. Notes by Syncellus were added to it, and together they appear in Jacob Goar's edition (Paris, 1652), and its successor, in the series *Corpus Scriptorum Historiae Byzantinae,* 2 vols. (Weber, Bonn, 1829) under the title ΕΚΛΟΓΗ ΧΡΟΝΟΓΡΑΦΙΑΣ (Chronographiae). I quote Goar's Latin translation from the latter.

Year as 15,000 years.[27] This is no easier to justify than most of the figures quoted above. The 490,000 for example, involving a square of 7, a 'week of weeks', or ten thousand jubilees, must appear numerologically inspired, to those who know the symptoms. The 473,000 might just possibly be an approximation to the product of the square of the *exeligmos* 223 lunations, about 18 years, and the Sothis period of 1461 Egyptian years, but this is very unlikely.[28] One certainly cannot rule out the use of the Sothis period in connection with Great Years, for the figure 1461 is the length of the Sun's 'maximum year' in the Great Introduction (*Kitāb al-ulūf*) of Abū Maʻshar.[29] Yet another explanation which has been offered for one of the figures quoted involves the 18-year cycle. According to William Hales, writing in the early nineteenth century, the period of 432,000 years is 'Chaldean', and is made up of cycles of 18 and 24,000 years.[30] Hales makes it clear that he takes the longer period to be a 'Hindu' parameter for the precession of the equinoxes—or, as it should be called in this historical context, the 'movement of the eighth sphere.' An explanation with better historical credentials will be given later (see the paragraph after the next).

This brings us, however, to an unavoidable clouding of the very conception of a Great Year. Even a modern editor of Plato's *Republic*, J. Adam, has suggested that when Hipparchus found a figure of one degree per century for the movement of the eighth sphere, thus implying that the stars circuit the sky in 36,000 years, he was influenced by Plato's perfect number of the *Republic*, which he interpreted as the square of 3600.[31] Quite apart from the fact that Hipparchus merely set one degree per century as a lower limit to the phenomenon he had discovered, the empirical character of his result is not in doubt. It would also be surprising if Hipparchus had understood the *Republic* passage in question, since it seems to have been beyond the resources of most of Plato's editors to do so.

Later writers did introduce the movement of the eighth sphere into the periodization of world history, as I shall have occasion to show, and there is nothing intrinsically improbable about a medieval 24,000-year cycle explained along these lines, since this corresponds closely to the data accepted by several

[27]For an English translation see Stahl,[12] pp. 220–2. Macrobius is arguing that Cicero's statement to the effect that a man's reputation cannot endure for a single year refers to the Great Year of the *Timaeus*. Cicero was obviously more realistic than Macrobius.

[28]The idea is to be found in Hales,[13] Vol. 1, pp. 40–41. The Sothis period is the time taken for the seasons (e.g., the summer solstice) to return to the same civil date in the Egyptian calendar. The shift of date is due to the discrepancy of about a quarter of a day between the actual year and the 365-day Egyptian year. (We have $1460 \times 365\frac{1}{4} = 1461 \times 365$).

[29]The table in which this figure occurs was copied out by Ashenden, as I shall mention later. For the original context of this material, see Pingree,[14] p. 64.

[30]Hales,[13] Vol. 1, pp. 41–42. cf. n. 67 below.

[31]See Heath,[9] pp. 171–2. A better commentary on the whole subject is to be found in Cornford.[15]

Islamic astronomers. Even so, the explanation of the 432,000 offered in the twelfth book of the Sanskrit *Mahābhārata* and in the first book of the laws of Manu (both stemming from a source earlier than the second century A.D.) is no doubt the original one, and is very different (Pingree,[14] p. 28; van der Waerden[16]): There, a 'year of the gods' is equal to 360 ordinary years, while 12,000 of these, that is, 432,000 years, make up a *yuga* of the gods—called by later astronomers in India the *Mahayūga*, or Great Year. This in turn was divided into ten parts, in the ratio $4:3:2:1$, and the last part, the historical period in which we are now, on this system, living, was thus to be of duration 432,000 years—the period we find in the early ninth-century Western writer Georgius Syncellus. It might well be the case that different explanations were found at an early date for one and the same number. Priority certainly seems to belong to the system as it was found in India, where a *Kalpa* or 'day of Brahman' (containing a thousand *yugas*) is mentioned, according to Pingree, more than two centuries before Christ.[17] That it is a Babylonian system in origin seems to follow from the statement reported by Syncellus from Berosus, that the total length of the reigns of Babylonian kings before the Flood was 120 saroi, where the (true) saros is 3600 years.[32] This is the earliest implied explanation of the figure of 432,000 years, although it is worth observing that, sexagesimally written, the number becomes 2,0,0,0, which in itself seems almost to guarantee a Babylonian origin. The interval's occurrence in Berosus as a historical period commencing with the Creation of the world makes it proper—if that is the right word—that it should have passed into Islamic, Byzantine, and Western chronological reckoning.

As an example of the way in which the bare bones of this now Indian and Iranian doctrine crept surreptitiously westward across half the world, we might first consider Pingree's masterly study of the *Kitāb al-ulūf* (Book of the Thousands)[33] by Abū Ma'shar (A.D. 787-886). The sources for this book were held to have gone back to antediluvian times, and it was supposedly written on the basis of written records of 'cycles of the Persians' discovered in a hole in

---

[32] In Goar's Latin version (*ed. cit.*, footnote 26 above), Syncellus writes: Sed et Berosus Saris, Neris et Sossis annorum numerum composuerit, quorum quidem Sarus ter millium et sexcentorum, Nerus sexcentorum, Sossus annorum sexaginta spatium complectitur. Saros autem centum et viginti, decem regum aetate, hoc est myriadum quadraginta trium et duorum millium annorum collegit summam Annos porro illos Historicorum nonnulli dies coniiciunt, ex quo Eusebius Pamphili causati, qui Sarrorum annos dies esse non animadvertit. Nullo tamen consilio inscitiae accusant. Qui namque quod non erat, vir alioquin eruditus, assereret? Vir ille, dico, qui Graecorum sententiam, plure saecula, annorumque portentosas myriadas ex fabulosa Zodiaci per partes adversas in idem signum conversione, ab arietis, inquam, termino ad eandem metam revolutione, iam a mundi natalibus praeteriisse asserentem probe dignosceret? Qua vero necessitate pressi mendacium veritati coacervare excogitaverunt?

[33] Pingree.[14] The work is lost, but summaries by other writers allowed Pingree to reconstruct most of the argument of the work, which is conjectured to belong to the period A.D. 840-860.

Isfahan, where they had been preserved from natural disaster from the time of the ancient kings of Persia (Pingree,[41] pp. 2-4). This is all reminiscent of Plato's reference in *Timaeus* to the antediluvian wisdom of the Egyptians.[34] In a *zīj* (a set of astronomical tables) by the same author, which again purported to transmit to posterity the astronomy of the prophetic age, some of the methods are of Indian origin, and introduce the *yuga*. Certain planetary parameters (in the *zīj*) were received indirectly from Persian sources, while the planetary models were Ptolemaic. On the score of antiquity, the most one can say of the writings supposedly recovered from the Isfahan hole is that they were certainly older than Piltdown man.

At the root of Abū Maʿshar's astronomical system, that is, of a system which was essentially Indian, was an assumption that the planets are in (mean) conjunction at the first point of Aries at regular intervals of time. Of four Indian methods he could have followed,[35] he takes one which assumes grand conjunctions of the mean planets at Aries 0° in 183,102 B.C.; at midnight of Thursday/Friday, 17/18 February 3102 B.C., the epoch of the Flood; and in A.D. 176,899. The *yuga* is now clearly one of 180,000 years. Most of the planetary parameters come from the *Sindhind*.

Of the three other Indian systems, two share with that followed by Abū Maʿshar a Flood date of 17 or 18 February 3102 B.C.[36] For the date of the Creation, al-Hāshimī tells us that Abū Maʿshar took this (in effect) as 4693 B.C., in his Book of the Thousands. In the *Kitāb al-qirānāt* (The Book of Conjunctions), however, Abū Maʿshar sets 2226 years, 1 month, 23 days, and 4 hours between the birth of Adam and the Flood, yielding a Creation date of 5328 B.C.[37] Dates coinciding with these, or near to them, appear in western writings, either borrowed or from a common source. As Pingree explains of the *Kitāb al-ulūf*, "there is an overwhelming mass of religious references which clearly belong to an Islamic milieu, not a Zoroastrian one," and that "though some of his astrological material may go back to a Sasanian source through the unknown scholar of the time of Hārun al-Rashīd, he has largely revised it to conform to the conditions of his own age" (Pingree, [14] p. 58). It is not surprising, therefore, that at least some Jewish and Christian authors should have come close to sharing epochs with Abū Maʿshar. From the Septuagint, Eusebius, Augustine, Isidore, and Bede we find from the Creation to the Flood 2242 years, only sixteen years or so removed from Abū Maʿshar's figure of the *Kitāb al-qirānat*. The discrepancy is

---

[34] See the long tale about ancient cultures by Critias near the beginning.

[35] Pingree,[14] pp. 28–9. The other three methods were known to the Arabs as those of the *Sindhind, al-Arkand,* and *al-Arjabhar.*

[36] There are problems over the precise day of the Flood. Most writers took the Thursday, and Abū Maʾshar is said to have been alone in taking Friday. See Pingree,[14] p. 38.

[37] Pingree,[14] p. 38. It seems that 4693 B.C. is the beginning of the second century of the millennium of the Sun, according to Māshāʾallāh's chronology. It is close to a figure for the Creation (4698 B.C.) abstracted by some commentators from Josephus' history of the Jews, and to a figure (4697 B.C.) in Magnus Aurelius Cassiodorus. But no one should ever pin an argument on Josephus, whose text is very corrupt and inconsistent.

easily explained in terms of the different approaches to the problem, the scriptural approach (by generations), and the astronomical (by conjunctions), which we must suppose was aiming to replicate it. As for the inconsistent Creation-to-Flood interval of 1591 years, from 4693 B.C. to 3102 B.C., this fits no well-known source very well. If a comparison with Josephus—of the early writers—seems profitable, this is probably for the wrong reason (see footnote 37). When we consider the Indian Flood-date of 3102 B.C., however, we find it frequently mentioned in Western astronomical works, which it apparently enters not only through the writings of Abū Ma'shar himself, but through the astronomical tables of al-Khwārizmī (c. A.D. 820) in the version of the Spanish astronomer Maslama al-Majrītī (c. A.D. 1000). Adelard of Bath put this into Latin early in the twelfth century.[38] Essentially the same table, slightly expanded but much more convenient to use, occurs in the astronomical tables produced for Alfonso X of Castile (c. 1272). There the intervals between eleven different eras are given in days (decimally as well as sexagesimally expressed).[39] The era of the flood is 17 February 3102 B.C. Time and again one finds this particular date quoted from the Alfonsine work, and combined with one of the several scriptural Creation-Flood intervals, thus resulting in a rather foolish blend of incompatible elements, but one unlikely to offend readers ignorant of its origins.

One author of such a blend was John Ashenden, the fourteenth-century astrologer of Merton College, Oxford. In an exceedingly long *summa* of judicial astrology, running to 375,000 words or so, Ashenden devotes perhaps 20,000 words to the age of the world and the time of year of its creation.[40] He is especially indebted to Abū Ma'shar's work on great conjunctions (but not *The Thousands*), and he is even conscious of the infusion of ideas from India. He quotes a work (no longer identifiable) by Walter of Odington (a monk of Evesham of the previous generation) on the age of the world, to the effect that philosophers, and Indians in particular, maintain that all planets are at the head of Aries at their creation.[41] Over the discrepancies between his principal authorities for the age of the world he is far from complacent, but rather than offer a critical account of his own he is content to produce a muddled digest of the works of others—

[38]The tables were edited by Suter.[18] See especially p. 109. Neugebauer[19] has since translated them into English and supplemented them. See especially pp. 82–4.

[39]Judging by the copy in the Bodleian Library,[20] the table is completely accurate, and (although in a different form) consistent with the Khwārizmī table. The only additional material is an era 'last of the Persian Kings,' 27 August 274, and of course the Alfonsine era, 1 June 1252.

[40]The first part was written in 1347, as is mentioned in the first chapter of the first *distinctio*. The whole was later printed under the name of Johannes Eschuid (1489).[21] All references given below to this work are to the better manuscript copy in Oriel College.[22] A general idea of the contents of the *Summa* can be had from Thorndike,[23] Vol. iii, Ch. 21, where references are also made to the MSS (Appendix 20 and pp. 329–34).

[41]MS Oriel 23, f. 3v. Ashenden goes on to report that the Jews and Christians differ in their interpretation of what comprises 'first moon.' Adam could not have seen its first conjunction, . . . , and so on.

among which may be mentioned the Septuagint and Hebrew Bible, Josephus, Orosius, Methodius, Eusebius, Jerome, Augustine, Julius Firmicus Maternus, Martinaus Capella, Isidore, Bede, Rabanus Maurus, Helperic, Gerlandus, and such 'moderns' as Grosseteste, Bacon, and Robert of Leicester.

Ashenden's chronology is inevitably cut down to essentials. There is no question of his listing the generations very systematically. He speaks in rather general terms when he divides history up into the 'seven ages of the world,' the first from the Creation to the Flood, the second from thence to the birth of Abraham, and so on. Touching lightly on the discrepancies among his many sources, he quotes Walter of Odington, who had said that there is nothing reprehensible about accusing an evangelist of lying.[42] This remark is a necessary preliminary to the act of collecting epochs from as many sources as possible, with a view to reconciling them. The Evesham monk's lost work clearly supplied Ashenden with the greater part of his material, and it is perhaps significant that Walter of Odington was, like him, an astronomer.(24) The work must have been a morass of genealogical comparisons and it is not difficult to find even among the products of that favorite nineteenth-century pastime of compiling from different authorities lists of biblical dates (of the Creation, the Flood, and so forth) signs of Walter of Odington's hand, doubtless known through Ashenden.[43] Some scores of dates are recorded, which would indeed have been more convenient if tabulated. Scriptural dates given for the Creation according to the authorities quoted vary from 3752 B.C. to 5530 B.C. When he comes to the astronomers, he complains that Albumazar is the only one whose book (on great conjunctions) he has seen which calculates the beginning of the world.[44] The others go no further back than the Flood. This is not surprising since events in their world histories were usually heralded by 'great conjunctions' of Saturn and Jupiter, conjunctions which could hardly take place before the Creation of the heavens. From Abū Ma'shar, Ashenden quotes the figure of 2226 years, 1 month, 23 days, 4 hours already mentioned for the Creation-Flood interval.[45] To this he adds the Indo-Alfonsine 3101 years, 318 days, and obtains for the period from the Creation to Christ's birth, 5328 years, 16 days, 8 hours, 30 minutes![46] This,

---

[42] MS Oriel 23, f. 6v (I.i.1): Sed dicit Odynton quos fas non est dicere vel existimare quemquam evangelistum fuisse mentitum et allegat hoc ab Augustino.

[43] In Hales,(13) the name is simply rendered 'Odeaton astrologus.' In the printed edition of Ashenden ('Eschuid') the name is almost unrecognizable as 'eduiconien'! The printed edition is sometimes described as accurate, but there are scores of mistakes on every page, many of them numerical.

[44] See the discussion of this work at p. 324 below.

[45] MS Oriel, 23 f. 9r and 9v (I.i.2), has both 22 days and 33 days, and it is possible that the discrepancy stems from the author's own carelessness.

[46] The month has 30 days. Carrying over $365\frac{1}{4}$ days as a year (as he seems to do) should surely leave 5 days, 22 hours. There is discussion of a calculation involving 6 days (rather than 16) in I.i.3, but this is scarcely worth discussing here. (See MS Oriel 23, f. 13r.) The precision to which Ashenden pretends is even more ludicrous in the light of the ten errors he warns against in I.i.3. There include the confusion of different calendars and years, as well as the acceptance of spurious precision in chronological statements.

he wistfully complains, differs from all authorities; and yet he sets great store by the precision and truth of the Alfonsine interval, while he is reasonably satisfied that the other interval is not far removed from the 2242 of Eusebius, Bede, and so many others.

If it is difficult for us now to feel any sympathy with works of synthesis like John Ashenden's, this is perhaps because the precise timing *per se* of a historical event does not seem to us to be a particularly significant characteristic of it. Accuracy is a minimum requirement of history. It may have prevented error in the ordering of events, but it did not provide an understanding of the pattern which some commentators, at least, were seeking. It was all very well for Robert Grosseteste to protest that men should stop querying *Genesis* on the subject of Creation.[47] (A bishop, Synesius, had argued for the eternity of the world and the preexistence of the soul, but that was in the fifth century.[48] Nearer home was Daniel of Morley, denying the eternity of the world, and yet in the next breath saying that God created everything simultaneously.) Grosseteste was no doubt right to protest. But in the protesting, as he also did, at doubts about the age of the world, he underestimated the obvious pleasure that was to be had in using mathematics in sacred chronology. It was the sort of pleasure some people now get from electronic calculators; and Roger Bacon had no doubts about the spiritual value of such mathematical activity—witness many passages of the *Opus maius*. This activity went wildly astray, however, once the groundplan of sacred history was established—not uniquely, but as one of a handful of possible systems, reconciled as far as possible in the Ashenden manner. It went astray because it fell into the hands of scholars with a taste for Great Years, Great Conjunctions, and a number of similar astronomical designs for the introduction of structure into history.

Perhaps the most pervasive of these designs was that which concerned conjunctions of the two most distant planets then known, Jupiter and Saturn. I have described at length elsewhere how the history of religions and sects was supposedly governed by so-called 'great conjunctions' of these planets, and how past religious change was thus interpreted, and how prophecies were derived.[(26)] I have touched upon some of the consequences of this doctrine already. It owed much to the philosopher al-Kindī, in the work of whose pupil Abū Ma'shar it was chiefly introduced to Christian Europe. Three signs of the zodiac symmetrically placed (i.e., each separated from the next by three signs) are called a *triplicity*. By chance the mean periods of Jupiter and Saturn are such that twelve successive conjunctions of those planets (conjunctions which occur at intervals

[47]Et sic desinant admirari cur mundus non sit antiquior, quam dicit scriptura, et cur non prius incepit quam dicit scriptura, quia non potest intelligi incepisse prius, quam incepit, ab intellectu comprehendente totum tempus praeteritum terminatum, sicut non potest intelligi mundum alibi esse, quam sit, ab intellectu, qui comprehendit extra mundum non esse spatium, cum tamen necessarium sit, ipsum posse esse alibi, quam sit, apud imaginationem ponentem spatium extra mundum.[(25)]

[48]There were writers who were not above suggesting that Aristotle himself had denied the eternity of the world.

of about 20 years) may all occur within a single triplicity. They will be in other triplicities for similar periods of time, about 240 years in each, and after approximately 960 years conjunctions will return to somewhere near the place of the first. There are certain differences in terminology between Abū Ma'shar and the second most important source of the doctrine, Māshallāh, but these are unimportant in comparison with the fact that chronologists were provided with three time intervals to conjure with: 20 years, 240 years, and 960 years.

A change of triplicity at a great conjunction might signify the end of the world, but, as I have already pointed out, there could be no such event to herald its creation. According to Māshallāh, the conjunction of 3321 B.C. (with change of triplicity) had indicated the Flood, and another conjunction of the same planets had taken place in 3301 B.C., the year in which he believed the Flood had occurred.[49] (And lest this date seem less important than 1776, I will remind you that in 1776 America was separated only from England, whereas at the time of the Flood, according to the Premonstratensian Francois Plaçet, America was separated from the rest of the world.[(28)]) Extrapolating back to the Creation from the Flood by astronomical means was done in various ways, all of them somewhat reminiscent of the Great Year scheme. The Indo-Persian versions, with their enormous time scales, did not find a ready acceptance in the West. An interval of 180,000 years from Creation to the Flood was, after all, utterly irreconcilable with the Old Testament.[50] From an era 180,000 years before the Flood-Kaliyuga date (17 February 3102 B.C.)[51] a variety of world-periods had been reckoned: the *tasyīrāt*, *intichā'āt*, and *fardārāt*. Kennedy lists eleven different sources, all Persian or having a strong Persian connection.[52] The periods of the maximum, great, medium, and small *tasyīrāt* are (for one revolution) 360,000, 36,000, 3600, and 360 years, respectively. The four grades of *intihā'āt* move through a zodiacal sign in 12,000, 1200, 120, and 12 years, respectively. There seem to be several variant systems of *fardārāt*. One *fardār* listed by Kennedy, for example, is of 75 years duration, and assigns a different number of years to each planet (including the lunar nodes). This system was known in the west through Abū Ma'shar's *Great Introduction*, and a table was copied by John Ashenden[53] from this work listing five sorts of period, all made up of lesser

[49] Kennedy.[(27)] This is an important contribution to a very extensive subject. Kennedy quotes several other suggested spacings between great conjunction and Flood (pp. 25–6). Ashenden (I.i.3; MS Oriel 23, f. 12*v*) discusses several different opinions on the timing of Flood and conjunction.

[50] According to Kennedy,[(27)] p. 24, this was ascribed to the people of Pārs in Persia, in a Persian zīj.

[51] Misprinted 1302 in Kennedy,[(27)] p. 27.

[52] See Kennedy,[(27)] pp. 26–30 for the source of this and most of the following information.

[53] MS Oriel 23, f. 7*v* (Ashenden's *Summa*, I.ii.3 and not I.ii.2, as in the printed edition). The table as included in the Latin Albumazar is accompanied by an utterly unintelligible explanation, and in consequence John Ashenden's version can make little sense to an uninformed reader. See Ref. 29. The table in the printed edition does not agree with the MSS I have seen, but I have not pursued the matter.

periods allocated to the planets, under the names *anni fardarie* (total 75 years), *anni maximi* (4588 years), *anni maiores* (588 years), *anni medii* (298½ years), and *anni minores* (129 years). The *anni maximi* listed correspond precisely with what Pingree calls the 'mighty years' of the *Kitāb al-ulūf*.[54]

There is not much consistency between the Ashenden–Albumasar table and the reports by Kennedy and Pingree, and there is little to be gained by pursuing the matter here. Western scholars were dimly aware of the Indo-Persian eras, but there are few signs, if any, that the underlying theory of planetary longitude was understood. The West had its own thousands, not overtly astronomical, but capable of being made so. The world had been created in six days, and this the apocalyptic writers took as an indication that the world would endure for six ages, each of a thousand years, before at last came the Day of Judgement, to be followed by a millennium of rest.[55] A thousand years is one day "in the testimony of the heavens" (*Jubilees*, 4:30); and one day "is with the Lord as a thousand years" (II *Peter*, 3:8). As a measure of the influence of this mode of interpreting, perhaps it is not without relevance that Creation dates tend to be brought forward so as to place the Day of Judgement in the future while preserving the assumption that the world's duration will be 6000 years. In other words, the later an author, the later he is inclined to date the Creation. (Most Jewish chronologists have been exceptions. They have always left an excessive margin of safety.)

Bouché-Leclerq[11] (p. 499) writes of a system, claimed as Tuscan but found again in Mazdean cosmogony, in which each sign of the zodiac was supposed to rule the world for a thousand years. Gayomart (the first man, in ancient Persian mythology) was born under Taurus, Adam and Eve under Cancer (horoscope of the world), and so on. Kennedy[27] (pp. 37–38) summarizes the evidence for the same system, which in Iran goes back to Sasanian times, and possibly the fifth century B.C. The 12,000 years is apparently never presented as a *cyclical* period. The 'mighty *intihā*' takes a thousand years to move through a sign, but there is considerable uncertainty about the point in time from which the Persian thousands were to be reckoned, and different chronologists took it in different ways. It is possible that several versions of this scheme, together with that rare thing, a comprehensible explanation, found their way into the European tradition. On the whole, the works of Abū Ma'shar, as they appear in Latin translation, are too obscure on the points at issue to have been very influential. The difficulty of recognizing the doctrine is that thousands, and high powers generally, are popular in all cultures as an expression of immensity in space and time alike. (Need I do more than recall the fascination of the number 10^{39}?) From Bardesanes, last of the great Gnostics, who put the duration of the world at 6000 years,[56] to

---

[54] Pingree,[14] p. 64; see footnote 29 above.

[55] For an elaboration of this point, see Robbins,[30] p. 27.

[56] He supposed that in 60 years the planets made an integral number of revolutions. Thorndike,[23] vol. i, p. 376.

the astrologer who either did or did not have to do with the dark lady of the Sonnets, namely Simon Forman,[57] a scheme with thousands was always at a premium. And of course it remains so in some religious circles to our own day.

If Christendom was not anxious to accept the Great Years of the Greek and Indo-Persian worlds, there were plenty of reasons. Many a writer touched on the 36,000-year period of the eighth sphere—and regarded it as a Great Year—with impunity. Alexander Neckham, William of Auvergne, Sacrobosco, and Bartholomaeus Anglicus all did so, for example, but not to make any doctrinal use of it. Others saw the same spiritual danger of repetition in history as Aguustine had written against, and the objection was codified in 1277, when Étienne Tempier, bishop of Paris, made as the sixth of the famous 219 condemned opinions:

> That when all the celestial bodies return to the same point, which happens every 36,000 years, the same effects will recur as now.[58]

There were subsequently many voices raised against the idea, with, for instance, Nicole Oresme—in the footsteps of Haly—suggesting its illogicality, and Pico della Mirandola that it was irreligious. This did not prevent the movement of the eighth sphere from being used as a guide to the age of the world. John Ashenden referred to its use. With his customary facility for introducing something of everything, he quoted a period of 300,000 years (from 'Hermes Book 2' and from 'Julius Firmicus Book 3') and also a period of 535 years, as a Great Year in the original Platonic sense, ascribing it to 'Johannes Rocensis' (John of Rupescissa, or Roquetaillade?)[59] This 535 is obviously an error for 532, the product of 19 and 28.

Two centuries later the game was still being played according to similar rules. Pierre Turrel was an influential French provincial astrologer of the early sixteenth century, who in a printed work of 1525, or thereabouts, considered the duration and end of the world, as reckoned in four different ways.[60] The first was by the motion known as the trepidation of the eighth sphere, a somewhat complex movement according to which the motion of the sphere of stars with respect to the quinoxes was doubly periodic. Without going into details, it should be enough to say that in the Alfonsine theory, which Turrel was clearly using, a secular movement—equivalent to a revolution of the sky in 49,000

[57]There is nothing very new or remarkable about Forman's system, to which I shall refer again below.

[58]For this and others of the articles with a bearing on astrology, see Thorndike,[(23)] vol. ii, pp. 710–12.

[59]The printed edition has 'Cretensis' and 525 years. The manuscript reads: Et mag. Johannes Rocensis dicit quod post 535 annos omnes stelle et omnes planete ad eundem punctum redeunt, et per easdem lineas ut prius vadunt (ibid. f. 15r).

[60]The work is of great rarity, and I follow Thorndike,[(23)] vol. v, pp. 310–11, who was himself describing the work at second hand. It is said to have been from a translation by Turrel of a Latin work in the monastery des trois Valées. On the history of its printing, see Thorndike,[(23)] vol. v, p. 310, n. 10.

years—was superimposed on a periodic (trepidational) movement with a period of 7000 years. Turrel maintained that there were four stations, one at the end of each quarter (i.e., at 1750-year intervals), marked by the Flood, Exodus, destruction of Jerusalem, and the end of the world. Although I have not seen the work, it seems that Turrel exchanged the Alfonsine radix of A.D. 15 for the date of the destruction of Jerusalem by Titus, namely A.D. 70. The date of Creation thus astronomically derived was presumably 5181 B.C., and that of the Flood 3431 B.C.[61] The first date was reasonably close to figures given by scholars of the early Church, while the Septuagint puts the Flood in or around 3246 B.C., so Turrel's sleight of hand was quite as good as that of his twentieth-century successors in cosmology.

According to Thorndike, Turrel's method had previously been advocated by Jean de Bruges, in a work written in 1444 under the title *De veritate astronomiae* (Thorndike,[23] Vol. V, p. 311, n. 11). There are reasons for thinking that this fifteenth-century work might stem in part from John Ashenden's *Summa*, and it is no surprise to find that Ashenden refers to the theory of trepidation, with a period which in his case was of 640 years, a period which certain unnamed individuals are said to have used as a Great Year. This period was supposedly derived from the theory of trepidation of Thābit ibn Qurra, an earlier and marginally simpler theory than the Alfonsine. The details are clearly ill understood by Ashenden, although interesting in themselves.[62] No one could possibly have used this passage of the *Summa astrologie* to learn how to apply Thābit's theory of trepidation. Ashenden, furthermore, fails to show how the anonymous 'others' to whom he refers actually arrived at a Creation date on the basis of the theory; but the Turrel method was obviously intended. Yet again we find, when looking for the source of the idea that trepidation could be applied to world history, that Albumasar is in the West almost always at the end of the trail. The references given in Ashenden are to his *De magnis conjunctionibus*, II.8 and to the work *De vetula*, lib. III, which Ashenden—like many, but not all, of his contemporaries—thought to have been written by Ovid. In fact, this work also alludes to the doctrine of great conjunctions in a form which we know stems from Abū Ma'shar.

When we compare Albumasar's work with Ashenden's we find that the relevant passage (about 700 words in all) has been copied out almost verbatim—if that is the right description of a copy in which almost every sentence is misconstrued, and in which the numerical data are almost totally at variance.[63] Two interesting points are worth noticing: Ashenden interpolates the name of Thābit,

---

[61] These figures do not tally with Thorndike's remark that Turrel (*c.* 1525) believed the world to have 270 years before its dissolution. This would put Creation at c. 5206 B.C.

[62] MS. Oriel 23, f. 15*r* and *v.*

[63] The printed edition I have used for the *De magnis conjunctionibus* (1515)[31] is none too perfect. See sig. C viii (*r* and *v*).

which does not appear in those copies of Albumasar I have seen. (The absence of the name is not surprising, since the parameters quoted do not appear to be Thābit's.) And second, the era named is not the Hejira, but that of Yazdigerd, about ten years later. The Latin work by Thābit on the movement of the eighth sphere is based on the former era, and if Ashenden had read more carefully one of his principal sources, Walter of Odington, he would have known as much.[24]

Pierre Turrel's borrowing from Albumasar, direct or indirect, did not stop at the theory of trepidation (access and recess). Not only does he reproduce the doctrine of great conjunctions (with a period of 240 years within a given triplicity), but he makes use of Albumasar's division of history into 300-year periods, one such interval being the approximate time for Saturn to circuit the sky ten times.[64] As a typical example of Albumasar's chronological use of this there is the interval between Alexander the Great and the 'son of Mary.' (The conventional era of Alexander was in 312 B.C.)

The idea of a Great Year, manifesting itself as a motion of the eighth sphere,[65] had in astrological quarters by the seventeenth century become a well-established dogma. Nicole Oresme, Pico della Mirandola, and Francesco Piccolomini were three powerful opponents of the idea, and yet none of the three can be considered a complete sceptic as to all astrological influence. A powerful spokesman on the other side had been Petro d'Abano, from whom, in all probability, Pierre Turrel took a fourth period—one of 354 years and four lunar months, for which time each of the planets was in turn to govern the world.[66] Petro d'Abano proposed a number of different astronomico-chronological schemes, and was particularly fond of a (steady) movement of the eighth sphere, to the periods of which he half-heartedly attempted to reconcile the classical chronologists.[67] Through his writings, some readers at least must have become dimly aware of the vast chronological cycles of the 'Indians' (*Indi*), for in one place he mentions "the beginning of the movement on Sunday, at sunrise, 1,974,346,290 Persian years having passed up to the present."[32] As an instance of Petro d'Abano's influence, we find the Elizabethan astrologer Simon Forman writing a short piece which comes to us among a collection of apocryphal tracts and genealogies, much concerning the book of *Genesis*. The item in question places the Creation at 3948 B.C.,[68] and the duration of the worlds as 6000 years, divided into twelve periods. During the course of world history, man's expecta-

---

[64] *De magnis conjunctionibus* II.8,[31] f. C.8r. There seems to have been no attempt to make the different world-periods commensurate.

[65] Only after Copernicus was this accepted as a precessional movement of the equinoxes, in the modern sense.

[66] A year of twelve lunar months, each 29 19/36 days, is equal to $354\frac{1}{3}$ days. The quoted period is, as it were, a lunar year of years.

[67] Bede, Septuagint, Abraham Judaeus, and Josephus, for example. As the duration of a complete cycle of the stars he cites al-Battānī (23,760 years) and 'Azolphi' (25,200 years).

[68] His great near-contemporary, Joseph Justus Scaliger, gave 3950 B.C.

tion of life was held to fall steadily, in a way which—however ludicrous—appears less arbitrary when located in Petro d'Abano's works *Lucidator* and *Conciliator*.[69] There the notion is linked with that of a golden age which gave way by degrees to a degenerate modern world, the diminishing life span of man playing its part as a symptom of the decline. There had always been those who, in comparing macrocosm and microcosm, wished to maintain that the ages of the world can parallel the ages of man, and who thought that when the senility of the world was evident in all secular affairs, then the last hour of the last age must be at hand. One did not have to be a seer or an astronomer to grasp the broad principle that the world had spent most of its force. In the words of C.W. Jones[7] (p. 133):

> the traditional chronology had combined in the popular mind with Augustine's doctrine of the Six Ages of the World, ably spread by Isidore, to create a millennial dogma which became a pseudo-scholarly fetish. To be sure, both Augustine and Isidore plainly stated that there was no predictable millennium. 'The end of the Sixth Age is known to God alone,' they repeated. But the pseudo-scientific saw that if five ages had lasted approximately five thousand years, the Sixth Age would last a thousand.

In the end, the habit died away as people tired of unfulfilled prophecies of doom, much as people tired of listening to the boy who cried wolf.

An argument concerning the age of the world highly reminiscent of that from the eighth sphere was put forward by a number of writers, the most notable of whom was Johannes Kepler. It was proposed that the solar apogee, at the time of the Creation, was at the head of Aries. Its rate of movement was known to Kepler, as well as its current position. From these two data he concluded that the Creation had taken place in 3993 B.C., at the summer solstice (the Sun then being at the head of Cancer).[70] By the same argument Longomontanus had five years previously found a Creation date of 3964 B.C.[71] Lesser intellects were grateful for yet another easy passage into the secrets of the divine strategy.

[69] Bodleian MS Ashmole 802, f.86, beginning "The wordle [*sic*] is divided into 3 partes. The firste from the Creation unto Noah his flod and after then untille 2000 years . . ." The age to which a man might live was supposed to reduce in steps from 1200 years ("and that begane in the head of [Sagittarius] ") to a mere 75. The top of the page has been torn away, and this might have given some clue as to what 'it' was which was supposed to move steadily round the zodiac, starting at the head of Sagittarius at the Creation, occupying each sign for 500 years, and finishing at the end of Scorpio at the end of the world. Anno Mundi 6000.

[70] Keplerus,[34] p. 42 (second pagination, that of the tables themselves). See the foot of the table 'Epochae seu radices': Ante Christum Anno 3993. die 24. Iulii, H.0.33'.26". Medius [Solis] 0.0'.0" [Canc.] Apog. 0.0'.0". [Arietis] . The clipped style (of which this is a sample) was later copied by Hevelius in his *Prodromus astronomiae*,[35] to the same effect. Kepler put the annual motion of the line of apsides as 1'2".

[71] Longomontanus,[36] put the annual motion of the line of apsides at 1'1" 54'" 5ᴵⱽ, a figure which would ensure that Christ's Passion was at 4000 A.M. and that Tycho Brahe's 'accuratissimi observationi' of the apsides for A.D. 1588 (95° 30') were correct.

Henry Power, for example, Yorkshire physician and early fellow of the Royal Society, was to write "An Essay, to prove the World's Duration, from the slow motion of the Sun's *Apogaeum*, or the Earth's *Aphelion*"[72] : but his proof was in reality that of Longomontanus, and occupied scarcely a dozen lines. He confessed that "We take it for granted, from the Scripture Account, that the world is about 5000 years old," and that the Sun's apogee at the Creation was at the head of Aries. Since these two assumptions were compatible with its present position and motion, that is, since the astronomical calculation "draws nigh to the Scripture Account," it was assumed that all were acceptable. Power now went further, and suggested that "in all likelihood, he that made this great Automaton of the world will not destroy it, till the slowest Motion therein has made one Revolution." This, he thought, would take about another 15,000 years. The word 'prove' in the title of the essay was obviously capable of many shades of meaning in Henry Power's time.

I have no idea where the Keplerian idea began, but there is little doubt that he and Longomontanus were writing in a tradition which took in Georg Joachim Rheticus, protégé and spokesman for Copernicus. A thoroughgoing astrologer, Rheticus had managed to work one or two astrological passages into the *Narratio Prima*. The Copernican system has a changing eccentricity and direction of solar apogee, and Rheticus related the rise and fall of kingdoms and faiths to the changes. "We look forward to the coming of our Lord Jesus Christ," he wrote, "when the center of the eccentric reaches the other boundary of mean value, for it was in that position at the creation of the world."[73] Rheticus went on to say that his calculation was in close agreement with Elijah's divinely inspired prophesy that the world would endure only 6000 years. He should perhaps have said, more precisely, that the implied 'Copernican' duration of the world was 6868 Egyptian years (of 365 days), assuming two full revolutions of the mean Sun on its small central circle.[74]

Bouché-Leclercq has shown that after the invention of eccentric and epicycle, in the ancient world, astrology adapted itself quickly to the concept of apogee at which a planet reached its peak of influence.[75] Nor for many centuries to come was it realized, however, that the apogees of the planets moved (apart from movement with the eighth sphere of stars as a whole), and when we en-

---

[72] Power[37] placed the apogee at "about the sixth degree of *Cancer*," and the motion per century as $1° \ 42' \ 33''$. There is a manuscript version among the Sloane MSS at the British Museum.

[73] Quoted from the Translation of the *Narratio Prima* (first printed in Danzig, 1540) by Rosen.[38]

[74] Rosen suggests the *Babylonian Talmud* as the source of the Elijah reference (Rosen,[38] n. 56). The equivalent period (two revolutions) in Julian reckoning is approximately 6863.3 years. Philip Lansberg, writing at the same time as Kepler, made the double period of the eccentric anomaly *exactly* 6000 Julian years, to conform with conventional chronology!

[75] Bouché-Leclercq,[11] p. 194, quotes Cleomedes (first century A.D.) and Theon of Smyrna (second century).

counter any doctrine of what might be called a variable 'physical' influence in medieval astrology, it usually involves the position of the planet on the epicycle, or the position of Mercury's deferent center on its small auxiliary circle. Perhaps this was the source of Rheticus' inspiration. At all events, his idea caught the imagination of sixteenth- and seventeenth-century astrologers, and it would not be difficult to name half a dozen contemporaries of Newton who subscribed to it. The last of any note, so far as I am aware, was Henri de Boulainvilliers, Comte de Saint-Saire, whose *Histoire du mouvement de l'apogée du soleil* was written in 1711.[39] Of Boulainvilliers, Feller's *Biographie Universelle* holds that, "il n'en voyait les événements qu'a travers le prisme de son imagination." While not quarreling with this judgement, I would like to point out that his imagination did not stretch to altering much the words Rheticus had written on the subject, 171 years earlier.

Throughout the long history of 'rational' attempts to solve the problem of the time of Creation, few writers appear to have had a very keen awareness of the difference between an argument based—even in part—on empirical premises and one based wholly on convention. The matter might have been clearer, had human and divine conventions been distinguished. The seventeenth-century Wittenberg theologian Aegidius Strauchius gives us an idea of what part a typical chronologist of the time thought God was playing in the arrangement of the cycles of the calendar. At the beginning of the fourth book of his *Breviarium chronologicum*,[40],[77] after paying his respects to Holy Writ and profane histories alike, he wrote that it seemed probable that God had deliberately so arranged matters that the first year of the world was the first also of the Sabbatic and Jubilean cycles (*anni Sabbatici et Jobelaei*); and that on the first day, the Sun was either at an equinoctial or solsticial point. (Strauchius himself favored the Autumn.) Another hypothesis proposed by the best chronologists, namely that the Moon was placed so as to show one of her principal phases immediately after the Creation, ought—said Strauchius—to be considered carefully; and the same attention should be given to the common belief that the hebdomadal cycle has continued since the first week of the world.

So much for God's design. But as for such arguments as those of Longomontanus, these were clearly now supposed figments of the minds of men. The apogee, Strauchius says, is a mere imaginary point, invented in order to save the celestial appearances, and the best astronomers are of the opinion that we know too little of it to make it the basis of a firm opinion (Strauchius,[40] p. 296). (This was a trifle disingenuous, since his contemporaries were quoting the movement, and by implication claiming an accuracy of one part in a quarter of a million.) Alternatively, as he might well have said, "let us stick to the week, the bare essentials of the astronomy of the sphere, and those lunae-solar cycles which allow us to correlate the phases of the Moon with the solar calendar." Of the voluminous effusions from the pens of biblical chronologists down the ages, by

[76] My list omits some of the points Strauchius raises.

far the greater part are uneasy with methods falling outside the terms of this description. For ourselves, I hope we are all of a mind in dismissing all inference to the Creation on the basis of 'cyclical' relationships, whether of apogee or Easter, as equally delusive.

For an early example of a way in which the precise date of the Creation became a matter for scholarly debate, we need only consider the problem of Easter, a problem which has obsessed the Christian Church almost from its foundation. Hippolytus, the third-century bishop and martyr, had mistakenly supposed that after 16 years the moons recurred in order in the Julian calendar.[77] In the year 243, under the name of Cyprian, a computistical doctrine appeared according to which the earliest Paschal full moon fell on March 18. The argument for this date is intricate, but depends on the premises that the first day of Creation was March 25, the supposed (Julian) spring equinox, and that the Moon was created at full, on March 28. The doctrine appears in every purely Roman computus, and in Bede's highly influential writings; and F. E. Robbins cites many medieval instances of its acceptance. In the words of C. W. Jones[7] (p. 13), "This constant recurrence of the notion shows how the [computists of the Roman church] struggled to support with confirmed Biblical doctrine the calculations which they could not easily support with astronomical knowledge." Inconsistent applications of these principles abound, but this fact seems to have been considered less important than that contemporary Hebrew usage be avoided, unless it be vouchsafed by divine sanction or revelation. (The main issue was "Could Nisan precede the equinox?")

It is difficult to comprehend the fervor with which scholars addressed themselves to the problem as to the season of Creation. They would cite the examples of the ancients, of the Chaldeans, Babylonians, Medes, Persians, Armenians, and Syrians, who were all said to have begun the year in spring; or, if they believed in an autumn Creation, they might refer to the Romans (before Numa's correction), the Egyptians, and those Jews who followed the Egyptian civil year. From Eusebius and Ambrose to Melanchton, Scaliger, and Kepler, the list of authorities in favor of spring was weighty; but so, too, was that stretching from Jerome and Josephus to Scaliger (who changed his mind for the second edition of his great work) and Ussher, all of whom favored autumn. There were those who favored one of the solstices. Vergil was quoted endlessly in favor of spring, and Solinus and Macrobius for summer. Arguments were culled from evidence of the slenderest sort. A discussion beginning with the testimony of *Genesis*, 8:11, that the dove returned to the ark with an olive leaf would end—after much horticultural discussion—with a precise day and month, and possibly even hour and minute for the Creation of the world.

No short summary can do justice to this enduring pseudoscholarly activity. John Ashenden, in the mid-fourteenth-century *Summa astrologie* to which I

[77] For further information on the subject raised in this paragraph, see Jones,[7] pp. 12–13, 63.

have already referred, thought fit to devote a large chapter (I.i.1) to its history up to his time. Most of his authorities I have listed in another connection. The arguments are, as one would expect, repetitive. Spring is a time of generation, said some, a time from which calendars begin, a time when the Sun is in the lamb, which came to take away the sins of the world.[78] The empirical, the conventional, and the religious. But what was the precise phase of the Moon at the Creation? And when were the days Creation reckoned to begin? On the answers to these questions hang the workings of the Easter *computus*, and certain more taxing arguments for an autumn Creation. There were biblical texts, for example, *Exodus* 23:16 and 34:22, concerning the ingathering of the fruits of labor, with which Roger Bacon, among others, thought to reinforce the choice.[79] And Adam, no doubt, needed food from the beginning. John Ashenden could not himself decide by reason, although he inclined to the vernal equinox, since Grosseteste had favored it, since the Church had asserted it, and since most astronomers were of the same opinion.

Throughout more than sixteen centuries of theological debate, this was the level of discourse. The most curious aspect of the entire situation was that almost invariably it was taken as axiomatic that Creation was with the Sun at one of the cardinal points of the ecliptic. Strauchius was often cited in the eighteenth and nineteenth centuries as one who could offer ten arguments in favor of autumn; and yet in not one does he explain why the *precise* moment of the equinox should have been chosen. John Ashenden brings Abū Muʿshar to the rescue, with an argument compounded from astrology and Aristotelian physics.[(42)] The argument was not calculated to find much favor with Christian computists.

Few medieval writers seem to have been conscious of the gross imperfections of the Julian calendar over four or five millennia, and of the shift in the equinoxes. Once the problem was widely recognized, however, there were no lengths to which a computist would not stretch his little astronomical knowledge in an endeavor to give a precise minute to the Creation. Examples based on the solar apogee we have already seen. For parallels to Ussher's famous "beginning of the night preceding 23 October 4004 B.C." we need only consult the books of Joseph Justus Scaliger, Seth Calvisius, Johannes Rodolphus Faber, Franciscus Allaeus,[80] Dionysius Petavius, or a hundred lesser writers from an age when the

[78] The third argument is unusual. It is ascribed to Walter of Odington. According to Robbins,[(30)] p. 59, the notion that the Creation took place in the spring (in fact on 25 March) can be traced back to Annianus, while Ambrose was the first to introduce it into the Hexaëmeral tradition.

[79] Richard Holdsworth (1590–1649) reproduced much the same arguments as Bacon's.[(41)]

[80] Allaeus, a 'Christian Arab,' traced history astrologically up to the birth of Christ and then to Calvin. His book, *Astrologiae nova methodus*,[(43)] was burnt at Nantes, as heretical. Since, in the copy I have seen, the diagrams have not been inserted, and the block for the grandiose initial capital has been cut in mirror image, there might have been more than one reason for the burning.

whole subject had run riot. What conceivable spiritual or historical enlightment could Petavius, for example, suppose that he was granting when he calculated that the mean full Moon at the Creation of the world was on 27 October at 9 hours, 5 minutes, 42 seconds after midnight, and that on the fourth day, when the Moon was eventually created, it was somewhat decreasing?

In Ashenden's *Summa* there are one or two statements which he might have reported less casually had he been adept at calculating precise planetary positions in the distant past. As already explained, he quotes Walter of Odington as saying that the Indians thought all the planets to be at the head of Aries at the Creation (Ref. 22, f 3*v*). He quotes Julius Firmicus and Macrobius as saying that at the beginning of the world the planets were in such and such specific positions—and each of the planet's was in fact said to be at the fifteenth degree of its domicile.[81] To extract a precise date from such information as this was to all intents and purposes beyond the mathematical and astronomical resources of the middle ages. Had things been different, God's universe would have suffered at the hands of chronologists in much the same way as the Pharoah's pyramids have suffered at the hands of astroarchaeology.

It is because the correlation of planetary longitudes with times past is so difficult that the method of Great Years was so well liked. A "Great Year," wrote Thomas Lydyat of New College, Oxford, in 1628, being "a period of the Sunne and Moone," is the "trew, right and onelie foundation of this business." The business here was that of calendar reform; but Lydyat did not refrain from fitting the chief points of history into integral multiples of 592 years, with occasional resort to half-periods.[(46),82] Two centuries later, Richard Greswell of the same university was to be found practicing a remarkably similar art. He observed that 129 mean solar years were just a day less than 129 Julian years. In $(31 \times 129 + 1)$ years, that is, 4000 years, the Sun will come back to its original position with respect to the meridian at the time of equinox (Greswell,[(8)] vol, i, pp. 234-36; vol. ii, pp. 32-34). The year 4004 B.C. having been accepted *a priori*, B.C. 4 is distinguished by this remarkable coincidence. Was this not the year of Christ's actual birth? Of the two, Lydyat's efforts were perhaps the more excusable, caught up as he was in the vortices of a fresh and strong scientific current. He set forth his chronologist's testament in these words, dedicated to King Charles I (16 September 1626):

[81] Firmicus in his *Matheseos*[(44)] simply states in which signs of the zodiac the planets have their domiciles. Macrobius in his *Commentarii in Somnium Scipionis*[(45)] says of the World Year that it begins when anyone wishes it to begin (as opposed to when the planets are at the head of Aries, or wherever). On the other hand, in I.21 he reported that the early Egyptians said that at the very hour of birth of the world Aries was in mid-heaven, the Moon was in Cancer, the Sun in Leo, and so on, and that each planet was considered lord of the sign in which it was then found.

[82] Lydyat set the Creation at 4004 B.C., and the end of the tenth period at A.D. 1916. The system is explained in his book *Solis et Lunae periodus eruditae Antiquitati appellatus Annus Magnus*, etc.[(47)] In part this is a defence against attacks by Scaliger.

But concerning those Reckonings and notes of mine in the end of mine Emendation of Times, I have not made them, nor these neither last set downe, as determining anything for certaine, or foresetting the time of any future event; which God hath put in hisowne power, and lockt up as it were in the closet of his Privie Councile: ordering his workes in so wonderfull proportion of time, rather that men should acknowledge and admire his Providence upon their so coming to pass, then upon any imaginarie proportion conceited by themselves, presume to foretell the time thereof, before the same come to passe. Although as by long and manyfold experience the learned and wise Physicians have found that there are indeed by Gods ordinance, certaine Criticall dayes in the diseases of mens bodies, and Criticall Climactericall years with distinct Characters of severall Ages, in the course of their lives; so experienced States men have noted the like in the bodies Politike of Kingdomes and Commonweals; and worthie Divines also in the state and ages of the Church; which rightlie to discerne, it behoves warilie and circumspectlie to consider the same, not in the mere speculative abstract, but alwayes with respect to the predominant or peccant humor of the body. As for Astrologicall or other like vaine predictions or abodes, I thank God, I was never addicted to them. . .

Lydyat's final disclaimer was largely superfluous. Chronology was too dull a subject for most practitioners of astrology. (Not many could bring themselves to enliven their tables by commencing with the Creation, and ending with such names as Ben Jonson, Shakespeare, Beaumont, and Fletcher—as did Philipp Kynder, a decade or two later.)[48] Chronology was fast becoming the province of dry mathematical astronomers with a firm religious purpose—the very men Roger Bacon had wanted for the government of the Republic of the Faithful. These men were generally too astute to rest their case on the flimsy foundations of other men's intuitions. Seth Calvisius, for example, was one of a new breed, who took extraordinary pains to regulate his chronology by computing more than 200 eclipses.[49],83

Isaac Newton struck out in a different direction. We may as well end with him as we began. Commencing with a loose description (by Hipparchus) of the way in which Eudoxus had made the colures of the solstices and equinoxes pass through the constellations, he derived what he considered a precise enough position for the equinoxes to allow the sphere of Eudoxus to be dated (at 939 B.C.) using an argument from precession. The sphere could not have been made by Eudoxus himself: It must be the first ever fashioned by the Greeks. Other evi-

---

83 On the use of eclipses for chronological purposes, Philipp Kynder has some apposite remarks which might be nailed to several famous doors[50]: "It is amazement to me to see how the most learned & wise should be carried away with tradition or receaved custome taking them upon trust without examination. As namely the Eclypses which generally they should to be the Bases, & muniments and infallible Characters of all cronologie, where you may see the simplicity of the Ancient, when with astonishment they take notice of an Eclypse in many yeers 10, 20, or 100. And tell us of praediction upon the tyme of Cyaxaris and think to astonish us, as Columbus did the ignorant Westindians at his discovery and conquest. [He continues, pointing out the high frequency of eclipses.] And how much, I pray, doe the most learned and acute computants differ? Josephus (lib. 17, 18) makes mention of an Eclypse A.M. 3949. Kepler, the greatest reformer, refers this to the yeere 3946, three yeeres praevention. A.D. 238 an Eclypse Bunting thinks this to be refer'd to the former yeere."

dence suggested to him that it was the time of the Argonautic expedition.[84]
With these fanciful ideas he did a little to turn scholar's thoughts away from the
Creation, even though some were deflected only as far as the book of *Job*.[85]
"Canst thou bind the sweet influences of Pleiades, or loose the bonds of Orion?"
A more difficult problem than *Genesis*, perhaps, but one which offered a refuge
safe from free-thinking evolutionists and the new astrophysics in the centuries to
come.

REFERENCES

1. Isaac Newton, *The Chronology of Ancient Kingdoms Amended*, new ed. (T. Cadell,
   London, 1770), p. 45.
2. James Ussher, *Annales veteris et novi testamenti a prima mundi origine deducti*, etc.
   (Gab. De Tournes, Geneva, 1722), p. 3.
3. Francis C. Haber, *The Age of the World: Moses to Darwin* (The Johns Hopkins Press,
   Baltimore, 1959).
4. D. R. Dicks, *Early Greek Astronomy to Aristotle* (Thames and Hudson, London, 1970),
   pp. 75-6, 88-9.
5. R. A. Parker and W. H. Dubberstein, *Babylonian Chronology, 626 B.C.-A.D. 75* (Brown
   Univ. Press, Providence, 1956), p. 2.
6. Geminus, *Isagoge*, cap. 8, ed. by K. Manitius (Teubner, 1898).
7. C. W. Jones, *Bedae, Opera de temporibus* (Publication 41 of the Medieval Academy of
   America, 1943), p. 38.
8. Edward Greswell, *Introduction to the Tables of the Fasti Catholici* (Oxford University
   Press, 1852), p. 149.
9. Thomas Heath, *Aristarchus of Samos* (Clarendon Press, Oxford, 1913), pp. 171-3.
10. Otto Neugebauer, *The Exact Sciences in Antiquity* (Brown Univ. Press, Providence,
    1957), pp. 141-3.
11. A. Bouché-Leclercq, *L'Astrologie grecque* (Paris, 1899), pp. 38-9.
12. W. H. Stahl (transl.), *Macrobius: Commentary on The Dream of Scipio*, (Columbia
    Univ. Press, New York & London, 1952).
13. William Hales, *A New Analysis of Chronology and Geography* (London, 1830).
14. David Pingree, *The Thousands of Abu Ma'shar* (Warburg Institute, London, 1968).
15. F. M. Cornford, *Plato's Cosmology* (London, 1937).
16. B. L. van der Waerden, *Science Awakening* (Noordhoff, Leiden, and Oxford Univ. Press,
    New York, 1974), Vol. II, pp. 306-8.
17. David Pingree, Astronomy and Astrology in India and Iran, *Isis* lxiv, 238 (1963).
18. H. Suter, *Kgl. Danske Vidensk. Selsk. Skrifter*, Raekke, *hist. og filos. Afd.* iii, no. 1
    (Copenhagen, 1914).
19. Otto Neugebauer, *Kgl. Danske Vidensk. Selsk. Skrifter, Raekke, hist. og filos. Afd.* iv,
    no. 2 (Copenhagen, 1962).
20. Bodleian Library MS. Canon. Misc. 499, f. 2*r*. the Khwārizmī table. The only additional

[84] See, for an extended account, Manuel.[51]

[85] William Hales,[13] tells that with the help of Dr. Brinkley, Professor of Astronomy at
Dublin, he calculated that the vernal equinox was in the constellation of Taurus in 2337
B.C., the date he had assigned to the trial of Job. (See *Job* 9:9, and 38:31-2.) He published his results in the *Orthodox Churchman's Magazine* (1802),[52] only to find that he
had been anticipated by one Ducoutant in a Sorbonne Thesis of 1765!

material is an era "last of the Persian kings," 27 August 274, and of course the Alfonsine era, 1 June, 1252.

21. Johannes Eschuid, *Summa astrologiae judicialis de accidentibus mundi* (Joh. Lucili. Santritter, Venice, 1489).

22. Oriel College, Oxford, MS 23, complete.

23. L. Thorndike, *A History of Magic and Experimental Science* (Columbia Univ. Press, New York, 1934).

24. John D. North, *Richard of Wallingford* (Clarendon Press, Oxford, to appear), Vol. iii, Appendix 38.

25. Robert Grosseteste, *De ordine emanandi causatorum a Deo,* ed. by Ludwig Baur, *Beiträge zur Gesch. der Philosophie* (1912), vol. ix, p. 147–50.

26. J. D. North, Astrology and the Fortunes of Churches, in proceedings of the International Colloquium in Ecclesiastical History (CIHEC) held in Oxford, 1974, to be published.

27. E. S. Kennedy, Ramifications of the World Year concept in Islamic astrology, *Proceedings of the Tenth International Congress of the History of Science, Ithaca, 1962* (Hermann, Paris, 1962), p. 34.

28. Francois Plaçet, *La Corruption du grand et petit monde* (G. A. & G. Alliot, Paris, 1668).

29. *Introductorium in astronomiam Albumasaris Abalachi* (Erhard Ratdolt, 1489), sig. h.3.

30. F. E. Robbins, *The Hexaemeral Literature* (Chicago, 1912).

31. John Ashenden, *De magnis conjunctionibus* (Melcior Sessa, Venice, 1515).

32. Bodleian Library MS. Canon. misc. 190, f. 83*v*. The *explicit* announces that this work is *Tractatus de motu octave spere secundum Petrum Padubanensem.*

33. Bodleian Ms Ashmole 802.

34. Joh. Keplerus, *Tabulae Rudolphinae* (Jona Saurius, Ulm, 1627).

35. Hevelius, *Prodromus astronomiae* (Danzig, 1690), p. 88.

36. Christianus S. Longomontanus, *Astronomia Danica* (Caesius, Amsterdam, 1622), I.2, pp. 46–7.

37. Henry Power, *Experimental Philosophy* (T. Roycroft, London, 1664), pp. 188–93.

38. Georg Joachim Rheticus, *Narratio Prima* (first printed in Danzig, 1540), transl. by Edward Rosen, in *Three Copernican Treatises*, 2nd ed. (Dover, New York, 1959), p. 122.

39. Henri de Boulainvilliers, *Histoire de mouvement de l'apogée du soleil* (1711), ed. by Renée Simon (1949).

40. Aegidius Strauchius, *Breviarium chronologicum* (Wittenberg and Frankfurt am Main, 1686), pp. 264–5.

41. Bodleian Library Ms Sancroft 129, f. 1*v*.

42. John Ashenden, *Introductio maior*, II.5.

43. Franciscus Allaeus, *Astrologiae nova methodus* (Rennes, 1654).

44. Firmicus, *Matheseos*, ed. by W. Kroll and F. Skutsch (Leipzig, 1897), II.2.

45. Macrobius, *Commentarii in Somnium Scipionis* ed. by F. Eyssenhardt (Leipzig, 1893), II.11.

46. Bodleian Library, MS Bodley 662, f. 2. Manuscript of a book to be printed, the prefaced petition dated February 1628 (O.S.), "Five years before mine enlargement" (f. 1*v*).

47. Thomas Lydat, *Solis et Lunae periodus eruditae Antiquitati appellatus Annus Magnus,* etc. (G. S., London, 1620).

48. Bodleian Library, MS Ashmole 788, f. 175*r*.

49. Seth Calvisius, *Opus chronologicum* (Frankfurt, 3rd ed. 1629; 4th ed. 1650).

50. Bodleian MS Ashmole 788, f. 175*r*.

51. F. E. Manuel, *Isaac Newton, Historian* (University Press, Cambridge, 1963), chapters IV and V.

52. William Hales, *Orthodox Churchman's Magazine* ii, 241 (1802).

Chapter XXI

Cosmic Order and Human Disorder

Robert S. Cohen

Boston University

Being neither cosmologist nor theologian nor historian, I may be the complete expert generalist for this conference. I had better initiate my remarks with a bit of a confession—which I take to be acceptable for this audience and our readers. I was a physicist and continued my work as a philosopher. I add that I am a Jewish atheist, which should set the context of my remarks as clearly as those of our colleagues who spoke from their religious positions. My being a Jewish atheist may raise a question, since I do mean I am a Jew and not merely by origin. Is to be a Jew inconsistent as a religious tradition with being an atheist? Not so, as we shall see. Now these autobiographical remarks are relevant, so let me go on a little more.

I have several heroes whose names also may set these remarks in context: Aristotle, Spinoza, Karl Marx, Bertrand Russell, and Paul Tillich. And a heroic sentence from Ernst Mach,[1] to read as an emotional theme: "The highest philosophy of the scientific investigator is to bear an incomplete conception of the world and to prefer it to any apparently complete but inadequate conception." Now, I think I am a Jew because there is a difference between the Jewish conception of God and the Greek conception of the order of the world. The Hebrew God increasingly stood for a notion of responsibility, not of first cause. The Hebrew set forth, not a cosmological doctrine that we should be deeply concerned with the model of a potter or a craftsman, but a different sort of doctrine, and a model of a different type, an ethical model. The ethical model was related to a source which is all-powerful; and there is a problem of personality involved with the notion of responsibility. For the moment, I just want to say that the Hebrew God is not God the mathematician, and James Jeans was not a Jew, but a Greek. Another preliminary remark. Despite these appreciative comments on the Hebrew God, religion for me has come to seem merely another object of study, another human phenomenon, like systems of law. Religion is at times terribly evil, at other times good, not consistently good, to my regret; and yet among the finest people I know are those who believe that the admirable human qualities are also religious qualities. Religion has a full moral spectrum; I have sympathy for those who feel horror at the history of religion, and I have sympathy for those who identify with religion, and feel peaceful toward it.

But having said that, it seems to me that toward religions as toward other aspects of human culture, as toward philosophies and scientific theories and legal systems, it is appropriate to take an attitude not of skepticism alone but of rational reconstruction. Perhaps this logical positivist phrase tells what reinterpretation means in theology, for we want to know what the doctrine has cognitively to say; or, a reasonable reinterpretation. I have no trouble with empirical and cognitive reinterpretation of myths, nor with accepting the clear historical evidence that mysticism has at times been very positive in human affairs, as well as at times very destructive. So, on balance, an open mind toward religious people, and toward religious doctrine. But when one talks of 'Cosmic Order and Human Disorder,' everything depends upon religion as well as upon science.

Cosmic order? One has to wonder, cosmos and order—what is the meaning of cosmos and what of order? One orderly entirety or whole is a little difficult to accept boldly in any *a priori* sense, but it is to be thought about in contrast to certain other notions. Not just with the notion of chaos, although chaos is interesting, too. Rather, cosmos is an orderly whole in contrast with anarchy, an orderly whole where there are links among the parts, whereas a cosmic anarchy (not social anarchism!), would be one in which the parts are not linked, and go off on their own. A cosmos could be partly ordered in that respect of linkage, and partly disordered, somewhat disconnected. There could be, so far as we now know, several parts to the Universe, each by themselves now, maybe always, maybe only sometimes. There might be several orders, spatially distinct, causally distinct. There might be some ordered parts, some disordered parts. More important than the option of a unified cosmos is that of an *ordered* cosmos, and of the kinds of order that there might be. And even more interesting is that the several orders, the plurality of orders, the nests of ordering, may be such that they, too, are not fixed. There may be a dialectic of novelties, which are not only fracturing breaks in cosmic order but the cosmic mode of introducing complexities of order.

Now I may wish a reductive explanation of such a natural history of emerging orders. I may think it possible to say that the necessary conditions for the arising of a new complex system of order should be found in the previously existing order, but for now that is just a prejudice, or shall we say a methodological heuristic concept? No such conclusions derive from the philosophy of science nor from the evolution of chemistry, physics, and biology, but we have metaphors, and among these, nested levels or envelopes. Needham uses the image of envelopes of order. As these levels of complexity arise, then, they provide further order in the Universe; and this is so, whether the circumstances of arising orders are reductive or nonreductive. Objectively there may be more and more order. Despite entropy, there may once have been a disordered cosmos.

What provides the order or the linkage among events? Roughly and briefly, a causal structure; the linkages are causal connections, which suggest the causal theory of time in philosophy of science. This ancient theory (which was revived 40 years ago in Mehlberg's classic essay),[2] construes time not as a Platonic reflection in actuality of an eternal order, but rather as the causal structure of events. Structure and order are one, because causation is the temporal relation. We may observe that it is not inconsistent with an orderly cosmos that the cosmic order be statistical, nor inconsistent with a causal theory of orderliness that the causality be statistical causality. Not only in the metaphorical sense of Peacocke's random sweeping search radar,[3] but beyond that, statistical constraints upon physical, chemical, and biological events provide orderliness with some flux otherwise accounted for. The fact that it may be incompletely deterministic with respect to every single event while being deterministic with respect to the sequence of properly defined, physical, chemical, and biological states, preserves quite sufficient orderliness for philosophical (or religious) provocation and struggle.

Thus, if an orderly cosmos is terribly important for the human spirit, neither the opportunity nor the threat from quantum mechanics seems very great. If there is religious implication in full determinism, then the determinism of state functions is sufficient to imply much. And if not, not. And so the relationship between the religious notion that the world is orderly (and hence responsible) and the modern physical notion of causal anomalies, that there may be apparently disorderly factors in the statistics of the natural cosmos, is less important than at first glance might seem to be the case.

Of course, the rational standing of generative factual conditions which set in motion the causal connections among events, that is, the ontological character of initial conditions and natural boundaries, is not itself a matter of orderliness. How shall we interpret the role of initial conditions? It would seem to be consistent with modern science, if not required by it, that there are possible and equally describable worlds, just as equally describable as the sample student problems which are equally consistent with a particular mathematical or physical theory. Possible worlds are differentiated by factors which the theories themselves apparently cannot specify, unless we offer an Eddingtonian or Hegelian view of nature and critique of science by which we demand of theories that ultimately they necessarily specify the initial and boundary conditions for a unique world. But then we are confronted by Eddington's subjectivist neo-Kantian trap, or by Hegel's endless and regressive mere hope to avoid it. To scientist or religious thinker, it would certainly appear that the initial conditions are so far contingent that a variety of worlds are consistent with the theories.

Perhaps our idea of causal order should be none of these, then, neither the statistical nor fully determined lawful regularities, nor the arbitrary uncaused initial conditions that instantiate the lawful regularities. Orderliness in the world may be a bit more local, and its extension to the Universe at large not necessarily something we need to settle. That is to say, orderliness may be more similar to the metaphor of harmony than it is to the metaphor of mechanism. Then cosmic order has balance in it, and the idea of cosmic order is the simile of cosmic justice: what is in balance, what is suitable, what is right. Kelsen persuasively argued that the idea of causality in Western thought has originated along with the notion of justice,[4] but it is not justice but retribution which is the common Babylonian origin for causal and judicial understanding at the very earliest times. In this static vision of universal structure, whatever happens does so because of a disturbance to the stable harmony, and this is the ancient metaphorical origin of a dynamics, in which effect and cause are in some respect equal, while not identical. Equal in that the cause is disturbance, and effect is the corresponding balance, restoring equilibrium, stability, harmony. The restoration of equilibrium is the historical source of a cosmic order, and of the causal relations which make it orderly. Actual political stability—order through law—is metaphor for natural stability—cosmos through lawlike causality. Now, equilibrium may be genuinely balanced or just as genuinely unstable, a house of cards rather than a marble in a bowl. And if the cosmic equilibrium is so disturbed that there is no restoration of the previous equilibrium, then we look either to anarchy and chaos or to a new cosmos, to a revolution with a new equilibrium. Indeed, the political theory and the natural philosophy of ancient times seem on this matter to have stemmed from the same source.

That cosmic orderliness is not static but a matter of harmony in process, is a rather simple notion, deeply elaborated by Hegel. The notion is of natural evolution, natural fulfillment to the things, organisms, and cultural entities of this world, self-generating rather than externally disequilibriated, and self-generating disequilibrium which restores the natural entity within itself by its own actions. (I read Hegel in a naturalistic mode, but naturalistic or religiously supernatural hardly matters.) The major causal evolution is evolvement of inner nature, Hegel building on Aristotle. Now, *pace* Leibniz, things *may* be isolated, but we had better think the natural and social world through otherwise. Only the Universe as a whole may be an isolated entity (or may not be, and Hegel does not speak to that, nor need he). Entities are not normally isolated, so that a civilization, marketplace, tribe, individual, system of laws or art form, whatever they may be, each may have external forces impinging. The mechanical problem of external force relations is there, along with the organismic problem of change through both natural evolution of that marketplace or tribe, and drastic, perhaps eternal, disruption by an external empire—social or natural. Then, the same entity enters a larger enveloping order, a primitive tribe becoming a new entity of a world marketplace, a new entity with its own inner dynamic; but whether fulfilled or drastically transformed and unfulfilled, the notion of an inner dynamic, of a fulfillment of that which is to be fulfilled within the terms of reference and values of the entity itself, is Hegelian orderliness. There may be incompatible orders and fundamental conflicts, perhaps inherent inner orderliness along with inconsistent orderlinesses of different natural entities; there may be local order and cosmic (or at least Earthly) disorder.

Let us look at disorder. At least to start, disorder is morally neutral. Disorder may bring something better. What is ordered may be morally rotten. Then disorder is not equivalent to undetermined, nor equivalent to uncaused. Disturbance of stable equilibrium is due to an external suitably caused and causal factor. So disorder need not mean chaos, not even partial.

There are four disorders to concern us here.

First, there are disorders of nature—disasters such as floods and drought, lightning, and earthquakes. There are natural disorders by which a species is destroyed. Is there a moral to this old tale? The dinosaurs could not have taken the point of view that their elimination by natural procedures was not a disaster, a disturbance to equilibrium which was disorderly and met by disharmony, certainly nonrestorative on any view. And there are artificial disorders of the natural order—human actions which are ecologically disastrous.

Second, there are disturbances and disorders of the social order. The tribe invaded by an alien power, warfare in general, warfare not only of the great against the small but of the great against the great—Carthage versus Rome—the interminable record of external disturbance to social order, these are a grave human disorder; and they may be internal, too. Slave revolts, class wars, civil strife open or covert—all are disturbances from the viewpoint of an existing order but orderly from another view, if one takes the evolution of a different entity, the entire society as such. Which entity is scrutinized decides whether or not the phenomenon is orderly process or in itself disorder.

Third, there are the disturbances and disorders in the personal order. We individuals introject, taking within us the standards, customs, and values of our societies and its parts; when the societies are disturbed, we are, too. But not only are there introjections of social, religious, racial, and class conflicts, and whatever other conflicts may be of a social nature: There is also private internal disorder. Just for the moment take the human individual as the natural entity, the person who—cliché or not—dies alone, the individual aspect of the human species. The individual's disorder which is so widely prevalent is neurosis, and the neurotic character is inherently self-frustrating, taking actions which are just such as to prevent accomplishment of the person's goals. Whatever that individual human being may have introjected, and without judging whether what is internalized from society at large is healthy, wise, good, or none of these, just merely taking introjected standards, the neurotic is the self-frustrating person, a disturbed entity with inherent and chaotic disorder.

A fourth form of disorder disturbs the larger moral order. This comprises all the uses of inappropriate and immoral behaviors, choices of action which bring pain, and prevent human fulfillment: the inappropriate unwarranted use of force. I say inappropriate to avoid absolute moral pacifism, but also to suggest the quality of sadism. The point is not only cruelty and its enjoyment, but deliberately more force than necessary for the attainment of the particular desired power (without judging whether what was desired was good or bad). If there is more force than is necessary, if there is an enjoyment of the infliction even of the necessary, then there is a gross disturbance of the moral order; but individual sadisms in human relations deserve separate mention. The socially structured disorder of mass sadism is far worse than merely personal neurosis.

These are the four disorders of the human sort.

Parenthetically, we need not claim that full cosmic order is incompatible with a human freedom. Human disorder in a *positive* interpretation has a very simple, appealing meaning. Postulate a cosmic order, impersonal, mechanical, forced, in which it appears that the human individual has no private, humanly individual, place; then human disorder may seem a positive move in such a mechanized universe. I think the postulate quite wrong because incomplete, but we have here the ancient Epicurean interpretation in which the human being introduces a swerve into the determinisms of nature, and thereby a disturbance, a disorderliness. I contend that the matter is quite different: Determinism is consistent with free choice of human actions, and freedom follows what Spinoza, Schopenhauer, and Einstein have described. To be free is to be able to do what we wish, with the footnote: Freedom is not to wish what we wish. That regress introduces the errors of the free-will antideterminists. Now we may wish this, or that; and we may be imposed upon from without; but we may also be imposed upon by forces within. We may be neurotic and hence unable to be free. We "escape from freedom." This may appear to be an escape to order and law, but it is only to the externally imagined totalitarian project of complete order. Thereby it shifts the causal initiative outward, and abandons free causative choice inward. Neurosis then bears upon responsibility; a deterministic freedom is required for responsibility. If we can predict what another will do in a given social, personal, natural situation, then we can hold that person responsible. If what is done is random, is not under personal control, not subject to personal choosing, and *if* thereby it is chosen irrationally, then we still may be able to predict it. What is at stake is the causal link of responsible choice. If our own actions could cause or determine another person's choice, and thereby action, to be otherwise, in a repeated, quite similar situation, then we may hold that person responsible. Or, in general, if additional causal factors in his consciousness brought from outside, or brought from within, could have made him act differently, then he is responsible. Therefore, the net response, the net result, the vector sum of all the causal determinisms upon a person bring him to choose to act, and then, if he can do it, he is free. The test of personal responsibility is a change in action in response to changed causal vectors and consequently different choice of action. In contrast, freedom without determinism is not freedom but chaos. The alternatives are responsible freedom and irresponsible chaos. Whether this view of responsibility is right or not, it is the cosmic human order which may, at a suitable level of human development, generate human freedom. Disorder, then, is not the sign of freedom, not positive but negative.[1]

If, then, human order is better than chaotic disturbance, we must take care to understand the special circumstances of freedom. The human species may be, indeed very likely is at times, the creator of new order, but otherwise a source of disorder, in ecological disturbances, sadisms, and surplus repressive violence. The network of circumstances is unclear; species-wide, there may be an inherent quality of destructive conflicts, but it may historically turn out the other way. We do not know. We do know that the human species has at one time or another been the source of new political, aesthetic, cultural order, or the destroyer of previous human order, of the natural order, or both intertwined.

Mankind as the creator? That is not inconsistent with the cosmic order. A naturalist thinks the human species is a part of nature, and nature works through the human species to make changes in nature; and these self-reflective action metaphors are both banal and simple. They are thoroughly homocentric. To speak of human creativity is to speak of new species-centered orderlinesses within a vast physical, chemical, and biological environment. What might humanity create? Dare we say that cosmic order may be seen as rational order? In reconstructing the history of the ideas of rationality and order, it is plausible to say that a reasonable society, and a reasonably saved and fulfilling life, are those not deviated from their self-determining natures by alien and subjugating factors. Then, the demand for genuine individual happiness is the legitimate paraphrase of the religious and metaphysical demand for a rational world order.[8]

To guarantee genuine individual happiness requires far deeper understanding, but this notion suggests the role of the human species, added to the cosmic order (and cosmic rationality?), to be the following: The world is not sufficiently rational, from this human point of view. Our task is to make the world rational, to realize reason; and thus it is a positive, and possible, achievement. But another viewpoint is quite negative. The likeable optimism in the history of idealist philosophy is the sober judgment that there could be a world which would be reasonable, not that there is one; and the fact that such a world must be made, rather than discovered or revealed, is very important. It means that men and things are, in a positivistic sense, all that they have been, all that they actually are, and yet they can be more. Now is this true? There may be cosmic disorder on a grand scale, with an imposed order. The orderliness we see may have been imposed by us, as our neo-Kantian filter for looking at the world as

---

[1] See the noted essay on "The Freedom of the Will" by a group of philosophers at the University of California.[5] Compare it with Carl Ginet's "Can the Will be Caused?," which offers a negative judgment and suggests that there may be no acts of will.[6] See also Erich Fromm.[7]

it comes to us through senses, instruments, and culture—epistemologically imposed and not cosmic orderliness. Or we may say that the order is placed upon the inchoate world by God. Alternatively, the orderliness may be a developing, growing quality in which the human species plays its own small but locally crucial part.

Let me summarize the disorders. There may be local disharmonies, local incompleteness, causes of local disorder; or there may be local violations of order, and they might be seen as good or bad. Miracles, I suppose, would be violations of order which some people would regard as good, and there may be devilish miracles which some would regard as bad. The disorder may be imposed by constraints upon mankind or upon individuals, such as to frustrate the inner nature and deny that which we wish. Hegel's optimism or a neo-Hegelian pessimism will depend upon such empirical factors as whether or not the external constraints upon the human species are such as inherently to frustrate achievement of human goals. Or, alternatively again, the disorder may be the interaction of two entities whose natures are such that only one may predominate. And there may be simply the normal disasters of nature in this disorderliness of pain rather than pleasure, the normal inextricable evils of human life.

Now I think of these disorders—incompletenesses, disharmonies, inexorable conflicts, inexorable frustrations, inexorable lacks, and dumbly terrible sources of pain—as met by three escapes, which each have the characteristic of imposing a kind of novel order upon the world.

One is serenity: acceptance of the world via an impersonal interpretation of what cognition is all about. This acceptance may be learned from the way that scientists see themselves, as very personal and yet learning something that transcends their and all personality. If there is any transcendence, it is through knowledge of that which is not within one's own self. Mach[9] again, talking about scientific knowledge:

> How great and comprehensive does the self become and how insignificant the person. Egoistical systems, both of optimism and pessimism, perish with their narrow standard of the import of intellectual life. I feel that the real pearls and values of life lie in the ever-changing contents of consciousness and the person is merely an indifferent symbolic thread on which those pearls are strung. The arm of man reaches far beyond the immediate. What an immense portion of the life of other men is reflected within ourselves, their joys, their affections, their happiness, their misery.

Now, this impersonal nature of positivistically defined elements, even of sensations, was in Schrödinger's mind, too:[10] "Shared thoughts really are thoughts in common." So the escape of serenity is an escape which accepts the Universe whatever it is, a mode which is particularly appropriate for the scientist. There is a cosmic religious dimension to learning that which appears to be true in itself, far beyond any merely personal enjoyment or personal instrumentality, neither a pragmatic view nor a libidinal view of science, and paradoxically not so much enjoyment as an enjoyment which transcends the life of the individual. Is it sublimation? Yet it is fulfilling through another being. Many scientists have had that.

A second escape from chaos is the good in itself, and it is sufficient. To seek eternal reality for the good is to place a spurious demand on this particular form of escape from cosmic disorders—Spinoza, for instance, and Russell, too. Let us recall a little tale from Freud's essay on transient phenomena:[11]

> Not long ago I went on a Summer walk through a smiling countryside in the company of a taciturn friend and of a young but already famous poet. The poet admired the beauty of the scene around us but felt no joy in it. He was disturbed by the thought that all this beauty was fated to extinction, that it would vanish when winter came, like all human beauty and all the beauty and splendor that men have created or may create. All that he would have otherwise loved and admired seemed to him to be shorn of its worth by the transience which was its doom.

> The proneness to decay of all that is beautiful and perfect can, as we know, give rise to two different impulses in the mind. The one leads to the aching despondency felt by the young poet, while the other leads to rebellion against the fact asserted. No! it is impossible that all this loveliness of nature and art, of the world of our sensations and the world outside, will really fade away into nothing. It would be too senseless and too presumptuous to believe it. Somehow or other this loveliness must be able to persist and to escape all the powers of destruction.

> But this demand for immortality is a product of our wishes too unmistakable to lay claim to reality: what is painful may nonetheless be true. I could not see my way to dispute the transience of all things, nor could I insist upon an exception in favor of what is beautiful and perfect. But I did dispute the pessimistic poet's view that the transience of what is beautiful involves any loss in its worth.

> On the contrary, an increase!

The third is a footnote to the acceptance of transience: What is good is good for itself and that is enough, a life judged by moral cliché, "Virtue is its own reward." Not heaven hereafter but virtue, love, and goodness when they exist, acceptance of moments of order or beauty or goodness, all moments of the moral order. Whatever may be the disorderliness or the uncaringness of the larger world, order and care arise in their human occurrence. So let the cosmic order or disorder be cosmos or chaos—it is normatively irrelevant.[2]

Now, religion may either be personal or impersonal about the nature of God. It may be that the meaning of God as person is crucial for reaching a religious response to cosmically local as well as cosmically large order or disorder. For me, the orderliness and the disorderliness of the cosmos are many-structured, many-leveled, and it is neither clear nor required that we can draw any conclusion from a cosmic order concerning what can save the human soul, nor from a cosmic disorder either. It may be either way, with either variety of cosmos. What does one do then about faith and reason, religion and nature? Can we make cosmic demands for religious faith? There is a well-established American scepti-

---

[2] Perhaps the young Russell put this in its most appealing romantic form; see his essay on "A Free Man's Worship," first published in 1903.

cism derived from Yankee atheists. Every town in America was once reputed to have at least one village atheist, whose classic remark asserted that faith is when you believe something you know ain't true. That surely is the first step in the scepticism about cosmic demands on behalf of a religious belief.

It is not cosmology which is important for a religious response to the structure and struggles within the world. What is important is the practical notion of a way of living which will achieve something of the particular goals and values of a human being within the amoral constraints of human existence. Are we to believe that you need a cosmic understanding in order to be a good person? Here is St Augustine: "Reason would never submit if it did not judge that there are some occasions on which it ought to submit. It is then right for it to submit when it judges that it ought to." Shall we then judge reasonably that we ought to abandon reason? And here is Pascal:[3] "If we submit everything to reason our religion will have no mysterious and no supernatural element. If we offend the principles of reason, our religion will be absurd and ridiculous." But finally his famous ambivalence: "The heart has its reasons which reason does not know." It may be that we *need* a mysterious element, but the argument for a mystery which is supernatural—that is different. The mystery of feelings and goodness, the mystery of the inner soul, may be entirely natural. Let me approach this cosmic element first with a remark about an ancient but continuing Jewish theme. The greatest repeated commandment from *Deuteronomy* is the famous *Sh'ma:* 'You shall love the Lord, your God, with all your heart and with all your soul and with all your might.' This is a statement of immense symbolic usefulness, and open to many interpretations. But of course the sharp point to all of this is in the last three phrases. The statement is not that there is a Lord your God. The statement is the command that you devote all your heart, all your soul, all your might, to whatever it is that you conceive of as this, your God. The interpretations of the Lord then are many and wide open, but what is not open to doubt is that whatever you hold of ultimate value is what you should devote yourself to. This is your commandment and your order, and it receives its obligatory quality from within. Do not be the neurotic character who is spread out, frittering, acting in conflict with itself, self-frustrating.

Your God is what God represents and this is what you are to seek, as I read that theme; and I see nothing which one needs to call either religious or atheist about it, either way. The theme of cosmic order and human disorder, then, although it can be made real by genuine religious actions and devotion, does not require religion. The devoted and committed and healthy individual, whether he be called atheist, communist, Confucian, Maoist, Catholic, Jew, or Buddhist, whatever he or she may be, brings order and responsibility into the character of his life experience, brings orderliness where there may have been only careless random disorderliness. But in any case, a cosmic order brings neither salvation

---

[3] *Pensées*, paragraphs 273, 277.

nor success. Only struggling with the natural disorders, and only the human struggles against human disorders, against racism, sadism, neglect, ignorance, and every exploitation, may bring some kind of order; or they may bring none. There is nothing *a priori* certain about the success of the struggles. But if there is any order on the human scale, it was and will be made by human beings.

REFERENCES

1. Ernst Mach, *Science of Mechanics* (Open Court, La Salle, Ill.), p. 560, slightly revised in accord with R. V. Mises, *Positivism*, p. 201.
2. Henry Mehlberg, "Essai sur la théorie causale du temps," translated in Henry Mehlberg, *Time, Causality, Indeterminism: A Study in the Philosophy of Science (Boston Studies in the Philosophy of Science*, Volume XIX) (D. Reidel, Dordrecht, Holland and Boston, forthcoming).
3. A Peacocke, this volume, Chapter XXIII.
4. Hans Kelsen, *Kausalität und Vergeltung*, translated in *Society and Nature* (University of Chicago Press, Chicago, and Routledge and Kegan Paul, London, 1946).
5. The Freedom of the Will, in *Readings in Philosophical Analysis*, ed. by Feigl and Sellars (Appleton-Century-Crofts, New York, 1949), pp. 594–618.
6. Carl Ginet, Can the Will be Caused?, *Phil. Rev.* 71, 49–55 (1962).
7. Erich Fromm, *Escape from Freedom* (Rinehart, New York, 1939).
8. Herbert Marcuse, *Reason and Revolution* (Humanities Press, New York, 1954), p. 26.
9. Ernst Mach, *Popular-Scientific Lectures*, 3rd English ed. (Open Court, La Salle, Ill.), p. 235.
10. Erwin Schrödinger, *My View of the World*, trans. C. Hashings (Cambridge University Press, London, 1964), p. 17.
11. Sigmund Freud, *On Transience* (1915), translated in *Complete Psychological Works* (Macmillan, New York, 1964), Vol. 14, p. 305.

Chapter XXII

Basic Christian Assumptions about the Cosmos

Philip Hefner

Lutheran School of Theology at Chicago

A recent reviewer in the Sunday New York *Times* set down a maxim for literature that applies equally well, I believe, to our reflections about the cosmos. "A novel," he wrote, "must stand on its own feet. We may tolerate its being pedestrian, but it must be ambulatory." Ambulatory, I would suggest, because it must go somewhere, be provocative and fruitful for further thinking, else it is useless. I write as a theologian, species Christian, subspecies Lutheran, and I intend to represent my species as well as I can. In the process, however, I am not interested simply in reciting what has been believed and reflected upon in my theological traditions, but rather to state what I believe is ambulatory in those traditions. Since, for better or for worse, I have set myself as the judge of what moves and what constitutes significant and interesting movement, I shall render the tradition in a highly selective manner and interpret it from a definite stance that corresponds to the place I want to go with it and to the direction in which I want to head. I shall not be able to avoid being pedestrian, but I do wish to be ambulatory, and I invite you to walk with me.

This presentation unfolds in several parts. In a relatively long introductory section, I discuss the nature of religious myths of the cosmos. Then I move on to compare scientific and religious cosmologies, setting forth criteria for interpreting both. In a third section, which is the main one, I elaborate the basic Christian assumptions about the cosmos, as I understand them. Finally, I speak about the significance of these assumptions and suggest criteria for assessing them.

QUESTIONS OF LANGUAGE; THE NATURE OF MYTH

At our present stage of highly developed self-awareness, we cannot avoid discussing questions of language. We cannot simply talk about the cosmos as if

347

all our talk were univocal with respect to the cosmos or as if it were all of the same type or order. I cannot embark here upon a full-scale technical discussion of language, but I will set forth some minimal statements that are pertinent to what I have to say.

Our self-awareness touches four issues relevant to language about the cosmos: (1) our relative placement on a tiny planet in the Universe; (2) the groping, tentative character of our scientific cosmologies; (3) the mythic character of our religious cosmologies; (4) the consonances and dissonances between scientific and religious cosmologies. Our relative placement in the Universe accounts for both the groping character of our scientific statements and the mythic character of our religious affirmations. The differences between scientific propositions and religious myth account for the apparent dissonances and consonances between scientific and theological statements.

It is not my primary responsibility or competence to rule on the linguistic character of scientific cosmologies, but I do wish to comment on the nature of religious cosmologies: They are mythical in character. By *myth* I mean that these cosmologies are symbolic narratives that purport to be rooted in a transcendent order and to describe the "way things really are," i.e., they claim to describe an objective transcendent order, even as they imply an imperative as to how we should comport ourselves in order to be in harmony with that objective transcendent order. These myths claim to describe both what *is*, with respect to the cosmos, and also what humans *ought* to be doing in the cosmos. Myth, in summary, has these components in its definition: (1) symbolic narrative, (2) description of a transcendent order, and (3) elaboration of what human action is commensurate with that order.

Myth, we know, is neither fanciful, nor is it arbitrary fabrication. Rather, it emerges out of the concourse between human beings and their world. Myth, in other words, is not most importantly a product of human autonomous creativity— although it is also that. More significantly, it is a rendering, in the human consciousness and by means of human modes of imagery and words, of what humans have encountered in their adventure in the world that surrounds them. No matter how great the human component is in myth, the entire mythic enterprise is record of an objectivity—the environing world and its impact on men and women.

Myth is a special type of rendering; it is a testimony concerning what human beings have learned from their concourse with their world about "what things are really like," and what human response is called for in the light of "how things really are." Myth, in other words, is the attempt to record that very wisdom and imperative which defies articulation. Myth is always operating at the boundary of the relationships between humans and their world, putting into image and language what can be only intimated and that only in part. We do not appreciate myth if we fail to underscore this boundary-line character as of the essence of mythic expression: The significance of myth lies precisely in its insistence on imaging what defies expression, grasping what refuses to be grasped, and making outrageous absolutistic claims about that which is irresistably ambiguous.

Despite frequent earlier claims to the contrary, myth is not dispensable in human existence, neither is it avoidable. Myth is necessary for human beings, because their encounter with the world is so serious in its actuality that they cannot refrain from bearing their testimony to what that encounter unveils to them and means for them in their living. The nature and content of this encounter determine that this testimony includes the mythic. The nature of this encounter is such that it presses upon humans at the point of their surviving or not surviving at all, at the point of their being stretched until their extremities ache. Sometimes the ache is one of pain, sometimes of joy. At times they feel the hollowness of melancholy and meaninglessness, while at other times, the fullness of incomparable richness. If the encounter with the world were more prosaic, less momentous, farther removed from the essentials of life and death, then myth would not be necessary. The content of the encounter consists of an unveiling to the human creatures of the lineaments of a foundational reality and the meaning of that foundational reality for the existence of the creature. The modality of the encounter with reality is marked by seriousness and ultimacy; the content of the encounter is the taste of the foundational depth. Because of this modality and this content, myth is forced upon the human creature, if that creature is to be faithful in bearing his testimony to what he has seen and touched and heard and felt. It is man's sense of responsibility to what he has himself perceived that makes him irrepressibly a myth-sayer.

RELIGIOUS AND SCIENTIFIC COSMOLOGIES.
THE SURVIVAL-INTEREST

There are those who would say that scientific cosmologies share much with what I have described under the rubric "myth." Scientific cosmological theory also insists on imaging what defies expression, grasping what refuses to be grasped, and making claims about that which is intrinsically ambiguous. Such hypotheses also emerge from a serious encounter with the world, and the basic dynamics of thought which underlie the scientific enterprise are of a piece with the human perceiving and thinking processes that surface in religious myth and theological proposition. Epistemologists have made this point often enough, and I have no desire to dispute it now. In fact, my own thinking is monistic enough to put a premium on the thesis that all human perceiving and thinking—whether scientific, artistic, philosophical, or theological—are of a piece, cut from the same cloth.

Religious mythic cosmologies and scientific cosmologies share a number of important characteristics. Both attempt to describe cosmology in a unified and comprehensive manner. We judge each of these cosmologies by such criteria as these: whether they take into account as many of the known data concerning the Universe as possible, their simplicity and elegance, their plausibility and consequent persuasiveness. If there be any doubt that even myth may be judged by these criteria, that doubt should be dealt with summarily. Humans in their concourse with their world, on the boundary we have already described between human sensibilities and ultimacy, take these criteria with utmost seriousness. In the myth-forming stance, the known data are not always identical with those that command the attention of the scientist, but nevertheless, if the myth could serve no explanatory function by taking into account a host of significant publicly recognized "known data," and when the myth does not bestow meaningfulness on those data, the myth collapses. Furthermore, since myth is crucial for the formation of human identity, functioning to organize the core of the human being, serving at times as a vehicle of adoration and worship, simplicity and elegance are built-in as positive characteristics. Theologians and philosophers may elaborate the complexities inherent in myth, but the myths themselves are simple and elegant, else they prove to be artificial and transient. Persuasiveness, rooted in plausibility, is essential to myth, because it would be a contradiction in terms to assert that a myth is an articulation of the order of things and the imperative of human response to that order, if that myth did not seem to be plausible and thereby persuasive. As Whitehead has said of all statements of truth, we too say of myth: If its persuasiveness does not figure so prominently that men can say, "It expresses what I already believed but did not know how to articulate," then it will not stand as truth.

These characteristics are common ground for religious mythic and scientific cosmologies—the intention to describe the Universe comprehensively, judged by the criteria I have mentioned. What, then, is the *proprium* of religious mythic cosmology? It is this, I believe: *That religious mythic cosmological statements focus primarily on describing the cosmos from the point of view of what assumptions are necessary if human beings are to live optimally in the world.* There is an *ex parte* character to religious cosmology that cannot be overlooked. The cosmological myths of the religions recognize that in the encounter with ultimacy, man's perspective is indelibly shaped by his desire to live optimally, to survive well—and surviving well means living in harmony with the invariant order, this-worldly and transcendent, that is set forth in the myth. This is not the only perspective from which to view the Universe. My impression is that it is not so central to the scientific cosmologists. The sciences have focused on clarifying reality apart from its implications for human concerns.

What I have called the *proprium* of religious mythic cosmologies may be called the survival-interest, provided that we do not set our sights too low when we define survival. We speak of survival as living optimally and as obedience to the invariant order that is revealed in both scientific statement and religious myth. This survival-interest is *not* the same thing as relativity, the recognition that the human knower is inextricably related to that which he seeks to know.

There is an element of subjectivity, or better, participatory knowledge, that is shared by all disciplines, from physics to the social sciences to theology. The survival-interest that I refer to implies that the Universe *per se*, as an end in itself, must without doubt figure more prominently for the scientist in his cosmological speculations than it does in religious myths *about* the Universe. This does not mean that the scientist is more objective in his speculations, but rather that his interest in his speculations is different from that of myth and the theologies, which expound myth and uncover its complexities.

While I would insist, at least at this moment, in the survival-interest as the *proprium* of religious myth, it may well be that this *proprium* should not be distinguished radically from scientific cosmologies. The physicist John Wheeler has been quoted as saying that, "No theory of physics that deals only with physics will ever explain physics. I believe that as we go on trying to understand the universe, we are at the same time trying to understand man. . . . Only as we recognize that tie will we be able to make headway into some of the most difficult issues that confront us. . . . Man, the start of the analysis, man the end of the analysis—because the physical world is in some deep sense tied to the human being" (quoted by Helitzer[1]). In the context of the article in which Wheeler makes this statement, he seems to be referring to the mode of quantum theory which maintains that observer and observed are indissolubly linked. Nevertheless, his language could be interpreted to refer to something like what I have called the survival-interest—"as we go on trying to understand the universe, we are at the same time trying to understand man."

Whatever the scientists say about themselves and their statements, this survival-interest is essential for understanding religious mythic cosmologies. A myth of the cosmos is not comprehensive enough for religion, its simplicity and elegance are empty, it simply is not plausible and persuasive, except as it speaks of the Universe with reference to what assumptions are necessary for optimal human living. If we fail to understand this proper interest, we cannot assess adequately religious myths of the cosmos. We fail to perceive the content accurately, we fail to comprehend what genre these myths represent, we miss their point and purpose. Teilhard de Chardin pointed in this direction when he said, "The truth of man is the truth of the universe for man, that is to say, the truth, pure and simple."[2] Religious cosmological myths, then, do not really answer directly the question, "What is the Universe like?" Rather, they speak to the question, "What must we believe about the Universe—as we know it to be from our science—in order for us to live optimally?" This is not an antiscientific question, nor even a nonscientific question. The question is not, "What must we believe in our fancy about the Universe of our dreams?" Rather, it is, "What must we believe about the Universe that we have come to understand through our very best, hardest scientific knowledge?" Nevertheless, even if the question that concerns myth-making man is neither anti- or nonscientific, it is a question that is not in the centerstage for the scientist *qua* scientist.

RELIGION AND SURVIVAL

Underlying what I have said about myth and religious cosmologies is an assumption that dare not go unstated, even though it cannot preoccupy us here. I accept the premise that religion and its myths are a part of the ongoing evolution of the ecosystem, more particularly the evolution of life and the human species. Religions and their myths, therefore, are caught up in the process of survival within the mechanisms of natural selection. For many persons, including leading theologians, such a view of religious faith and myth is repugnant. For many, it smacks of a demeaning utilitarianism to place religion within the context of evolution and survival. I can understand this dismay. After all, we hold also that our religious faith and its myths are vehicles of what we call divine revelation, precious vessels of transcendent truth. While I believe that I can understand the objection, I cannot share it. It has never seemed to me that God or his revelation is demeaned by suggesting that they work within the processes of physical and biological and cultural evolution, or by insisting that they have a survival-value. To admit the biological significance of morality, for example, has never seemed to denigrate either my belief systems or my morality. On the contrary, it has always seemed to me that religion, its faith and its myths and its morality, acquire all the more powerful force when they are intimately related to the process of evolution and natural selection in which I believe God himself is involved.

Now I have suggested that the interest that governs mythic cosmologies of religion is that of discerning the assumptions about the Universe that are necessary for optimal human life—survival. The question immediately arises, "Why should we even put the matter in that way?" Why should we ask "What assumptions must be accepted if humans are to survive?" It seems an extraordinarily subjective perspective. It is a perspective that invites fantasy, projection, and self-delusion. Are we not wholly in the box in which 19th century atheistic philosophers put us, and which Freud nailed up tight: That religion and its myths project what humans want to believe? That these myths insulate man from harsh reality and thereby pacify him? Provide him with an opiate, as the radical Marxists would claim?

Obviously, I do not believe that I am fitting myself or religion for a place in the coffin that Marx and Freud prepared for us. On the contrary, I would argue that religious traditions, including their myths, are part of the necessary, albeit fallible, wisdom that has been encoded in the cultural patterns of the human species during the course of its evolution, as part of the adaptive mechanisms (both biological and cultural) that have led up until now to survival. In the contemporary jargon, I might say that religious traditions and myths are part of the information, which in the human species supplements the genotype and takes the place of the instincts that are programmed into lower species. As a part of the sociocultural evolution that has followed upon biological evolution and is its extension in the human species, religious myths and other traditions are part of the millennia-long human project in which the human species has had to learn what other species knew instinctively. Human learning and culture, defined as learned patterns of behavior, are what make human survival significantly different in style, quality, and significance from survival processes in other species.

Religion and its traditions, it has been argued by some significant scientists (Goodenough, Katz, Wallace, Williams, Campbell) possess survival significance.[3],[1] Ralph Burhoe writes that "the technological function of religion includes the accomplishment of several things necessary for making man into a social animal and a highly intelligent and self-conscious animal. . . . Religions are special sociocultural environmental boundary conditions that modify the expression of genetic information in populations" (Burhoe,[3] p. 2). This author has devoted considerable attention to charting the implications of this scientific research. He maintains: "That traditional religious belief systems as they have evolved under preconscious as well as conscious selection processes have accumulated a basic and still essentially valid salvatory or sacred wisdom is attested by some recent findings in the psychobiological and anthropological sciences" (Burhoe,[3] p. 10).

If this view of religion be true, and I am assuming that it is, and if it be also true that religious mythic cosmologies focus upon the assumptions about the Universe that are deemed essential for human survival—then we must revise our assessment of the projective and hypothetical character of religion and myth. Such projection in myth is not necessarily deceptive and fantastical. Rather, it emerged through the millennia as part of the information that enabled the human being to respond adequately to his environment. And since we believe that God created the environment and created *Homo sapiens* as adapted for survival in that environment, such myths have both human and divine significance for us.

The task, then, is to ask, "What is the significance of the fact that what the religious myths of the Universe convey was considered to be essential for human living?" To put it a bit too crassly, "What survival-value did these assumptions about the Universe have? What value do they have now?" In other words, we begin to recognize that religious myth is not necessarily a tranquilizer or an opiate, but rather part of the mechanism by which men and women face up to the realities of evolution, the demands of environmental natural selection, and their own survival.

Precisely because the myths of the Universe do pose the question of human projection so sharply, they provide an excellent field upon which to approach the survival significance of religion and its myths. Our task in the remaining parts of this paper, then, is to examine the central Christial assumptions about

---

[1]For extended discussion of this idea, see Burhoe,[4] where references to Goodenough, Katz, Wallace, Williams, and Campbell are also presented in detail.

the cosmos, and then to ask after their status and significance, against the background of this rather long introduction.

BASIC CHRISTIAN ASSUMPTIONS ABOUT THE UNIVERSE

What is the basic Christian mythic description of the Universe? Insofar as there is such a description that carries normative weight, I suggest that it is gleaned from certain passages in the Niceno-Constantinopolitan Creed of A.D. 381—as follows: "I believe in one God, the Father Almighty, Maker of heaven and earth . . . and in one Lord Jesus Christ . . . who shall come again with glory to judge the living and the dead, whose kingdom shall have no end. . . . I look for the resurrection of the dead, and the life of the world to come." The fundamental myth of Christianity, with respect to the Universe, then, is that (1) God is the maker of the Universe, (2) that he continues to be active in it (at least on this planet Earth), and (3) that God will bring his creation to a fulfillment or consumation. (God's creation has everlasting significance for man's life, which does not end with death.)

This fundamental creedal statement I have chosen as my basic text precisely because it is a brief and universally accepted text. There are richer and more beautiful expressions of the Christian myth, however. I cite two, one from Martin Luther, one from Scripture: From Luther:

> I believe that God has created me and all that exists; that he has given me and still sustains my body and soul, all my limbs and senses, my reason and all the faculties of my mind, together with food and clothing, house and home, family and property; that he provides me daily and abundantly with all the necessities of life, protects me from all danger, and preserves me from all evil. All this he does out of his pure, fatherly, and divine goodness and mercy, without any merit or worthiness on my part. For all of this I am bound to thank, priase, serve, and obey him.[5]

From Psalm 104:

> Thou dost cause the grass to grow for the cattle, and plants for man to cultivate, that he may bring forth food from the earth, and wine to gladden the heart of man, oil to make his face shine, and bread to strengthen man's heart. These all look to thee, to give them their food in due season. When thou givest to them, they gather it up; when thou openest thy hand, they are filled with good things. When thou hidest thy face, they are dismayed; when thou takest away their breath, they die and return to their dust. When thou sendest forth thy Spirit, they are created; and thou renewest the face of the ground.

To put it very plainly, the Christian mythic cosmology portrays a beginning, a middle, and a vision of the distant future. It is worth our while, I believe, to discuss (unfold) Christian belief about each of these moments—beginning, middle, and the distant future.

The Beginnings

In its conception of the beginning, of cosmic origins, the Christian myth, which is also coincident with the Hebrew myth, since Christians and Jews share the same ancient Scriptures, can be classified with other myths that represent the "creation from nothing" conception. Historians of religion have made such classifications, and one of these historians, Charles Long, has conveniently summarized four salient characteristics of such myths that apply fully to the Hebrew–Christian myth: "First of all, the Creator deity is all-powerful. He does not share his power with any other deity or structure of reality. Secondly, and a correlate of the first point, the deity exists by himself alone, in a void, or space. There is no material or reality prior to him in time or power. He creates the cosmos out of the void or nothingness in which he exists. Thirdly, the mode of creation is conscious, ordered, and deliberate; it reveals a definite plan of action. Finally, the Creator is free since he is not bound by the inertia of a prior reality."[6]

Now there are two crucial considerations that we must keep in mind, if we are adequately to understand this creation-from-nothing conception of the Christian tradition. The first is the question of how this myth came to be a creation-from-nothing affirmation, and the second is the basic content that is poured into that affirmation.

The Genesis of the Creation-from-Nothing Myth. We must consider how this myth came to be a creation-from-nothing affirmation, because it has not always been such, and it did not originate as such. In other words, if we may indulge a *bon mot*, the genesis of the creation-from-nothing myth is not in *Genesis*! Let us backtrack for a moment. Earlier I said that religious myths of cosmology focus on those assumptions that are necessary if human life is to be lived optimally. In other words, the cosmos as such is not the chief concern of these myths. As Long has put it, "The myth is a symbolic ordering which makes clear how the world is present for man" (Long,[6] p. 14). The world was present for Hebrew man, first of all as a historical reality in which he had to carve out an existence, and *not as a cosmos about whose origins he speculated*. For at least 2600 years of literary record, the Hebrew–Christian myth of creation, of cosmic beginnings, was a brief prologue statement to the history of the Hebrew nation and its Christian offspring. The account of beginnings in *Genesis* is primarily (at least until the time to Theophilus of Antioch, who died in the 180s A.D.), a statement, not about beginnings as such, but about the *dependence* of the world, and the nation of Israel, and the Christian Church, upon God. It was a myth of dependence before it was a myth of beginnings. This dependence motif came out clearly in the passage from Luther and the 104th Psalm. As the historian of Christian ideas, Jaroslav Pelikan, has written:

> Therefore, the story or stories of creation in Genesis are not chiefly cosmogony but the preface to the history that begins with the calling of Abraham. *Genesis* is not world history but the history of the covenant people of God. And as the Book of *Exodus* is interested in Pharaoh only for his part in the Exodus of Israel and otherwise cares so little about him that the Pharoah of the Exodus is still difficult to identify historically, so the Book of *Genesis* is interested in "the heavens and the earth" as the stage for the essentially historical, rather than cosmic, drama it sets out to recount.[7]

Even the Nicene Creed bears the clear trace of this original character, since the "Maker of Heaven and Earth" affirmation cannot be separated from the redemptive concern of Jesus' coming again to judge the living and the dead and the final resurrection of all humankind to live eternally in God's bosom.

I shall return later to this historical concern of the cosmic myth, but let us now ask why and how the myth of dependence became an explicit myth of beginnings. Very simply: The myth was recast in the second century A.D., because the philosophical challenge of the Hellenistic culture made it necessary. That challenge persisted in one form or another for centuries, so that the great medieval thinkers, like Thomas Aquinas, reinforced it. In other words, the myth of creation—which originally included both *continuing creation* or dependence and *original creation* or beginnings and which emphasized continuing creation over original creation—was inverted, so that origin loomed larger than dependence. This emphasis upon original creation engendered the technical phrase, *creatio ex nihilo*, which became synonymous with the creation myth.

The two challenges that Christian theologians faced in the Hellenistic world were dualism and pantheism. Dualism took two forms, metaphysical and moral. Metaphysical dualism asserted that there were several *archē* or first principles of reality. The monotheistic faith of Israel and the Church, which had proven itself over two millennia of historical existence, at least, for those two groups, would not tolerate a plurality of first principles. Therefore, creation-from-nothing was asserted. There was nothing before God, and he made all that is from that nothing. Moral dualism asserted that this earthly realm was inferior to a heavenly realm, and therefore it was to be escaped from. This violated the experience of the Hebrews and the Christians that this created order was "in essence and so in possibility good" (Gilkey,[8] p. 51). Creation from nothing nailed down these two polemical points. But, as commentators have pointed out, the negative polemic was also a powerful positive argument: that God is the sole source of all existence and of the entire Universe, and that nothing is intrinsically evil. This question of evil could consume our entire attention. Let it suffice here that the fundamental assertion of the Christian doctrine of creation is that the world is intrinsically good. Precisely because of this assertion (as Langdon Gilkey has pointed out) Christians, of all religious groups and philosophies, are most embarrassed by the presence of evil, and we have our own ways of dealing with that embarrassment (Gilkey,[8] p. 60). But we do not deal with it by placing evil either in God or in the process of creation.

Pantheism tried to identify God and the world in a univocal manner—something that once again violated the wisdom accumulated in two millennia of Hebrew and Christian experience. Ironically, even though pantheism has sought to identify the world and God, it has denigrated this world. Whether the pantheism be Mahayana Buddhism, Neoplatonism, or 19th century idealism, it has "represented a denial of the reality and value of individual creaturely existence" (Gilkey,[8] p. 59). This denigration of worldliness was also antithetical to the traditional wisdom. *Creatio ex nihilo* seemed useful as a counterpoise to both pantheistic propositions. It established the unmistakable difference between God and the world, and yet it also affirmed the goodness of the world, because it was created by the good God, and therefore it was not to be denigrated.

The Content of the Creation-from-Nothing Myth. We have actually already set forth the content of this Hebrew–Christian myth in the course of our describing how it came to be. Its content centers on (1) the sovereign God, who is absolutely preexistent to all matter and beings, (2) the free and uncoerced creation of the Universe by God, and (3) the finite world of created things, which has a being or existence "which is at one and the same time dependent upon God, and yet is real, coherent and good" (Gilkey,[8] p. 47). With this in mind, let us move to the second moment of the myth.

The Middle

The middle is, in some ways, as we have already hinted, the heart of the cosmological myth, since it is the arena of continuing creation, the dependence of the world upon God, which is at the center of the original myth and which has never ceased to be important for Christians. When I use terms like "continuing creation" and "dependence," I do not mean to say that the tradition of Christian belief has been indifferent toward speculation about the origins of the world, nor that such speculation was foreign to the belief system of Christians. Rather, I mean to emphasize what Teilhard and Long have said in the citations I have included, namely that the Christian reflection upon origins has always been refracted through a lens that relates those origins and the God of origins to the ongoing history of this world and the community of belief. Through this lens, the tradition of Christian reflection has always seen that this community and this world are beholden to the God of origins, and this affirmation of being beholden to the Creator has seldom been separate from any reflection upon origins.

Hebrews and Christians have been passionately involved in the historical and social processes of this world. Precisely because of this passionate involvement and because as a people their first memories were of history, their relationship to God as ongoing creative power at the foundations of their historical existence was central. Both the Hebrews and Christians remembered historical events as the matrix of their existence. Their memories were of Abraham called to the west from Ur, of Moses and the Exodus from Egypt, of the entry into Canaan, of the carpenter from Nazareth who taught, healed, and died on the cross. Religiously, this is of enormous importance. The distinguished historian of religion, Mircea Eliade has elaborated upon this unique emphasis upon history:

> . . . for the first time, we find affirmed, and increasingly accepted, the idea that historical events have a value in themselves, insofar as they are determined by the will of God. This God of the Jewish people is no longer an Oriental divinity, creator of archetypal

gestures, but a personality who ceaselessly intervenes in history, who reveals his will through events. Historical facts thus become "situations" of man in respect to God, and as such acquire a religious value that nothing had previously been able to confer on them (quoted by Long,[6] p. 160).

The point is not that the Hebrews and Christians are unique because they portray their God as being active in history. Several of the other Near Eastern religions did the same. Rather, the point of importance is that because the Hebrews looked to this-worldly history as the matrix of their coming to be, rather than some primordial time inhabited by a pantheon of divinities, and because their monotheism precluded such a pretemporal pantheon—for these reasons, the world of history and nature had to bear an incredible weight for them. Their cosmological glances, sketchily elaborated in myth, functioned to link this historical order to the very foundations of being.

The dialectic between nature and history, this world, on the other hand, and origins and the originating God, on the other hand, is exceedingly rich. The Hebrews could not permit themselves the luxury of being concerned primarily with origins and cosmology, except as that concern was relevant to their life in the world. And yet they had to be concerned with primordial matters, because unless the primordial creator God was indeed the undergirding power of the world and their history, their historical and natural existence would have collapsed into meaninglessness and insignificance. They were concerned more with dependence upon God than with origins, but dependence amounts to very little, if its ground is not the God of origins. G. van der Leeuw puts it squarely: Since there is only one God and he has created out of nothing, "God's act of creation is the only reality" (quoted by Long,[6] p. 161).

As a consequence, the Hebrew (Jewish) and Christian religions stand under an excruciating burden to find meaning and value in the worldly realm that is not surpassed, or even matched, by any other religion or philosophy of life. The philosophies that bear the same burden—Marxism, for example—probably have taken up this burden, because of their western origins as bastard offspring of Judaism and Christianity. There is no time before creation that is of any significance to Jews and Christians, and therefore this created order, which includes the entire Universe, must have meaning and value, else the belief-system and the life of faith are shipwrecked.

I choose to speak of these concerns as the questions of the meaningfulness and the trustworthiness of the created order. Both of these terms—meaningfulness and trustworthiness—are specifications of what it might mean to say that the creation is "good." Meaning implies "that our life exists within an environment of events coherent enough, so that what we do can be regarded as an effective means for what we want; unless there is this coherent relation of means and ends, our life seems to us pointless" (Gilkey,[8] p. 144). Trustworthiness is related closely to meaningfulness. The question of the trustworthiness of the processes of creation is at stake here, i.e., whether we can give ourselves to the world processes (including evolution, for example) which bear us along in the confidence that they will not destroy us. This is not only the question of our survival as a species, but also the question whether there is ultimately a resonance between man and his world or a dissonance—whether man is fundamentally at home in his world or out of phase with it.[9]

In these questions of the meaningfulness and the trustworthiness of world processes we come upon the inextricable relationship between questions existential and cosmological. At the point of these two questions, everyone of us must feel uncomfortable, because—while the appropriateness of these questions is existentially beyond all dispute—the answers (which our experience cries out for) are intellectually and scientifically unavailable, at least with any high degree of certainty. Teilhard de Chardin put the question of meaningfulness into cosmological terms very directly in a 1945 essay entitled, "Life and the Planets."[10] He was responding to Sir James Jeans' assertion that life on the planet Earth is incidental to a Universe that is indifferent and even hostile to it. Teilhard could not tolerate this axiom of meaninglessness, and he insisted that it was fundamentally inhuman to accept it. He dealt with Jeans' position by developing a counterposition of his own, in which he argued that planets and their developments are incidental if we consider things from the point of view of the infinitely large and the infinitesimally small. If, however, we choose another measure of significance, that of complexity, which eventually evolves into consciousness and self-consciousness, then life on the planet Earth is in the avantgarde of cosmic evolution.

Whether one agrees with Teilhard or not, he is a representative Christian in the sense that he insists upon the meaningfulness and essential coherence of the Universe and its processes. In recent times, particularly in America, it seems that Christian tradition is associated with the view that the created order is intrinsically inferior and evil. This is a profound misunderstanding, and Teilhard is a convincing witness to its falsity. Other contemporary theologians have stated the same thing by saying that for Christians, "evil is given moral status, but never ontological status" (Tillich), or that "evil is always the intruder, it never belongs essentially to creation" (Gilkey).

The Christian assertion of the goodness and meaningfulness of the processes of creation is not of recent origin, however. In the first centuries of its existence, Christianity faced a succession of foes who insisted that materiality was inferior, evil, and to be escaped. Christian faith fought a life and death battle against each of these foes and emerged with a definite bias in favor of the created world. At one stage, it was Gnosticism, which was a widespread religious and philosophical movement that stretched from Asia to the Mediterranean. Its basic message was that creation is the fruit of a defect, the primordially tainted offspring of two rebellious divinities who copulated in pretemporal bliss. Obviously, to be worldly was to be defective, and gnosis, or true knowledge, was an instrument for escaping this fleshly, material world. The struggle against Gnosticism was bitter.

Theologically, morally, and church-politically, the Church energetically fought to rid itself of Gnostic influences. Montanism was a moralistic, rigoristic, world-demeaning schismatic movement that was also rejected. Manicheism was an outright dualistic philosophy, more at home in Zoroastrianism than in Christianity. By the fourth century A.D., we may say that Christianity became what it was in a very large part because of its struggle against these foes. The great 19th century historian, Adolf von Harnack, roots the rise of theology as an activity in the struggle with Gnosticism, in which the Church was forced to develop a philosophy that could match the Gnostic thinking. Other scholars observe that the present Christian Bible was put together in order to counter an abridged version of Scriptures put forth by the Gnostic, Marcion, in an attempt to provide Christian foundation for his views. The episcopal structure of the Catholic Church and the dogma of the primacy of Rome were in part responses to the Gnostics, who had set up rival bishops.

The point I want to make with these historical references is that the Church did not affirm the goodness, meaningfulness, and trustworthiness of created processes in an incidental manner, neither as a peripheral issue. On the contrary, the Christian faith recognized over the span of several centuries that these world-denying and creation-distrusting philosophies and pieties were antithetical to the very core of Christianity. The Chalcedonian Formulation of A.D. 451, which sets forth in classic and normative fashion the incarnation of God in Jesus Christ, both as God and as man, seals the book, in a way, on this question. Once and for all, it was impossible to deny the possibilities of the created order and its potential goodness and still remain orthodox. And—to repeat what I said earlier—since Christianity is monotheistic and also denies preexistent matter and rejects the notion that pretemporal events are of any significance aside from creation itself—in light of these assumptions, Christians have put a very great investment in the cosmos and its intrinsic possibilities for meaningfulness and trustworthiness. To quote van der Leeuw again, "God's act of creation is the only reality" for the Christian.

Resorting to this conception, we can understand the statements of Teilhard, when he claimed that belief in God and belief in the world go together. In my own opinion, as I have stated it in an article entitled "The Relocation of the God Question"[9] (the title summarizes my point), the challenge of belief in God hits us precisely at this point, where we must decide whether the cosmic and planetary processes of evolution are meaningful and trustworthy or not. If we answer "Yes," I would argue that we have uttered an affirmation of God. If we say "No," then I believe we have declared ourselves atheistic in the only meaningful sense of that word.

The Distant Future

You will notice that I have not spoken of the "end" in my discussion of the Christian myth of cosmology. I do not believe that Christians hold to a view of the end. Christians believe in a fulfillment or consummation, and there is a great difference between "end" and "consummation." The creedal statements speak of the kingdom of Christ that is without end, of the resurrection of the dead, and of the life of the world to come. These, I would suggest, are authentically Christian terms, and they bespeak consummation, not termination. Consummation implies that things will be brought together, brought to fulfillment or perfection, brought to completeness. Consummation is end in the sense of goal or meaning (as in the statement, "To this end I have disciplined myself for these years"), but not necessarily in the sense of termination. If termination means the obstruction of or abortion of fulfillment, perfection, and completedness, then termination is antithetical to Christian belief about the cosmos.

CONCLUDING CONSIDERATIONS

There are several considerations that flow from my ruminations in this paper that might be discussed further:

(a) Although Christian cosmological myth can scarcely validate or invalidate scientific cosmologies, that myth does have a great stake in the work of scientists, and theologians gladly look over the shoulder of the scientists at work coveting every opportunity to converse with them. Some Christian theologians are eager to learn what may be said about purpose or meaning and trustworthiness in the scientific findings about cosmic evolutionary processes, and negative findings are naturally disturbing. Christians would probe with the cosmologists several theories and their consequences. For example, does the "Big-Bang" hypothesis imply, as the writer in the newest edition of the *Encyclopedia Britannica* suggests, that "the present dark and relatively empty universe is doomed to greater darkness and emptiness"? He goes on to say, "If the cosmos must forever expand, the glory of the early universe has departed forever, and an eternal future lies gripped in a frozen state of meaningless death."[11] Or does the steady-state theory preclude a Creator? And what is the status of planet Earth and human life? Is it but a match struck in the dark, to burn only an instant? What are the implications of Wheeler's comment (see Helitzer,[1] p. 32) that man is participant in the Universe, that "the physical world is in some deep sense tied to the human being"?

(b) Let us return for a moment to the first section of my paper, in which I insisted that the *proprium* of religious mythic cosmology is its focus on the assumptions about the Universe that are necessary for optimal living. In other words, that mythic religious cosmologies are part of the encoded information that forms the human counterpart to instinct, which is part and parcel of the human enterprise to adapt to environment and survive through the mechanisms of natural selection.

My discussion has implied that there are differences between religious mythic cosmologies. The Hebrew and Christian myths emphasize the possibilities of the cosmic processes for meaningfulness and trustworthiness, leading to consummation. What is the survival status of such a belief? Can this rank as adaptively helpful or even necessary information that humans have gleaned from their encounter with their world? How would one go about assessing such a question and what criteria would be appropriate for judging the answer?

Such a belief as I have outlined as the Christian focus on cosmology is clearly myth, and as such clearly human projection or analogy. It is very much in the interest of the human being to project the hypothesis of meaningfulness and trustworthiness—there is no illusion on this score, neither among theologians nor scientists. But the question—as the sociologist Peter Berger has put it—is not whether or not this is projection, but whether or not it is correlated to an objective reality. This may be the most pertinent question of all: Does the success of theoretical physics and mathematics reinforce our confidence in human processes of projection or making analogies? Western philosophy and Christian theology have put a high premium on the *analogia entis*, the assertion that there is a proportionality between human thinking and objective reality, particularly as that reality involves the Universe and God. The most important method of thinking in the West has been that of proportionality or analogy. Unfortunately, the theologian is seldom able to test out immediately his analogies in the way that the physicist can. The theologian can rarely say (with Remo Ruffini of Princeton), "Often we are both shocked and surprised by the correspondence between mathematics and nature, especially when the experiment confirms that our mathematical model describes nature perfectly" (quoted by Helitzer,[1] p. 27). It is only in the historical testing of the life-enhancing power of the religious myth that the theologian will be able to begin to discover whether the hypothesis of meaning and consummation is true!

The survival significance of the Christian cosmological myth is argued well in the sharp differences between the Frenchmen, contemporaries yet unacquainted with other personally, Pierre Teilhard de Chardin, the Jesuit paleontologist, and the writer and philosopher Albert Camus (whose pupil was Jacques Monod). Teilhard insisted that the Universe must be able to be thought, to be conceived, and that anything less than a thought-out Universe was less than human beings could settle for. Furthermore, he insisted that humans could not survive apart from this thought-outness, this meaningfulness, because the nerve of their action would be cut by despair. Thus, Teilhard wrote passionately, in "How I Believe,"

> If, as a result of some interior revolution, I were to lose in succession my faith in Christ, my faith in a personal God, and my faith in spirit, I feel that I should continue *to believe* invincibly *in the world*. The world (its value, its infallibility and its goodness)— that, when all is said and done, is the first, the last, and the only thing in which I believe. It is by this faith that I live. And it is to this faith, I feel, that at the moment of death, rising above all doubts, I shall surrender myself.[12]

He goes on to justify this belief in the world by affirming a cosmic interrelationship of processes that will bring him to wholeness. He insisted that to espouse ultimate pluralism and atomism was intellectually indefensible and a weakness. As a consequence, he developed his ethics of "building the world," he argued that only in building the world could one survive, and only thus would one be able to know God and serve him.

Albert Camus wrote the contrary. In his "The Myth of Sisyphus," he argues as follows:

> I said that the world is absurd, but I was too hasty. This world in itself is not reasonable, that is all that can be said. But what is absurd is the confrontation of this irrational and the wild longing for clarity whose call echoes in the human heart. For the moment it is all that links them together ... man stands face to face with the irrational. He feels within him his longing for happiness and for reason. The absurd is born of this confrontation between the human need and the unreasonable silence of the world. This must not be forgotten. This must be clung to because the whole consequence of a life can depend on it. The irrational, the human nostalgia, and the absurd that is born of their encounter—these are the three characters in the drama that must necessarily end with all the logic of which an existence is capable.[13]

Camus, like his figure Dr. Rieux in the novel *The Plague*, chose to live courageously and energetically to build up the world, but he expected neither meaningfulness nor success in his efforts because he was caught up in an absurd relationship to the world, fighting the plague in Algeria as in a "war without armistices."

For Camus, Sisyphus was prototypical man, because in his fated efforts to push a boulder up a hill, only to have it roll down again, he arrived at truth. He was fated to carry out this exercise eternally. But every time he walked down the hill after the boulder, he gained a moment of insight—this was authentic existence—energetic work, but no meaning. For Teilhard, Sisyphus is the epitome of the hopeless, antlike slave, who sees no hope, and whose position is therefore untenable.

For both Teilhard and Camus, the most significant question facing humans is whether or not to commit suicide. Teilhard decided against it, because there was meaning in the "infallible world." Camus decided the same, because it is human to fight in wars without armistices. Teilhard insisted that Camus' position was ultimately untenable, because it was maladaptive; one could not survive with such a belief system. We cannot judge men by their lives—even though Camus died tragically in his middle years in an auto accident that some believe was a form of suicide; Teilhard died at peace on Easter morning, 1955. Whether their lives and deaths are more than coincidence, we cannot know. But the two men present the options to us, and they are the options that Christian cosmic myth finds pertinent. Both men were "Christian" in their seriousness with regard to the world. The belief system of only one was in the tradition of Christian faith.

Camus and Teilhard symbolize very personal, existential questions that confront us. They seem, in some sense, to be far removed from cosmology, from the intergalactic spaces and the evolution of stars and planets. But in another sense, I can think of no area other than cosmology in which the issues of the human project are more poignantly confronted than in the study of the Universe and reflection upon that study.

The question is one of survival—is one belief-system more adaptive than the other, and if so, does that mean that it is rooted in an objective reality?

The question is also, therefore, one of man's relationship to the cosmos. I cite Wheeler again, "No theory of physics that deals only with physics will ever explain physics. I believe that as we go on trying to understand the universe, we are at the same time trying to understand man . . . . Only as we recognize that tie will we be able to make headway into some of the most difficult issues that confront us." And I close with Teilhard: "The truth about man is the truth about the universe for man, that is to say, it is the truth, pure and simple."

REFERENCES

1. Florence Helitzer, The Princeton Galaxy, *Intellectual Digest* 1973 (June), p. 32.
2. Pierre Teilhard de Chardin, *Human Energy* (Harcourt Brace and Jovanovich, New York, 1969), p. 55.
3. Ralph W. Burhoe, The Relation of Belief to Behavior, an unpublished paper, p. 2.
4. Ralph W. Burhoe, The Human Prospect and the Lord of History, *Zygon: Journal of Religion and Science* **X**, 3 (1975).
5. Martin Luther, *The Small Catechism*, Explanation of the First Article of the Creed, in *The Book of Concord*, ed. by Theodore Tappert (Fortress Press, Philadelphia, 1959), p. 345.
6. Charles Long, *Alpha, The Myths of Creation* (George Braziller, New York, 1963), p. 149.
7. Jaroslav Pelikan, Creation and Causality in the History of Christian Thought, in *Evolution After Darwin*, ed. by Sol Tax (Univ. of Chicago Press, Chicago, 1960), p. 31.
8. Langdon Gilkey, *Maker of Heaven and Earth* (Doubleday, Garden City, N.Y. 1959).
9. Philip Hefner, The Relocation of the God Question, *Zygon: Journal of Religion and Science*, **V**, pp. 11-12, (1970).
10. Teilhard de Chardin, *The Future of Man* (Harper and Row, New York, 1964).
11. Edward R. Harrison, Universe, Origin and Evolution of, *Encyclopedia Britannica* (15th ed., 1974), "Macropaedia," Vol. 18, p. 1011.
12. Teilhard de Chardin, *Christianity and Evolution* (Harcourt Brace Jovanovich, New York, 1971), p. 99.
13. Albert Camus, *The Myth of Sisyphus and Other Essays* (Knopf, New York, 1955), pp. 21-28.

Chapter XXIII

Cosmos and Creation

Arthur Peacocke

Clare College, Cambridge

The activities of cosmologists and theologians have at least this in common: both are assuming that some sense can be made by man of the universe he inhabits, or, perhaps more precisely, of that which he constructs from his observations and experiences. To this extent both are assuming that the universe is a 'cosmos,' i.e., a general world order, a generally ordered entity. They also have it in common that both are aware of the limitations of their observations. The scientific cosmologist knows that the observable universe is only a finite population of galaxies and clusters of galaxies in a greater whole, a subsystem within the theoretical limits of observations (the distance at which the galaxies recede with velocities approaching that of light), limits which themselves need not exhaust a supposedly identifiable entity called "the universe." Similarly the theologian knows in principle that he cannot ever know or say all that might conceivably be said or known of God and his relation to the natural world. Nevertheless the intuition that the universe is ordered, which observation and experience tend to validate, encourages man *qua* cosmologist and *qua* religious thinker to construct models in which that order is made explicit and is given interpretative potentiality. Later on I shall be referring to the models of creation which theologians have found useful and illuminating.

CHANCE

Before we reach that point, it is worthwhile to examine interpretations of the universe which accord to *chance* a metaphysical role in these models, both cosmological and theological, a role which can be both confusing and disturbing.

The ancient and classical antinomy of Cosmos was Chaos—

Chaos umpire sits,
And by decision more embroils the fray
By which he reigns; next him high arbiter
Chance governs all[1]

Chaos was the mythical state of affairs which preceded the emergence of the world order, of Cosmos, which was thought to manifest itself in the totality of natural phenomena. Chaos was apparently a transitional stage, for although Hesiod says[2] that 'at the first it came to be' ($\gamma\epsilon\nu\epsilon\tau$ (o)), Ovid at least pictures it as containing the seed or potentialities of ordered matter: "Before the sea was, and the lands, and the sky that hangs over all, the face of Nature showed alike in her whole round, which state have men called chaos: a rough, unordered mass of things, nothing at all save lifeless bulk and warring seeds of ill-matched elements heaped in one . . . . No form of things remain the same: all objects were at odds . . . . God—or kindlier Nature—composed this strife, for he rent asunder land from sky, and sea from land, and separated the ethereal heavens from the dense atmosphere."[3] Chaos is superseded by Cosmos, though not entirely without some continuity. However, the victory now appears to be a Pyrrhic one and the celebrations premature, so some would argue. For Jacques Monod[4] has recently made chance, in the sense of the existence of events which are the concurrence of two independent causal chains, into the cornerstone of his understanding of the implications of recent molecular biology—and has thereby once again endowed "chance" with the ability to render man's existence meaningless. It is that abyss (and Chaos was, in one of the myths, a dark and windy chasm) into which many of Darwin's contemporaries peered. It was this daunting vision which induced the anguish which Tennyson had expressed in *In Memoriam* and which provoked the Bertrand Russell of the 1920s to this famous passage:

> That Man is the product of causes which had no prevision of the end they were achieving; that his origin, his growth, his hopes and fears, his loves and beliefs, are but the outcome of accidental collocations of atoms; . . . all these things, if not quite beyond dispute, are yet so nearly certain that no philosophy which rejects them can hope to stand. Only within the scaffolding of these truths, only on the firm foundation of unyielding despair, can the soul's habitation henceforth be safely built.[5]

Today Monod contrasts the "chance" processes which bring about mutations in the genetic material of an organism and the "necessity" of their consequences in the well-ordered, replicative, interlocking mechanisms which constitute that organism's continuity as a living form. More specifically, mutations in DNA are the results of chemical or physical events and their locations in the molecular apparatus carrying the genetic information are entirely random with respect to the biological needs of the organism. Those that are incorporated into this apparatus, the genome of the organism (if they are not lethal), are only permanently so incorporated if, in interacting with its environment, the differential reproduction rate of the mutated form is advantageous. So put, it cannot be said to add anything new in principle to the debates of the past 100 years. For the essential crux in these debates was, and is, that the mechanism of variation was causally entirely independent of the processes of selection, so that mutations were regarded as purely random with respect to the selective needs of the organism long before the molecular mechanism of transmission, and alteration, of genetic information was unraveled in the last two decades.

The general conclusion he draws is that man, and so all the works of his mind and culture, are the products of pure chance and therefore without any cosmic significance. The universe must be seen not as a cosmos (that is, a directionally ordered whole) but as a giant Monte Carlo saloon in which the dice have happened to fall out in the way which produced man. There is no general purpose in the universe and in the existence of life (and so none in the universe as a whole). It need not, it might not, have existed—nor might man. Therefore any system of philosophy or religion which presupposes any plan or intention in the universe is founded on a fallacy, now fully exposed by the molecular-biological account of DNA and its mutations.

Monod has been strongly criticized by philosophers in his account of much of Western thought and his labeling of it as "animism" (in his special sense: any view which attempts to show that there is any sort of harmony between man and the universe). But the attraction of his book, in addition to its lucid scientific exposition, is that it tries to understand man's significance in the world by starting from the most accurate view of the physical and biological world available to us—namely, that afforded by the natural sciences.

In the past 100 years, the perspective of the sciences concerned with the origin and development of the physical and biological worlds has, or should have, altered our attitude to the natural surroundings which human minds appear to transcend as subjects. For our familiar environment is seen no longer to be a kind of stage for the enactment of the human drama, but rather to share with man common molecular structures and to be steps in a common continuous development in time.

Although this continuity of man with the inorganic world had sometimes been accepted in principle (e.g., *Genesis* 2:7, "And the Lord God formed man out of the dust of the ground"), it was not before the last century that the scientific evidence for man's relation to other species began to appear. Only in the past few decades could the emergence of primitive living organisms from inorganic matter be outlined in any fashion which had a scientific basis in the new knowledge of biological evolution and of molecular biology and in new insights into the development of the physical cosmos. The broad picture is familiar enough (from atoms to complex molecules and cells and thence to living organisms), but there is one stage in this development which occupies a key position in Monod's thesis: the randomness of the molecular events on which natural selection is based.

The whole context of the fundamental idea of natural selection of living organisms has been amplified, since Darwin and Wallace, by our knowledge of the existence of genetic factors, "genes," which are molecular patterns in DNA and are subject to sudden changes (as a result of irradiation or chemical events) that are random with respect to the biological needs of the organisms. It is this which so impresses Monod that he regards all living forms, including man, as the products of "chance." For the complex and subtle processes of natural selection that favor the survival of particular mutated organisms can only operate among the spectrum of possibilities provided by the random chemical events at the level of DNA.

However, I see so reason why this randomness of molecular event in relation to biological consequence has to be raised to the level almost of a metaphysical principle in interpreting the cosmos. For in the behavior of matter on a larger scale many regularities, which have been raised to the peerage of "laws," arise from the combined effects of random microscopic events which constitute the macroscopic (e.g., Boyle's law and its dependence on molecular kinetics and all of statistical thermodynamics). It would be more accurate to say that the full gamut of possible forms of living matter could only be explored through the agency of the rapid and frequent randomization which is possible at the molecular level of the DNA.

This understanding of the role of chance, as the most effective means of exploring all the possibilities innate to matter-energy-space-time, also illuminates, I would urge, an interesting cosmological question. I understand, and I am open to correction on this point, that many cosmologists are taking seriously the possibility that, on the other side from us of the 'hot big bang,'[1] when "the universe is squeezed through a knot hole,"[6] all physical constants might be different— even though all such extrapolation is speculative and hazardous. So that the nuclear and electronic energy levels of the atoms of carbon, nitrogen, hydrogen, oxygen, and phosphorus on which living matter, as we know it, is so utterly dependent, would be different. Indeed these atoms, or even atoms as such, might not be constituent units of the universe on the 'other' side of the condensed gravitational mass from which our present universe expanded. If this were so, then we have to envisage the possibility that our universe (the one which has allowed the emergence of life, and so of man) is but one among a possibly infinite cycle of universes and perhaps even of an infinite set of cycles of universes. Our universe then, on this view, just happens to be one in which the physical constants have had values so that living matter, and thus man, could, in time, appear within it—man who could then argue about cosmology and the existence of God! A small change in the physical constants could result in an uninhabitable universe and, if there were various possible sets of values of these constants, then there could be a run of universes of which a number, possibly the great majority, would not produce forms of living and eventually conscious, matter.[2]

[1] The alternative cosmology, that of the continuous creation of matter, seems now to be generally regarded (e.g., by the astrophysicists at this Symposium) as lacking adequate support in the observations.

[2] I am indebted to a letter published in the *New Scientist,* 22 August, 1974, by Mr. I. Watson (in reply to an article of mine) for drawing my attention to this possibility.

If we are to take this suggestion seriously, and I believe we must, then we have to extend both the time scale and the ontological range over which 'chance' is thought to operate. It must now be regarded as not only operating to elicit the potentialities of matter-energy-space-time over the spatial and temporal scale of our present universe, but also over the ensemble of possible universes in most of which matter-energy-space-time are replaced by new entities consistent with other values of the physical constants and acting, presumably, according to quite different physical laws than those we can ascertain in principle in this universe. Even so, the point is that over the extension of space-time (or whatever replaces it) the potentialities of the ensemble of cosmoses is being run through, 'explored' if you like, and it is this which transforms 'chance' and randomness into creative agents. Or, to put it another way, if we see 'chance' operating in this way, we are no longer daunted by the fact that the existence of life, in general, and of man, in particular, and, indeed, of our actual universe, is the result of the operation of chance; we no longer have to apotheosize "chance" as a metaphysical principle destructive of any attribution of significance and meaning to the universe. For however long it may have taken on the time scale of our universe, or however many universes may have preceded (and might follow) it, the fact is that matter-energy has in space-time, in *this* universe, acquired the ability to adopt self-replicating living structures which have in fact acquired self-consciousness and the ability to *know* that they exist and have even now found ways of discovering how they have come to be.

With reference to our own universe, it can be confidently affirmed that the primordial,[3] and still existing, cloud of neutrinos, antineutrinos, helium nuclei, electrons, and protons (particularly the last) has developed in our corner of the universe into living organisms and into man, with all his special qualities, achievements, and potentialities for sublimity and degradation. Each transition (for example, the origin of life) within the development of our cosmos can be legitimately regarded as proceeding in accordance with regularities in parallel observations we can make in, or infer from, our present experimental and theoretical investigations of the world we know; that is, briefly, the development of our cosmos has proceeded by natural "law." It is important to stress that the cosmic development presents us with an ordered behavior of matter which is not abrogated, so it seems to me, by its depending on the random, chance character of the microevents which underlie the regularities of many kinds of macroobservation.

The cosmic development is apparently a process in which new forms of organization of matter emerge. The description of terrestrial, biological evolution as displaying "emergence" is often used to point to the difficulty of fully explaining the mode of being of the newly appearing form in terms of its immediate, and certainly of its distant, predecessors. It is important to realize that it is reasonable to affirm and recognize this emergent character of the cosmic development, as, for example, Michael Polanyi does,[8] without thereby intending

---

[3]Existing, according to current views, from 100 sec after the 'big bang' up to the present and preceded, in the 10^{-2} to 10^2 sec after the 'big bang,' by photons, electrons, positrons, neutrinos, antineutrinos, and a trace of protons and neutrons.[7]

to postulate in any sense any special superadded force or principle ("élan vital," "entelechy," "life force"), which somehow mysteriously distinguishes living organisms from their nonliving components. For the principle applies equally— as Polanyi rightly argues, it seems to me—to the logical (not contingent) impossibility of reducing the principles of operation of, for example, a steam engine to the physics and chemistry of each of its components considered separately. New properties, functions, and abilities have genuinely emerged in the successive stages of the development of our cosmos, and the laws, principles, and categories of thinking and vocabulary needed to describe each stage of this process will be particular to and characteristic of it. In this sense, biology is not "nothing-but" physics and chemistry; nor is human psychology and sociology "nothing-but" biology.

MAN

Even allowing for our natural anthropomorphism, there are purely biological criteria (which I have elaborated elsewhere,[9] following G. G. Simpson) whereby man is seen to represent a point of development in which many tendencies have reached a preeminently high level. Yet a full description of the "man" who has emerged in the universe goes beyond his purely biological features. One's assessment of the nature of man has a determinative influence at this point and the challenge of the presence of man in the universe as the outcome of evolution evokes various responses among scientists. To some, like Monod, it is a stark fact, but in itself not significant for the nature of the cosmos. But to others, such as Eccles, Dobzhansky, Hardy, Hinshelwood, and Thorpe (to name a few known to the author) it is a "false modesty, verging on intellectual perversity . . . to renounce, in the name of scientific objectivity, our position as the highest form of life on earth and our own advent by a process of evolution as the most important problem of evolution," as Polanyi[10] affirms.

For to take seriously, as scientists ought, the presence of man as the outcome of the evolution of matter-energy in space-time, in this cosmos, or in the ensemble of cosmoses, is to open up many questions which go far beyond the applicable range of languages, concepts, and modes of investigation developed by the natural sciences for describing and examining the less developed and less complex forms of matter which preceded the emergence of man. For if the stuff of the world, the primeval concourse of protons, etc., has, as a matter of fact and not conjecture, become man—man who possesses not only a social life and biological organization but also an "inner," self-conscious life in relation to others, which makes him creative and personal—then how are we properly to speak of the cosmological development (or the development of the cosmoses) if, after aeons of time, the atoms have become human beings, have evidenced that quality of life we call "personal"?

Moreover, paradoxically and significantly, knowledge of the process by which they have arrived in the world seems to be confined to human beings. We alone reflect on our atomic and simpler forbears and we alone adjust our behavior in the light of this perspective. To ignore the glory, the predicament, and the possibilities of man in assessing the trend and meaning of the cosmic development would be as unscientific as the former pre-Copernican account of the Universe, based as it was on the contrary prejudice.

Thus the perspective of science on the world raises acutely certain questions which by their very nature cannot be answered from within the realm of discourse of science alone.

I. What sort of cosmos is it, if the original primeval mass of protons, etc., or whatever earlier simpler stage one chooses as one's reference point, has eventually manifested[4] the potentiality of becoming organized in material forms such as ourselves—forms of matter which are conscious and even self-conscious, can reflect, and love and hate, and pray, have ideas, can discourse with each other, can exhibit the creative genius of a Mozart or Shakespeare, or display the personal qualities of a Socrates or Jesus of Nazareth?

II. How can we explain the existence of such a cosmos of this particular character, as outlined above? It seems to me that any explanation (not cause in the cause-effect sequence of our spacetime) of the existence of such a cosmos to be plausibly adequate must be one that grounds this existence in a mode of being which is other than the cosmos so described and which transcends mental activity as much as mental activity transcends physical processes. Such a cosmos-explaining entity must be not less than personal or mental in its nature. Its (his?) existence would make it more comprehensible how matter could possess the potentiality of the mental activity evidenced in man than would the designation of "chance" alone (à la Monod) as a sufficient explanation of the cosmos. The role of randomness in natural processes does not of itself preclude the possibility of the existence of such an entity which, as Aquinas would say, "men call God." So we come explicitly to the question of God and the cosmos and the idea of creation.

CREATION

The Christian[5] idea of creation is principally an answer to questions of the sort "Why is there anything at all?" "Why is there something rather than nothing?" "Is there a reason for the world's existence?" "Why are things the way

---

[4] Apparently by pure randomness, which is just the surest way of trying out all the possible permutations and combinations.

[5] Perhaps, more strictly, the Judeo-Christian idea of creation; but since the main cultural thrust of the idea has, in the West, been mediated by Christian thinkers, I will continue to use this adjective, with this qualification taken as understood.

they are?"—which overlap and are forms of the earlier two questions (I and II), that, it was argued above, are provoked by the intellectual content of the perspective of science on the cosmos and also by the sense of wonder and mystery and open-endedness which that perspective engenders.[11] But it is also evoked by man's awareness of the mystery of personality, of its birth and growth and death, and by the realization that there is nothing necessary about our existence. We are contingent; we might not *be* at all. Equally man becomes aware of the contingent character of all that surrounds him and extends this sense of contingency to the universe as a whole. In the individual this generates a search for meaningfulness; he must make sense of his own being in relation to the world he experiences and observes.[6]

The Christian doctrine of God as Creator is an intellectual and rational attempt to provide a suitable answer to these questions and to satisfy the desire for meaningfulness. The doctrine of creation also stems from a recognition of the natural human response to and attraction for the awesome and mysterious in both numinous and aesthetic experience. This response and attraction find their most satisfying focus, at least for the theist, in an Other who beckons as well as dazzles us. These aesthetic and mystical elements in Christian theism have provided the fertile soil of an integrating and culturally creative experience in art, architecture, painting, music, and poetry which has enriched mankind. However, in the present context we shall in the following concentrate on the more purely intellectual features of theism.

Christians have affirmed, in concord with Hebraic tradition, that the whole cosmos is other than God, but dependent on him and that, more metaphysically, the cosmos has derived being, whereas only God has underived being, that is, an existence not dependent on any other entity. The classical expression of this belief is to be found in the phrases of the Nicene Creed which attribute creativity to God: God the Father is believed in as 'Maker of heaven and earth, and of all things visible and invisible'; God the Son as he 'by whom all things were made'; and God the Holy Spirit as 'the Lord, the giver of Life.' Without going into Trinitarian doctrine at this point, it is notable that in this Creed one God is said to be creator, but that each *persona (hupostasis)*

---

[6] That the Christian idea of creation is a response to the intellectual and personal question of *meaningfulness* has been very well elaborated by Langdon Gilkey in his *Maker of Heaven and Earth*[12]" . . . there is another kind of question that involves the issue of 'origins'. This is a much more burning, personal sort of question than the scientific or metaphysical ones, which are rightly motivated by a serene curiosity about the universe we live in. Here we ask an ultimate question with a distinctly personal reference. We are raising the question of 'origins' because we are asking about the ultimate security, the meaning and destiny of our own existence" (p. 19). "Whenever (therefore) we become conscious of the essential contingency of our life and its fulfillment, and when the forces that we cannot control seem to thwart our every hope, then in desperation we ask 'what possible meaning can this dependent existence of mine have ?' The problem of meaning springs initially from the contingency and so insecurity of man as a finite creature, crucially subject to external forces beyond his power that determine his weal or woe" (p. 169).

of the Triune God—what is so unfortunately translated into English as 'person'—
is also explicitly involved with creation. In its developed form, the Christian doctrine of creation became an assertion that the world was created with time[7] and
that the cosmos continues to exist at all times by the sustaining creative will of
God without which it would simply not be at all.

The magnificent first chapter of *Genesis* is properly regarded by modern
Christians not as a literal record of what happened but as a way of expressing
a present situation, an ever-present truth, by telling a story, a 'religious myth.'
The *Genesis* story is in opposition to the myths, for example, of Babylon,[13]
which always postulated something other than God (or the gods) out of which
he (or they) shaped the world, and also gave accounts of the birth of the gods
themselves. In *Genesis*, and in Christian thinking, the world exists entirely by the
fiat of God 'And God said, Let there be . . . .' In more propositional and less
literary terms, *Genesis* is asserting that nothing comes into being but by the one
God. Thus in the Biblical teaching about creation the prime emphasis is on the
dependence on God's will for the existence of everything that is; and, correspondingly, that there was and is nothing whose existence is apart from God. We
recall the ascription to God in the *Revelation*: ' . . . thou didst create all things;
by thy will they were created, and have their being' (*Rev.* 4:11, N.E.B.).

Reflection on these passages, at a variety of levels of sophistication, has proceeded throughout Christian history. It early become clear that the word 'creator' and its implied analogy to 'maker' was ambiguous and needed very careful
handling. For example, in the extensive debates on Arianism, Athanasius[14]
had to insist that the term 'creator' was not literally meant when it referred to
God 'creating' the World in a much-disputed passage, *Proverbs* 8:22. The special
way in which the verb 'create' was predicated of God was emphasized in the
subsequently developed assertion that creation was *ex nihilo*, 'out of nothing,' which thereby denied pantheism, dualism, Manichaeism, and, of course,
atheism. This assertion may at first appear to be logically absurd, for only nothing can be made out of nothing, as has frequently been pointed out. But the
phrase was intended to emphasize that 'creation' itself is not an act at a point in
time, but an analogical word representing God's relation to the cosmos now, a
relation of absolute dependence of the cosmos on God's will for its very being.

There has been much discussion of these issues in modern times; for example,
D. MacKinnon[8] quotes William Temple as interpreting the situation pithily as: God
minus the world equals God; whereas the world minus God equals nothing at all.
Furthermore, I. T. Ramsey[16] has analyzed, in a contemporary idiom, the logical form of 'creation out of nothing' (*creatio ex nihilo*) into two components.
The first is the analogical model, 'creation' (*creatio*), which is a word used of human beings making paintings, symphonies, etc., out of something or by means of

---

[7] The relation of 'creation' to time is discussed in a later section.
[8] See the discussion between A. Flew and D. M. MacKinnon in *New Essays in Philosophical
Theology*.[15]

something and is here applied to the creation of all that is in the cosmos. The second component is the qualifier, 'out of nothing' (*ex nihilo*), which has, Ramsey suggested, the function of developing a sense of wonder with the universe as its focus and which, by eliciting an awareness of its logical oddity, shows how the word 'God' is being posited in relation to this situation. Ramsey[17] (p. 75) concludes, 'whereas creation out of nothing seems, on the face of it, and from its grammar, to be talking of a great occasion in the past, it is rather making a present claim about God, and its logical grammar must be understood appropriately.' He goes on to suggest that perhaps 'creator out of nothing' (*creator ex nihilo*) might have made the point better, but that then its connection with the empirical world around us, 'creation,' would be concealed.

Ramsey referred to 'analogical models' of creation, in the plural, and this reminds us of the rich variety of ways in which has been depicted the constant, but asymmetrical, relation subsisting between God and the world, whereby God maintains the world in being: asymmetrical because all-that-is is contingent and depends on God, who is not contingent and is not dependent on all that which is other than himself. Although the uniqueness and asymmetry of this relation underlies the whole Biblical tradition (e.g., the special Hebrew verb translated as 'create' in *Gen.* 1:1 is employed only with God as its subject), Christian thinkers have found it necessary to unfold further the content of the idea of creation by means of a variety of analogical models for God and his actions.

A number of complementary and reinforcing models have been necessary to avoid the extreme of a completely transcendent God (merging into deism) and of a completely immanent God (merging into pantheism). In deism God has no effective interaction with the world; and in pantheism God tends to be identified with the world—and the Christian idea of creation repudiates both extremes. The models which have been used in the past have been elaborated by contemporary theologians (e.g., MacQuarrie[17]) and range from those stressing transcendence and the distance between Creator and creation (such as God as Maker, closely related to God as King and Sovereign) to those stressing the close and intimate relation between Creation and Creator (e.g., the relation of breath to a body, or parts of a living organism to the whole). Some modern authors have furthermore seen relevant analogies in the relation of an artist to his work and the process-theologians have espoused the term "panentheism" to denote a doctrine which attempts to combine transcedence and immanence by asserting that the world is "in" God, but that his being is not exhausted by the world. Biblical images have included the thought of creation as like a garment worn by God; as like the work of a potter; as a kind of emanation of God's life-giving energy, or spirit; as the manifestation of God's Wisdom (*Sophia*) or Word (*Logos*), which are hypostatized so that they represent the outgoing being and action of God in creation from within his own inner and ineffable nature. This nature is believed, within Christian thought, to have been partly unveiled in the person of Jesus and thereby to have been revealed essentially as Love, insofar as man can understand God's being at all. Thus creation is regarded as proceeding from the inner life of God as Love and his creating and sustaining beings other than himself as in accord with that more-than-personal inner life.

The expanding scale of cosmology, which, as we have seen, now even embraces the possibility that this universe we observe might be but one of a 'run' of possible universes, demonstrates convincingly that vast tracks of matter-energy-space-time have, and probably will, exist without any man to observe them and with no apparent reference or relation to man (unless it is to provide a reservoir of potentiality vast enough to allow to emerge that improbable coalescence of matter we call man). The excessively anthropomorphic cosmic outlook of medieval, and even of Newtonian, man is thereby healthily restored to that more sober assessment which characterizes the Old Testament.[9] The book of *Job*, moreover, urges emphatically that man's whole attitude to the created order is wrongheadedly egoistic and anthropocentric, for it is clear, the writer says in effect, that the greater part of creation is of no relevance to man at all. John Baker has recently[18] summarized this aspect of Old Testament utterances about Nature in the following terms: "Nature [in the O.T.] is not to be evaluated simply in terms of Man's needs and interests; and to think that it is, is merely a mark of folly. God created the greater part of the world for its own sake; and wisdom consists in recognising this, and the limitations which this imposes on us, so that the truly 'wise' man will never imagine that he knows what God was 'at' in creation." Any model of God's activity in creation must, therefore, allow for what we can only call God's gratuitousness and joy in creation as a whole, apart from man. The created world is then seen as an expression of the overflow of divine generosity. The model then is almost of God 'playing' in delight and sheer exuberance in creation—as when the personification of God's creative Wisdom is depicted as a child playing in front of its Creator Father:

> "When he set the heavens in their place I [Wisdom] was there,
> when he girdled the ocean with the horizon,
> . . . .
> and knit together earth's foundations.
> Then I was at his side each day,
> his darling and delight,
> playing in his presence continually,
> playing on the earth, when he had finished it,
> while my delight was in mankind" (*Proverbs* 8:27–31, N.E.B.)

The list of models of creation can, and no doubt will, be extended. But enough has, I hope, been said to demonstrate the variety of models which Chris-

---

[9]Especially the Psalms (e.g., the famous "when I consider thy heavens, the work of thy fingers, the moon and the stars, which thou hast ordained: what is man that thou art mindful of him? and the son of man, that thou visitest him?" of *Psalm* 8:3,4 (A.V.), as well as 19:1–6; 65:9–13; 104; 136; 148) and the 'Wisdom' literature (*Job, Proverbs, Ecclesiastes,* and, in the Apocrypha, the *Wisdom of Solomon* and *Ecclesiasticus*), but also some of the prophets (e.g., *Jer.* 8:7).

tian thought has employed, a variety which testifies to the ineffability of its theme. At least it illustrates that the struggles of cosmologists have long been paralleled by those of Christian theologians: They share the same grandeur and scope in their themes, similar logical pitfalls in their schemata—and perhaps, similar opprobrium from their excessively empirically minded and pragmatic colleagues!

TIME

Theological models of the relation of God and the world and the cosmological models of modern astrophysics for the most part use entirely different resources for their construction. Nevertheless each has to face the problem of how *time* is incorporated into its models and I shall now turn to an examination of certain aspects of the understanding of time in the theological models. The principal stress in the Christian doctrine of creation, as hinted above, is on the dependence and contingency of all entities, and events, other than God himself: It is about a perennial relationship and not about the beginning of the Earth, or the whole universe at a point in time. The phrase "the whole universe" is notoriously ambiguous since it seems to imply a boundary or limit 'beyond' or outside which, in some sense, God exists—and this will not do at all, for God would then be in a 'beyond' or 'outside,' which would be but an extension of the same framework of reference as that in which the "whole universe" is conceived to exist. To avoid this conceptual impasse, the principal stress in this doctrine of creation is often (e.g., by Schleiermacher) an affirmation that any particular event or entity would not happen or would not be at all were it not for the sustaining creative will and activity of God.

One of the most influential attempts to understand the relation of God, time, and creation was that of St Augustine in the eleventh chapter of the *Confessions*, where he addresses himself to those who ask, "What was God doing before he made heaven and earth?," and who point out the paradox,

> "If he was at rest," they say, "and doing nothing, why did he not continue to do nothing for ever more, just as he had always done in the past? If the will to create something which he had never created before was new in him . . . how can we say that his is true eternity, when a new will, which had never been therefore, could arise in it? . . . The will of God, then, is part of his substance. Yet if something began to be in God's substance, something which had not existed beforehand, we could not rightly say that his substance was eternal. But if God's will that there should be a creation was there from all eternity, why is it that what he has created is not also eternal?" (*Confessions* xi. 10,[20] p. 261)

This provokes St Augustine to a profound analysis of our experience of time, from which he concludes, in our terms, that the world was created along with and not in time. Time itself is a feature of the created cosmos and therefore no 'act of creation' can be located at a point within created time itself. There is no time without events and God's eternity is not just endless *temporal* duration but a mode of existence which is qualitatively different from successive temporal experience (but whether or not this mode of God's existence, 'eternity,' can accurately be described as 'timelessness' is another question, to which I refer below). Augustine addresses God thus:

> How could those countless ages have elapsed when you, the Creator, in whom all ages have their origin, had not yet created them? What time could there have been that was not created by you? How could time elapse if it never was? You are the Maker of all time. If, then, there was any time before you made heaven and earth, how can anyone say that you were idle? You must have made time, for time could not elapse before you made it. But if there was no time before heaven and earth were created, how can anyone ask what you were doing 'then'? If there was no time, there was no 'then' (*Confessions* para. 13,[20] pp. 262–3).

> It is therefore true to say that when you had not made anything, there was no time, because time itself was of your making. And no time is co-eternal with you, because you never change; whereas, if time never changed, it would not be time (*Confessions* para. 14,[20] p. 263).

> Grant them [those who ask the question 'what was God doing before he made heaven and earth?' or 'How did it occur to God to create something, when he had never created anything before?'] O Lord, to think well what they say and to recognise that 'never' has no meaning when there is no time. If a man is said never to have made anything, it can only mean that he made nothing at any time. Let them see, then, that there cannot possibly be time without creation . . . Let them understand that before all time began you are the eternal Creator of all time, and that no time and no created thing is co-eternal with you, even if any created thing is outside time (*Confessions* para. 30,[20] p. 279).

These questions formed part of the later famous debate[21] between St Thomas Aquinas and St Bonaventure. Although both accepted that the world had a beginning in time, on the basis of revelation, the former thought that the world could have existed from eternity (that is, for an infinite time)—though, in fact, revelation shows it does not so exist—whereas the latter argued that, if the world is created, then time necessarily had a beginning. On this latter view, it is more accurate to say the world was created *with* time, which seems to be Augustine's position in the *Confessions*.

Such a close connection between time and 'things,' so that both are part of the created order, is entirely in accord with the outlook to which relativistic physics has accustomed us. For we can no longer regard time, as Newton did, as a flowing river of endless duration, as a mode of extension into which events are inserted as if time had an entirely discrete and unrelated kind of existence from the events themselves and from the matter participating in them. For it is certain distinctive features of space and time, or rather space-time, which are now seen as constituting the gravitational forces that act on matter, which is known to be interchangeable (under appropriate conditions) with energy, a concept which itself involves an intimate relation to the concept of time. Thus time, in modern relativistic physics, is an integral and basic constituent of nature; hence, on any theistic view, it has to be regarded as owing its existence to God, as Augustine perceived. It is this 'owing its existence to God' which is the essential core of the idea of Creation.

Scientific cosmology, in investigating and making theoretical deductions about the remote history of our universe, cannot, in principle, be doing anything which can contradict such a concept of creation. From our radiotelescopes and other analyses of the nature of the universe at great distances of space and time we may, or may not, discover if there was a point (the 'hot big bang' of p. 368) in space-time when the universe, as we can observe it, began. We may never be able to infer what happened on the other side of this critical point of the 'hot big bang,' but whatever we eventually do infer would have no effect on the argument for affirming that the cosmos (or succession of cosmoses) has (have) derived and contingent being and that God alone is Being in Himself. The doctrine of creation itself would be unaffected since that concerns the relationship of all the created order, including time itself, to their Creator, their Sustainer and Preserver.

Nevertheless there is an important feature which the scientific perspective inevitably reintroduces into the idea of creation in the form just described. This is the realization, now made explicit, that the cosmos which is sustained and held in being by God (this sustaining and holding itself constituting 'creation') is a cosmos which has always been in process of producing new emergent forms of matter. It is a world which is still being made and, on the surface of the Earth at least, man has emerged from biological life and his history is still developing. Any static conception of the way in which God sustains and holds the cosmos in being is therefore precluded, for the cosmos is in a dynamic state and, in the corner which man can observe, it has evolved conscious and self-conscious minds, who shape their environment and choose between ends.

That the world was in flux and change, with all its corollaries for the destiny of the individual man, has been reflected upon since the ancient Greeks. But that the matter of the world developed in a particular direction to more complex and ultimately thinking forms was not established knowledge. The people of Israel, and following them, the Christian Church, have always believed in the providential hand of God in human history, with the nonhuman world usually being regarded simply as the stage for that drama. Science now sees man as part of 'Nature' and both together as subject to continuous development. Any static conception of the relation of God and the world is therefore excluded, for if the emergence of new forms of matter in the world is in some way an activity of God, then it must be regarded as his perennial activity and not something already completed. The scientific perspective of a cosmos in development introduces a dynamic element into our understanding of God's relation to the cosmos which was previously obscured, although always implicit in the Hebrew conception of a 'living God,' dynamic in action, and in the formulations of (for example) the Nicene creed. If time itself is part of the created cosmos, there seems at first to be no more difficulty in regarding God as having an innovative relationship to the cosmos at all times than in postulating such a relationship only at some posited 'zero' time. Indeed there is less difficulty, for why should God have a relation to one point in time which is different from his relation to any other point, if his mode of being is not within the temporal process?

But if God's mode of being is not within the temporal process, does this not mean that God is 'timeless,' just as he is nonspatial and without parts? This is a particular form of the problem of the relation between transcendence and immanence to which we have already referred in discussing various models of the activity and nature of God as creator. How can God be thought to act *in* time and yet be the creator *of* time? Recent analyses of this question show that a number of important traditional attributes of God (e.g., his personhood,[22,23] his ability to act in the cosmos,[23] his ability to know the world as temporal and changing[22,24]) lose coherence and meaning if God is regarded as 'timeless' in the fully negative sense of that term, i.e., as 'outside' time altogether in a way which means time cannot be said to enter his nature at all, so that he can have no temporal succession in his experience. But similar remarks pertain to space in relation to God: How can God be thought to act *in* space, and to be the creator of space, and yet to be nonspatial, to have no spatial location? We appear to have less difficulty with space in relation to God than with time, presumably because the content of mental life includes the sense of temporal succession.[10] There seems no way out but to posit both (a) that God transcends space and time, for they owe their being to him, he is their creator, and (b) that space and time can exist 'within' God is such a way that he is not precluded from being present at particular points in space and time, indeed at all such points. It was such considerations, *inter alia*, which led to the idea of 'panentheism' to which I have already referred, and which attempts to ameliorate this conceptual impasse of trying to understand how God the transcendent, the Other, who comes to men from 'beyond' and discloses himself to man, can be related to time at all.[11] This all arises because of the classical problem that we cannot assert who or what God is in Himself: we can only speak of Him by analogy.

[10]Cf. Lucas[22] (p. 304) "(such) a view of God as pure mind leads us to regard God as nonspatial, because minds are not necessarily but only contingently located in space. It seems reasonable to say with Boethius that God is not in space but space is present to, or (with Newton), in God. But although we may in this sense say that God is non-spatial, we cannot analogously argue that he is non-temporal, since minds although only contingently located in space are necessarily 'located' in time."

[11]Lucas[22] (p. 305 ff.) distinguishes between the *interval* that God can regard as "the present," which is different from our "present"; and the present *instant*, which, he argues, must be the same for God as for us, else He would not be in time at all and so could not be a person. Moreover, our present is limited: an interval but only a finite interval. But God's present, which is also an interval, is an infinite one that includes all time—"the extremest bounds of any interval which we in any context call present" (p. 306). He goes on to (§ 56) to argue that since time is a concomitant of consciousness, and so of God, it is therefore 'eternal.' "His existence implies the existence of time. He made time only in the way in which love made Him suffer" (p. 310). His whole discussion is illuminating and should be consulted.

CONCLUSION

The convergence of the lines of Christian thought which see God as immanent in the cosmos in general, in man in particular, and as consummated in a
particular historical person and in the community expressing his spirit is, I would
suggest, peculiarly consonant with the scientific perspective. For in that scientific
perspective we see a cosmos in which creativity is ever-present, in which new
forms of matter emerge, and in which in the end there emerged man, mind, human society, human values—in brief, what people call the 'human spirit.' These
two perspectives form, on the one hand, the Hebrew and Christian experience
and, on the other, the gamut of the sciences, and mutually illuminate each
other. Each has its own autonomy and justification but, if both are recognized, a
combined insight into the cosmic development is then afforded in which, it
seems to me, the features elaborated by the sciences are in harmony with the
experiences which cluster around particular events in history and which theological language expounds.

The Christian theological interpretation complements and develops the
scientific account in the significance it attributes to these events in human history. Moreover, the theological perspective, if accepted, gives meaning to the
present and a sense of direction for the future to a world still regarded as in process and as the matrix of new emergent forms of human life. The theological
persepective itself is correspondingly reshaped by this consideration of the
scientific account of the cosmic development, as I have elaborated elsewhere.[9]
The two perspectives are complementary, for the scientific provides the necessary grounding in material reality which the theological requires, and the theological furnishes the means whereby contemporary man in his community can
consciously participate and find both personal and corporate meaning in a cosmic process which, without the Christian perspective, would appear impersonal
and even inimical.

However, the contemporary Christian theist in stressing the immanent
creativity of God in the cosmos must recognize that it is by the "laws" and
through regularities of nature that God must be presumed to be working. This
recognition is linked with the important understanding that matter is of such a
kind, and the "laws" which it obeys are of such a kind, that creativity, in the
sense of the emergence of new forms of matter, is a permanent potentiality
whose actualization depends on circumstances. This potentiality is not injected
into the cosmos from "outside" by God, for God sustains through his immanent
power and will the whole world-process of matter: he it is who allows its possibilities to be actualized. For he made it thus with these potentialities and not
otherwise—including the emergence of man with all *his* hope of glory and risk
of degradation.

I should not like it to be thought that I have been arguing that the Christian
understanding of the cosmos and man's place in it can be *deduced* from the cur-

rent scientific perspective. But what I am urging is that the Christian understanding of man and the cosmos is, at least, consonant with that scientific perspective. This consonance assures us of the meaningfulness of this Christian understanding of man and the cosmos and gives it a strong claim to be the framework which is to be tested in man's life today—a life in which man's personal consciousness and aspirations must now come to terms with his physical and cosmological origins and with his individual finitude.

REFERENCES

1. John Milton, *Paradise Lost*, Book ii, 1.907.
2. Hesiod, Theogony, 1.116 of the transl. of H. G. Evelyn-White (Heinemann, London, 1929).
3. Ovid, Metamorphoses, Book I, 1.5ff in the transl. of F. J. Miller, 2nd ed. (Heinemann, London, 1921).
4. Jacques Monod, *Le Hasard et la nécessité* (Editions du Seuil, Paris, 1970); English transl., *Chance and Necessity* (Collins, 1972).
5. Bertrand Russell, *Mysticism and Logic, and other essays* (Edward Arnold, London), pp. 47–48.
6. C. W. Misner, K. S. Thorne, and J. A. Wheeler, *Gravitation* (W. H. Freeman, San Francisco, 1973), Chapter 44.
7. A. Webster, *Sci. Am.* **231**, 26 (1974).
8. M. Polanyi, *Personal Knowledge* (Routledge and Kegan Paul, London, 1958).
9. A. R. Peacocke, *Science and the Christian Experiment* (Oxford Univ. Press, London, 1971), Chapter 3.
10. M. Polanyi, *The Tacit Dimension* (Routledge and Kegan Paul, London, 1967).
11. H. K. Schilling, *The New Consciousness in Science and Religion* (SCM, London, 1973).
12. L. Gilkey, *Maker of Heaven and Earth* (Anchor Doubleday, New York, 1965).
13. C. F. von Weizäcker, *The Relevance of Science* (London, 1964).
14. Athanasius, Orat. ii, 18–22.
15. A. Flew and A. MacIntyre (eds.), *New Essays in Philosophical Theology* (SCM, London, 1955), pp. 172–73.
16. I. T. Ramsey, *Religious Language* (London, 1957), Chapter II.
17. J. MacQuarrie, *Principles of Christian Theology* (SCM, London, 1966), pp. 200ff.
18. J. A. Baker, Biblical Attitudes to Nature, Appendix i to *Man and Nature* (Report of a Subgroup of the Doctrine Commission of the Church of England), ed. by H. Montefiore (Collins, London, 1975).
19. H. Cox, *The Feast of Fools* (London, 1969); J. Moltmann, *Theology and Joy* (London, 1973).
20. St Augustine, *Confessions*, transl. by R. S. Pinecoffin (Penguin).
21. F. Coplestone, *A History of Philosophy* (London, 1964), Vol. II, pp. 262–65 and 363ff.
22. J. R. Lucas, *Treatise on Time and Space* (London, 1973), paragraphs 55, 56.
23. Nelson Pike, *God and Timelessness* (London, 1970).
24. P. T. Geach, The Future, *New Blackfriars* **54**, 208 (1973).

Chapter XXIV

Creation and Redemption

David W. Peat

University of Cambridge

The concept of God which the people of Israel developed in the course of their history stemmed directly from their deliverance from Egypt. Yahweh was revealed in history, and the Israelites experienced Him in mighty, saving acts. Yahweh was Redeemer, and He was Redeemer both of the individuals of which the nation was composed, and also of the nation as a whole—for the Israelites did not distinguish, as sharply as does modern Western man, between individual personality and corporate personality. But the Israelites were not an isolated people; they lived in Canaan among many different tribes. And the religion of these tribes had little to do with historical experiences; for these tribes religion was concerned with Nature and its own peculiar rhythms. Their gods were concerned with the seasons, the natural cyclical periods of Sun and Moon. These gods had to be propitiated in order to ensure the continued fertility of the soil and a plentiful supply of rain and sunshine in order to water and fertilize the crops on which the survival of the tribes depended.

As a result of their meeting with the religions of Canaan, the Israelites steadily expanded their early ideas of the redemptive nature of God. Gradually Yahweh came to be seen as Lord of Creation as well as the Lord who had redeemed His people from the hands of the Egyptians. And so, two strands of thought became fused in the later Yahwism—God as Redeemer and God as Lord of Creation. The primitive cosmological statements became intertwined with personal statements of faith—Creation and Nature on the one hand, personal and national redemption on the other. Even the historical tradition of Israel reflects the polarity of Creation and Redemption, as typified, for example, in the Priestly history known as Chronicles, and in the Deuteronomic history known as Kings-Samuel. It is the priestly tradition which gives rise to the Creation narrative in *Genesis*, but it is the Deuteronomist who is concerned with the obedient relation of the people to God. The two traditions are partially brought together by Deutero-Isaiah, who links Creation and Redemption by thinking of the Redemption from Egypt as the Creation of the Israelite nation.

The early Christian writings show the same polarity between Creation and Redemption. The cosmological aspect of faith is clearly shown in, for example, the writings we know as *Ephesians* and *Colossians*, and in the *Epistle to Romans*, an intellectual statement of the meaning and nature in faith in a personal Redeemer. Paul talks of "The whole of creation groaning and waiting for the glory to be revealed." And the use of the word 'Kyrios' by the early church implied both Lord of Creation as well as Lord the Redeemer.

The tradition of closely relating Creation and Redemption was followed in the early centuries particularly in the East, and it reoccurs many times in, for example, the writings of the Cappadocian Fathers, Basil of Caesarea and Gregory of Nyssa. But in the West, the development was rather different. Here the theologians were heirs to the logical, moral, and juridical traditions of Rome rather than to the speculative, mystical tradition of the East. The West was more concerned, for example, with Pelagius and the nature of moral endeavor rather than with Nestorius and Apollinarius and the nature of the Person of Christ. Nor is the Roman tradition particularly concerned with Nature at all. Practical, concerned with the administration of the state, it has much to say about the relationship between man and man, but much less to say about the relationship between man and Nature. It was the African lawyer trained in the Roman tradition, Terhillian, who uttered the famous "What has Athens to do with Jerusalem?," and later in the second half of the fourth century Ambrose, the ecclesiastical statesman concerned with the relationship between Church and State, who discourages his congregation from interest in natural phenomena.

It was perhaps natural therefore that, when in the fifteenth and sixteenth centuries the study of natural phenomena—in the sense in which modern science understands the study—was renewed, it grew up increasingly outside a Church which by centuries of tradition was not particularly interested in it. An organic view of the cosmos concerned with the totality and unity of all things was replaced by a mechanistic deterministic view of the Universe seeing relationships entirely in terms of cause and effect. A view of the whole was replaced by a view of the interacting local parts.

Now it is just this view that some cosmologists would now wish to question. They would suggest that microphysics and macrophysics—the world of the elementary particle and the Universe as a whole—may somehow be interdependent. Not only do atoms determine the Universe as a whole, but also the totality of the Universe determines the nature of matter. Indeed such a notion is only an extension of Mach's principle asserting that the inertial properties of all objects derive from distant matter. And this possibility arises against the background of a dynamic Universe in which new forms and species arise out of the extinction and re-creation of older forms, and in which it is not possible to say with certainty that the cosmos or created order had a beginning in time. Eternity, evolution through destruction and re-creation, and mutual interdependence and interaction, are of the nature of the created Universe.

MAXIMUS THE CONFESSOR AND TRACES OF THE TRINITY

The notion of attempting to discern the Holy Trinity in ordinary human experience is a commonplace in early Christian theology. In the West, for example, Augustine discusses it extensively, and earlier in the East, Origen had speculated in terms of Father, Son, and Holy Spirit corresponding to maker, provider, and discerner. But in the seventh century the monk Maximus takes this principle a stage further. "From a wise consideration of creation we perceive the idea concerning the Holy Trinity, I mean the Father, Son, and Holy Spirit." There is no doubt here that Maximus is developing his thought from the famous verse of St Paul about the invisible things of God being knowable through the visible creation. Just how Maximus does this need not concern us here; the point is that by arguing from the nature of the Universe, as understood by Greek cosmology, Maximus seeks to illuminate classical Trinitarian doctrine as developed three centuries earlier by the Cappadocians. Maximus moves from creation to final consummation; he allows his cosmology to provide the framework for the whole of his anthropology, and thereby maintains the largeness of that cosmic view seen in Origen.

Closely related to the cosmic view is deification, as seen by Maximus. Deification is the ultimate fulfilling of the capacity of human nature for God. Deification is brought about by the Incarnation; the "logos" of our nature has its powers enkindled so that it attains to God. The potentiality of human nature is elicited in the context of its relationship to God, made possible through the Incarnation. The last end of the whole of the created order is the deification of man, and this deification, although not implanted in our nature, lies within the possibility of the renewed power of man. Thus, according to Maximus, creation may point the soul to God in that it is such that is capable of being divinized as a consequence of the Incarnation.

In contemporary terms, therefore, can an actual (or real) cosmological world picture be seen as containing within it vestiges of the Divine Trinity? Can it in any way be used to illustrate the Divine Nature? In the categories described here, then, let us draw the parallel Father:Son:Holy Spirit as eternity, evolution through destruction and re-creation, and mutual interaction. God as Father represents the eternity of the created Universe, God as Son personifies within Himself through Crucifixion and Resurrection the destruction and re-creation within the Universe, and God as Holy Spirit represents the mutual interaction and sharing between the one and the whole, the individual and the many.

CONCLUSION

Our aim here is not to follow these parallels in detail, but rather to reiterate the importance of seeing within creation itself the expression of the Divine Nature. Western theology with its redemptive approach has placed much emphasis on the nature of man. But man is continuous with his natural environment, and the workings of God, if they are seen in man, must also be seen in Nature. Man can only understand his responsibility for Nature if he is willing to accept that Nature is the scene of the expression of God Incarnate.

Appendix

An English Translation of the Third Part of Kant's *Universal Natural History and Theory of the Heavens*

Stanley L. Jaki

Seton Hall University

Introductory Remarks

In reading about Kant one is almost inevitably exposed to the claim that he was a potentially great scientist, a philosopher imbued with the spirit of Newton, a remarkably original cosmologist, and the like.[1] All these claims are ultimately based on Kant's *Allgemeine Naturgeschichte und Theorie des Himmels*,[2] a cosmology which he published in 1755, at the age of thirty-one.[1] The publication on which he staked his academic hopes reached only a few hands as the publisher went bankrupt and his holdings were impounded and scattered. The first two Parts of the book have been available in English since 1900 in the translation by W. Hastie, professor of divinity at the University of Glasgow.[3] Hastie certainly shared the late nineteenth-century admiration for Kant; with his translation and especially with his introduction to it, he wanted to give further impetus to the cause of neo-Kantianism in the English-speaking world.

An avowed admirer of Kant's philosophy and personality, Hastie chose not to translate the Third Part of Kant's cosmology. This choice was retained when his translation was republished in recent years with new introductions.[4,5],[2] Thus, while the French have been able to read for almost a hundred years the

---

[1] According to its subtitle the book was "an essay on the constitution and mechanical origin of the whole universe treated according to Newtonian principles." Consisting of 200 small octavo pages, the book was published by Johann Friederich Petersen in Königsberg and Leipzig.

[2] Ley's introduction is considerably shorter than Whitrow's.

full text of Kant's cosmology[6],[3]—to say nothing of the Germans, who have it now available in a paperback edition,[7] in addition to many other editions of Kant's collected works—the English-speaking public is still deprived of that Third Part. It is, of course, available to anyone who had mastered German to the extent of being able to read Kant, whose style appeared even to his eighteenth-century countrymen a hard nut to break.

The tacit excuse for omitting the Third Part is that its topic has, on a cursory look at least, nothing in common with the hierarchical organization of galaxies and with the evolution of the planetary system, the topics of the first two Parts. Clearly, it can be made to appear plausible that planetary denizens cavorting from Mercury to Saturn, as described in the Third Part, form no integral part of scientific cosmology. Indeed, until about a few years ago speculations about intelligent beings on other planets were carefully kept out of respectable scientific literature published since the early part of the nineteenth century. Now that Mars has been found as desolate as the Moon, and Venus as uninhabitable as a cauldron, curiosity has shifted to planetarians inhabiting other solar systems. In view of recent efforts to establish radio contact with civilizations around some nearby stars and in view of the equipping of the space probe Pioneer X with a hieroglyphic plaque containing information about us earthlings,[4] there would have been enough extrinsic reasons to give the English-speaking world an accurate and full glimpse of Kant's own speculations on denizens of other planets. Moreover, there would have been some intrinsic reasons as well. These reasons, unlike the foregoing ones, are not rooted in the fleeting fashions of the day. They should rather seem to be of permanent importance to anyone interested in the true history of cosmology and in the correct mental physiognomy of Kant, the philosopher, who, as it is usually claimed, was not only deeply imbued with the spirit of Newtonian science, but might even have become, given better circumstances, a truly great scientist.

The critical sense of a reader, not specialized in physical science and its history, can easily be disarmed by the glowing introductions to the incomplete English text of Kant's cosmology. By the time he begins to read Kant's own discourse, he most likely will have no second thoughts on finding that Kant presents himself as a second Newton. By then he was already told in those introductions that Kant, although never a professional student of physics and mathematics, had taught himself through reading books lent to him by Martin Knutzen, a sympathetic and progressive professor at the University of Königsberg. Although Knutzen's tenuous connections with physics had been revealed (though somewhat indirectly) a hundred years ago,[9],[5] the myth still lingers on that Kant learned enough about Newton and physics from Knutzen, who

---

[3] See pp. 237–255 for the translation of the Third Part.

[4] For the diagram, explanation, and justification of that plaque, see Sagan and Drake.[8]

[5] The fourth and fifth chapters of this work deal with Knutzen's philosophical works, the sixth with his theological publications, and the seventh with his scientific writings, of which Erdmann found worth mentioning only one, a little treatise of the famous comet of 1744.

(and this is what is invariably ignored) did not lecture and write on physics, let alone on Newton's *Principia*, but on a wide variety of topics relating to Wolffian metaphysics.

In Kant's own evaluation of his cosmology, it offered the definitive physical part of a topic of which the definitive mathematical part had already been created by Newton. To this he added that to furnish the physics of cosmology was actually the easier task, and that he could without great effort furnish its specific mathematics, if so requested (Ref 3, pp. 36 and 73). An extraordinary boast indeed on the part of one whose writings and school record do not suggest that he had mastered even the elements of differential and integral calculus either in its geometrical or Newtonian form, or in its more recent Eulerian or algebraic formalism. In particular, Kant claimed that his cosmology dispensed with the need of resorting, à la Newton, to the Creator's arm to explain the orbiting of planets in their almost circular paths around the sun (Ref. 3, p. 72). Let it suffice to remark here that the solution of this problem, closely tied to the distribution of angular momentum in the solar system, is still awaiting, if not the Creator's arm, at least a purely scientific explanation, recent claims about its having been achieved notwithstanding.[6] Concerning Kant's inconsistency in recognizing the impossibility of having an absolute center in Euclidean infinite space and then reintroducing that center through the back door, even the nonspecialist reader might recognize something which is hardly the hallmark of a genuinely Newtonian or of a potentially and rigorously critical philosopher.

To a careful reader of Kant's cosmology it should be clear that apart from his explanation of the visual appearance of the Milky Way—the first correct one to appear in print[7]—each and every step in Kant's explanation of the evolution of the planetary system is patently *a priori* and invariably wrong. C. V. L. Charlier, himself an enthusiastic advocate, like Kant, of a hierarchical organization of galaxies, but also a first-rate mathematical physicist, became a lonely voice when about half a century ago he boldly challenged what he called "the high place" which Kant's cosmology "has obtained in the popular treatises on astronomy," a place which "it not at all deserves." As Charlier explained himself in his Hitchcock Lectures at the University of California in April 1924: "I mean that the 'Naturgeschichte des Himmels' is, scientifically, of very small value; that the comparison of it with the planetary cosmogony of Laplace is highly unjust and misleading; also that it cannot be used as a working hypothesis, which, however, may be the case with the atom-theory of Democritus. As a *popular* treatise on cosmogony I consider the 'Naturgeschichte' of Kant unsuitable and even dangerous as inviting feeble minds to vain and fruitless speculations."[12]

---

[6] For a detailed discussion of this point, see Ref. 10.

[7] For details, see Ref. 11.

[8] This point has been argued, and with specific reference to Kant's scientific publications postdating the *Critique*, in the eighth of the first series of my Gifford Lectures given under the general title, "The Road of Science and the Ways to God," at the University of Edinburgh in 1974–75 and 1975–76 (to be published by the University of Chicago Press).

This devastating judgment will not appear too extreme to those familiar with the age-old connection between *a priorism* and falsehood. This connection should for any student of Kant's cosmology make its Third Part appear an instructive piece indeed. Kant's speculations on planetarians are based on the very same *a priori* principles on which he based his *dicta* on the evolution of planets. While the obscurantism of the latter might remain hidden to a nonspecialist, the obscurantism of the former should be strikingly obvious. Once this obscurantism is seen for what it is, it will not sound too harsh to learn that as far as science is concerned, Kant never transcended his precritical stage.[8] When his ill-fated book on cosmology was republished half a dozen times in the 1790's,[9] admirers of Kant were eager to show that the celebrated spokesman of a critical, or Copernican, turn in philosophy, had even anticipated with the 'eyes of his mind' what the famous Herschel saw through his telescopes. Revealingly, Kant did not disavow his *a priori* inferences about the existence and specific characteristics of denizens on the inferior as well as on the superior planets.[10] His reluctance to do so should cast further light on the *Critique of Pure Reason*, the basis of Kant's reputation as a critically-minded philosopher, a work in which science was treated somewhat shabbily. The author of the *Critique* was, however, aware of the fact that the *a priori* theory of knowledge and especially scientific dicta in the *Critique* had to be applied to the physical sciences to prove its validity. In other words, on the basis of the *Critique* it was not only possible but simply imperative to outline in an *a priori* fashion the main structure and details of physical science, and in a definitive manner at that.

This application of the *Critique* to physics was accomplished by Kant mainly[11] in a thousand-sheet long manuscript, which became known, once it had been found and published in the 1880s, as the *Opus postumum*. It is a work into which Kant scholars, to say nothing of his unreconstructed admirers, can look only with horror. For as E. Adickes, one of its editors and the author of a massive study on Kant, the scientist, concluded half a century ago, the science of the *Opus postumum* was equivalent to Schelling's Naturphilosophie,[15] an evaluation which would have thrown Kant into a frenzy. Those who had a glimpse of the maddeningly *a prioristic* and obscurantist discourse of Schelling on science will find no difficulty in recognizing the true weight of Adickes' conclusion. If the Third Part of Kant's cosmology is read in that broader perspective, it will not only titillate by its bizarre details and reasoning, but will also instruct as few pieces from the writings of major philosophers ever will.

[9] For a full listing of those editions together with their brief description, see Ref. 13.

[10] Kant's critical sense was not strong enough to suspect that general acceptance of the principle of plenitude (Herschel himself placed people on the Sun!) does not guarantee its soundness. The hollowness of that principle is well portrayed by Lovejoy.[14]

[11] Some smaller publications of the post-critical Kant are equally revealing in this respect. Particular mention is deserved by his essay published in 1794 on the Moon's influence on weather, in which a scientific topic is discussed in the thesis–antithesis–synthesis framework of the *Critique* with an affectation of profundity but with no true scientific merit.

Text

[171]

Universal Natural History and Theory of the Heavens[12]

Third Part, Which Contains an Essay on Comparing the Inhabitants of the Various Planets on the Basis of the Analogies of Nature

> *He, who thro' vast immensity can pierce,*
> *See worlds on worlds compose one universe,*
> *Observe how system into system runs,*
> *What other planets circle other suns,*
> *What vary'd being peoples ev'ry star,*
> *May tell why Heav'n has made us as we are.*

Pope

[173] Appendix on the Inhabitants of Planets

I feel strongly that the status of learnedness is degraded when used to give with sophisticated expressions plausibility to flights of fancy with the simultaneous claim that entertaining was one's [174] sole purpose; therefore, in the present essay I wish to submit only such propositions that can really contribute to the advancement of knowledge, propositions whose probability is so well founded that one can hardly refuse to admit their validity.

It may seem though that in such a project there is no proper limit to unbridled imagination and that in judging the characteristics of the inhabitants of far-away worlds one might give free rein to phantasy with greater abandon than would a painter in depicting the plants and animals of undiscovered lands, and that such speculations are incapable of being proved or disproved; nevertheless, one must admit that the distances of the celestial bodies [planets] from the Sun embody specific relationships, which in turn entail a decisive influence on the

---

[12]Unfortunately, there is no facsimile edition of the very rare original. The text in the *Theorie-Werkausgabe*, Vol. I, *Vorkritische Schriften bis 1768*[(16)] contains the original pagination (indicated in the present translation by numerals in square brackets). The title page [171] of the Third Part is adorned with a passage from Pope's *An Essay on Man* (Epistle I, lines 23-28), which Kant quoted in B. H. Brocke's translation (1740). Other passages quoted from it are lines 31-34 of Epistle II [p. 188] and lines 237-241 of Epistle I [p. 196]. The quotation on p. [197] is from Albrecht von Haller's "Über den Ursprung des Übels" (1734), Book III, lines 197-198. The "witty author from The Hague" quoted on p. [176] is still to be identified. He is not the author (Huygens) of the *Cosmotheoros*. In the same quotation reference is being made to [Bernard de] Fontenelle, author of *Entretiens sur la pluralité des mondes* (1686), possibly the most influential science popularization of all time. The Philip mentioned on p. [179] is, of course, the father of Alexander the Great. In the concluding phrase Kant seems to allude to the new heaven and earth resulting from the final resurrection of all.

various characteristics of thinking beings located there; their manner of operating and feeling is bound to the condition of the surrounding material world and depends on the intensity of impressions which nature evokes in them according to the modalities of the relation of their habitat to [the Sun,] the center of attraction and heat.

I am of the opinion that it is not even necessary to claim that all planets should be inhabited, although it would be well-nigh absurd [175] to deny this in respect to all, or even to most of them. In the richness of nature, where worlds and world-systems are but specks of dust compared with the whole of creation, there may very well exist barren and uninhabited regions that are not useful in the slightest for the purpose of nature, namely, for contemplation by intelligent beings. To deny this would be tantamount to refusing—with a reference to God's wisdom—to admit that sandy and uninhabited deserts cover large tracts of the Earth's surface and that there are in the ocean abandoned islands where no man can be found. At any rate, a planet is much smaller in respect to the whole of creation than a desert or island in respect to the Earth's surface.

It may well be that not all celestial bodies are yet fully developed; hundreds or perhaps thousands of years are needed before a great celestial body obtains a firm structure of its material. Jupiter seems still to be in that phase. The notable changes in its form at different times have long ago made astronomers suspect that it must undergo great convulsions and that its surface is not nearly as undisturbed as the case should be with a habitable planet. Should it not have, or should it never have any inhabitant, would this not be still an infinitely smaller waste [176] of nature compared with the immensity of the whole creation? And would it not be a sign of nature's poverty rather than an evidence of her abundance, if she were to display with diligence all her richness at every point of space?

But it is more satisfactory to imagine that, although at this time Jupiter is not inhabited, it will be at a time when the period of its development is completed. Perhaps our earth had been around a thousand or even more years before it found itself in the condition to support men, animals, and plants. It does not disrupt the purposefulness of a planet's existence that it should reach such a stage of perfection in only a few thousand years. In fact, precisely because of this a planet will stay longer in its state of perfection once it has reached it; for there is a basic law in nature: Everything that has a beginning steadily approaches its decline and is all the closer to it, the farther it gets from its starting point.

One cannot indeed help agreeing with the satirical portrayal by that witty author from The Hague who after listing the general news from the realm of [the various] sciences knew how to present the humorous side of the idea about all celestial bodies being [177] necessarily inhabited: "Those creatures," says he, "which live in the forests on the head of a beggar, had long since considered their location as an immense ball, and themselves as the masterpiece of creation, when one of them, endowed by Heaven with a more refined spirit, a small

Fontenelle of his species, unexpectedly spotted the head of a nobleman. Immediately he called together all the witty heads of his quarters and told them ecstatically: We are not the only living beings in nature; see, here, that new land, there live other lice."

Let us judge without prejudice. These insects, which both in respect to their manners and insignificance represent most men, can with good reason be used for such a comparison. Since in man's belief nature is infinitely adapted to his existence, he holds negligible the rest of creation which does not embody a direct reference to his species as the central point of her purpose. Man, standing [178] immensely removed from the uppermost rank of beings, is indeed bold to flatter himself in a similar delusion about the necessity of his own existence. With the same necessity the infinity of nature includes within itself all beings which display her overwhelming richness. From the highest class of thinking beings to the most abject insect, there is nothing indifferent to nature; and nothing can be missing without breaking up the beauty of the whole, which consists in interconnectedness. There everything is determined by universal laws which nature makes operative through the connection of their originally implanted forces. Because in her procedure nature displays aptitude and order, no particular purpose should disturb and interrupt her course. In its primordial stage the formation of a planet was but an infinitely small effect of nature's fruitfulness; and it would indeed now be senseless that her well-established laws should be subservient to the particular aims of that atom. If the condition of a celestial body sets obstacles to its being inhabited, then it will not be inhabited, although in and for itself it would be more beautiful that it should have denizens of its own. The splendor of creation loses nothing thereby: For the infinite is among all magnitudes the one which by the subtraction of a finite part is not diminished. One would imply this by complaining that the space between [179] Jupiter and Mars is unnecessarily empty, and that there are comets with no denizens. In fact, that insect may appear to us as insignificant as possible, still nature is more interested in maintaining its whole species than in supporting a small number of prominent creatures of which an infinite number would remain, even though an entire region or locality would be deprived of them. Because nature is inexhaustible in producing both, one sees both of them in their preservation and decay mercilessly abandoned to general laws. Did the owner of those populated forests on the beggar's head ever make greater devastation among the numbers of that colony than did the son of Philip in the ranks of his fellow citizens when his evil genius made him think that the world existed solely for his sake?

At any rate, most planets are certainly inhabited, and those that are not, will be one day. What relationship will then obtain among the different kinds of those inhabitants in relation to their place in the world-edifice to the center, out of which diffuses the heat which keeps all alive? For it is certain that this heat produces specific relationships in the properties of the substances of those celestial bodies in proportion to their distances [from the Sun]. Man, who [180] among all rational beings is the best known to us, although at the same time his inner nature remains for us an unsolved problem, should in this respect serve as the foundation and general reference point. Here we wish to consider him neither in his moral traits, nor in the physical structure of his body: We merely wish to investigate as to what limitations would devolve on his ability to think and on the mobility of his body, which obeys the former, from the properties of matter with which he is linked and which are proportionate to the distance from the Sun. Whatever the infinite distance between the ability to think and the motion of matter, between the rational mind and the body, it is still certain that man—who obtains all his notions and representations through the impressions which the universe through the mediation of bodies evokes in his soul, both in respect of their meaning and of the faculty to connect and compare them, which man calls the ability to think—is wholly dependent on the properties of that matter to which the Creator joined him.

Man is so constructed as to receive the impressions and emotions which the [external] world must evoke in him through that body which is the visible part of his being, and the material of which serves not only to impress on the invisible soul that dwells [181] in it the first notions of external objects, but also to recall and connect them interiorly, in short, [that body] is indispensable for thinking.* In the measure in which his body develops, the faculties of his thinking nature also obtain the corresponding degree of perfection, and they reach a definite and mature status only when the fibers of his body-instrument achieve the strength and endurance which is the completion of their development. Those faculties develop in him early enough through which he can satisfy the needs to which he is subject through his dependence on external things. In some men the development stops at that level. The ability to combine abstract notions and to master the bent of passions through a free application of considerations makes its entry later; in some never in their whole lives; in others it is rather weak: it serves the lower instincts which it rather should dominate and in whose mastering consists [182] the excellence of his nature. When we consider the lives of most men, this creature seems to have been created to absorb fluids, as does a plant, and to grow, to propagate his species, and finally to age and die. He of all creatures least achieves the goal of his existence, because he uses his outstanding faculties for such purposes which other creatures accomplish more reliably and surely with far less excellent faculties. He would hardly become among all the most worthy of attention in the eyes of true wisdom, if the hope of a future life did not inspire him and if a period of complete development were not in store for the faculties enclosed in him.

*It is clear from the principles of psychology that in virtue of the actual arrangement along which the creation made soul and body dependent on one another, the former not only must obtain all concepts about the universe through union with the latter and under its influence, but that also the exercise of the faculty of thinking depends on the latter's disposition and borrows from its support the needed ability.

If one looks for the cause of impediments which keep human nature in such a deep debasement, it will be found in the crudeness of matter into which his spiritual part is lowered, and in the unbending of the fibers, and in the sluggishness and immobility of fluids which should obey its stirrings. The nerves and fluids of his brain mediate to him only gross and unclear concepts, and because he cannot counterbalance in the interior of his thinking activity the impact of sensory impressions with sufficiently powerful ideas, he will be carried away by his passions, confused and overwhelmed by the turmoil of the elements that maintain his bodily machine. The efforts of reason to rise in opposition [183] and to dissipate this confusion with the light of judgement will be like the flashes of sunlight when thick clouds continually obstruct and darken its cheerful brightness.

The grossness of the material and of the texture in the make-up of human nature is the cause of that sluggishness which keeps the faculties of the soul in perennial dullness and feebleness. The handling of reflections and of representations enlightened by reason is a tiresome process in which the soul cannot endure without effort, and out of which the soul would, through the natural inclination of the bodily machine, soon fall back into the status of passions, as the sensory impressions determine and rule all its activities.

The sluggishness of his thinking, which is a consequence of its dependence on gross and rigid matter, is the source not only of depravity but also of error. Because of the difficulty which is connected with the effort to dissipate the cloud of confused notions and to distingush and separate the universal knowledge obtained through the comparison of ideas from the sensory impressions, one's thinking readily yields to overhasty approval and acquiesces in the possession of a view which, because of the sluggishness of its nature and because of the resistance of matter, could hardly be given a close look.

Because of this dependence, the spiritual faculties disappear together with the vigor of the body: When due to the slackened flow of fluids advanced age [184] produces only thick fluid in the body, when the suppleness of the fibers and the nimbleness decrease in all motions, then the forces of the spirit, too, stiffen into a similar dullness. The agility of thought, the clarity of representation, the vivacity of wit, and the ability to remember lose their strength and grow frigid. Concepts ingrained through experience offset somewhat the disappearance of these forces, and reason would even more effectively betray its incapacity, should not the strength of passions that need its rein also diminish simultaneously and even sooner.

It becomes evident from all this, that the forces of the human soul become hemmed in and impeded by the obstacles of a crude matter to which they are most intimately bound; but it is even more noteworthy that this specific condition of matter has a fundamental relation to the degree of influence by which the Sun in the measure of its distance enlivens them and renders them adapted to the maintenance of animal life. This necessary relation to the fire, which spreads out from the center of the world-system to keep matter in the necessary excitation, is the basis of an analogy which will be firmly stated in respect to the different inhabitants of planets; and in virtue of that relationship each and any class of theirs is tied through the necessary structure of its nature to [185] the place which has been assigned to it in the universe.

The denizens of the Earth and of Venus cannot exchange their habitats without mutual destruction. The former, whose constitutive substance is adapted to the measure of heat of his distance [from the Sun] and is therefore too light and volatile for a greater heat, would in a hotter sphere suffer enormous upheavals and a collapse of his nature, which would arise from the dissipation and evaporation of his fluids and from the violent tension of his elastic fibers; the latter, whose constitutive substance has in its elements a grosser structure and sluggishness, and is in need of a stronger influence of the Sun, would in the cooler regions of the celestial space grow numb and perish in lifelessness. By the same token, there ought to be much lighter and more volatile materials of which the body of Jupiter's inhabitant is composed, so that by the weak excitation which the Sun can produce at that distance, those [bodily] machines might move as powerfully as is the case with those closer to the Sun; and thus I would sum up all this in a general form: *The substance, out of which the inhabitants of different planets as well as the animals and plants grow there are built, should in general be all the lighter and of finer texture, and the elasticity of the fibers together with the principal disposition of their build should be all the more perfect, the farther they are removed from the Sun.*

[186] This relationship is so natural and so well grounded that not only the considerations of final purpose—which in natural science should be utilized sparingly and only as secondary reasons—lead to it, but also the proportions of the specific conditions of matter composing the planets (which are established from Newton's calculations as well as from the foundations of cosmogony) confirm the same relationship according to which the material composing the celestial bodies is always lighter in those that are more distant than in those which are closer, and this should entail a similar relationship in the creatures that are produced and sustained on them.

We have established a comparison among the conditions of the material with which rational creatures on the planets are essentially united; and it may easily be seen also from the introduction to this consideration that these relations would entail a sequence also in respect of the spiritual faculties of those creatures. For if these spiritual faculties necessarily depend on the substance of the [bodily] machines which they inhabit, then we can conclude with more than probable assurance: *That the excellence of thinking beings, the promptness in their reflections, the clarity and strength of their notions that are theirs through external impressions, together with their ability to connect them, finally also the skill* [187] *in their actual use, in short, the whole range of their perfection, follow one specific rule according to which these become more excellent and perfect in proportion to the distance of their habitat from the Sun.*

Since this relationship has a measure of credibility which is not far from demonstrated certainty, we find an open field for pleasant speculations that stem from the comparison of the characteristics of these various [planetary] denizens. Human nature, which on the ladder of beings occupies exactly the middle rung, finds itself between the two extreme limits of perfection, standing equally distant from both endpoints. When the thought of the most elevated classes of rational creatures which inhabit Jupiter or Saturn hurts the pride of human nature and humiliates it through the knowledge of its lowliness, then a look at the lower rungs would bring it satisfaction and peace, for those on the planets Venus and Mercury are lowered far beneath the perfection of human nature. What an outlook worthy of wonderment! From one side we saw thinking creatures compared with whom a man from Greenland or a Hottentot would be a Newton, and on the other side some others who would stare at him as if he were an ape:

> [188] Superior beings, when of late they saw
> A mortal Man unfold all Nature's Law,
> Admir'd such wisdom in an earthly shape,
> And shew'd a NEWTON as we shew an Ape.
>
> Pope

What advances in knowledge should not be achieved by the insight of those happy beings of the uppermost spheres of the heavens! What beautiful consequences would not this brightness of insights have on their ethical disposition! The insights of intellect, when they achieve the proper degree of perfection and clarity, have much more vivid stirrings than do sensory allurements, and are able to overcome these and hold them underfoot. With what majesty would not God, who depicts Himself in all creatures, portray Himself in these thinking beings, which as an ocean undisturbed by the storms of passions would receive and reflect His image! We do not wish to stretch such considerations beyond the limits prescribed to a physical treatise; we merely wish to recall once more the analogy set forth above: *That the perfection of the world of souls increases and progresses in a straightforward gradation according to the measure of their distance from the Sun exactly as does the perfection of the material world in the realm of planets from Mercury to Saturn, or perhaps even beyond* [189] *(insofar as there are still other planets).*

Since this rule follows to an extent naturally from the consequences of the physical relation of their habitats to the center of the world, to that extent it will conveniently be admitted; on the other hand, a serious consideration of those magnificent habitats, which are fully adapted from the perfection of those [intelligent] natures in the higher regions [of the heavens], confirms that rule to such an extent that it can make a claim to being fully convincing. The agility of activities that are connected with the characteristics of a highly elevated nature matches much better the rapidly changing time periods of those spheres than does the slowness of slothful and less perfect creatures.

The telescopes teach us that the alternation of day and night on Jupiter takes place in ten hours. What would the inhabitant of the Earth do with that period if transported to that planet? Ten hours would hardly suffice for that amount of rest which that crude machine [of man's body] needs for recuperation through sleep. Would not the whole daytime be taken up with the business of awakening, dressing, and preparing food, and how would a creature, whose activities take place with such slowness, not be confused and incapacitated [190] for anything productive, as his five hours of work would be suddenly interrupted by the onset of a night of the same length? However, if Jupiter is inhabited by more perfect beings which combine more elastic forces and a greater agility in execution with a more refined build, then one may believe that those five hours are as much or even more for them than what is offered by the twelve hours of day for the lower class of humans. We know that the need for time is something relative, which can be recognized and understood in no other way than from the magnitude of that which is to be done, including in the comparison the speed of execution. Therefore the same time, which for one class of creatures is but a moment, might very well be for another a long period in which a long sequence of changes unfolds through a fast chain of efficiency. Saturn has, according to the probable calculation of its rotation which we presented earlier, an even far shorter division of day and night, and this prompts us to presume even more excellent traits in the nature of its inhabitants.

In the final count all ties together for a confirmation of the foregoing law. In all evidence nature has prodigiously spread out her provisions to the farthest regions of the world. The moons, which compensate for the industrious beings of those happy regions the absence of daylight through an adequate substitute, are in the greatest number [191] present there, and nature seems to have been careful to give all the aid to the effectiveness of those beings, so that at no time would they be deprived of utilizing such a help. In respect to moons Jupiter possesses an obvious advantage over all other lower planets, and Saturn again over Jupiter; Saturn's outfitting with a beautiful and useful ring that surrounds it makes even more probable the still greater excellence of its conditions; on the contrary, the lower planets, in whose case such a provision would be uselessly wasted and whose class rather closely borders on [the realm of] unreason, do not share in such advantages or only to a very small extent.

But one cannot consider (here I anticipate an objection that could foil all the foregoing harmony) the greater distance from the Sun, this source of light and life, as a drawback against which the special features of abodes in the case of the more distant planets would appear a mere compensation, and then object that in fact the superior planets had in the world-edifice a less distinguished location and a position which was disadvantageous to the perfection of those abodes, because they were subject to a weaker influence from the Sun. For we know that the influence of light and heat is determined not through their absolute intensity but through the ability of the material substance which absorbs them and more or less also resists their impact, [192] and that therefore the very same distance, which for a cruder material can be called a proper climate, would destroy more subtle fluids and would be for them of disastrous intensity; consequently, it takes a more refined substance composed of more mobile elements to turn the distance of Jupiter or of Saturn from the Sun into a felicitous location.

Finally, the excellence of beings in these higher celestial regions seems to be tied through a physical connection to a durability which is most proper to it. Decay and death cannot affect those excellent creatures to the extent to which they affect us lower beings. For the very sluggishness of matter and crudeness of substance, which are in the lower echelons the specific principle of debasement, are also the cause of that propensity that leads to decay. When the fluids, which nourish and make grow the animal, or man, incorporate themselves amidst its small fibers and add to its mass, can no longer enlarge those vessels and canals in volume once the growth has been completed, these additional fluids of nourishment constrict—through the mechanical drive which is expended for the nourishment of the animal—the cavities of its vessels, block them, and destroy the structure of the whole machine through a gradually increasing numbness. It is likely that although decay affects even the most perfect natures, [193] nevertheless the advantage in the refinement of substance, in the elasticity of vessels, and in the lightness and efficiency of fluids (of which those perfect beings that inhabit the more distant planets are composed) check far longer this frailty which is a consequence of the sluggishness of the cruder matter and secure for those creatures an endurance the length of which is proportional to their perfection, just as the frailty of the lives of men has a direct relation to their unworthiness.

I cannot leave these considerations without anticipating a doubt which may naturally arise from the comparison of these ideas with our previous statements. In respect to the abodes in the world-structure, we have recognized in the great number of satellites which illuminate the planets of the most distant spheres, in the speed of their rotations, and in the composition of their substances which resist the influence of the Sun, the wisdom of God that disposed so fittingly everything for their inhabitants. But how would now man reconcile with the doctrine of purposiveness a mechanistic philosophy that what the Highest Wisdom itself planned is entrusted to raw matter and that the course of Providence was to be implemented by a nature left to herself? Is the former not rather an understanding that the orderly disposition of the world-structure could not have developed through the general laws of the latter?

[194] One will quickly dissipate this doubt if one only recalls what in a similar connection was set forth in a previous section. Should not the mechanism of all natural processes have a fundamental propensity toward such consequences that fittingly correspond to the plans of the Highest Reason in the whole realm of interconnections? How could those processes have erroneous trends and an uncontrolled dissipation in their origin when all their properties, from which these consequences follow, have been determined by the eternal idea of the

Divine Intellect in which all things must necessarily be related to one another and fit together? If man reflects properly, how can one justify that kind of thinking in which nature is considered a rebellious subject who only through some harness that sets limits to her free movements can be kept in the tracks of order and of common harmony, unless one thinks that nature is her own sufficient principle whose properties know no cause, and whom God strives, as well as this can be done, to force into the plans of His intentions? The better man learns to know nature, the better will he realize that the general properties of things are not alien to and separate from one another. One will be sufficiently convinced that things have essential affinities through which they are in harmony and support one another in achieving more perfect [195] dispositions, namely, the mutual influence of elements for the beauty of the material and even for the advantage of the spiritual world, and that in general the single natures of things in the realm of eternal truths already form, so to speak, within themselves a system in which one is related to the other; one will also forthwith recognize that those affinities are proper to the things through the common unity of origin out of which they together obtained their essential properties.

And now to apply this familiar consideration to the present purpose. These general laws of motion, which in the world-system assigned to the superior planets a distant place from the center of attraction and inertia, have placed them in a regular relation to the influence of heat which also emanates from the very central point according to a similar law. And it is precisely these regularities that make the development of celestial bodies in these faraway regions more unimpeded and the generation of motions depending on those bodies much faster, in short, the whole system better established, so that finally the spiritual entities will have a necessary dependence on matter to which they are tied in person; therefore it is no wonder that [196] the perfection of nature is implemented from both sides in a single connection of causes and from the same foundations. This harmony, on closer reflection, is not something sudden and unexpected, and because the latter [spiritual] entities through a similar principle are embedded into the general disposition of material nature, the spiritual realm will be more perfect in the faraway spheres due to the same reasons by which the material world is more perfect there.

Thus in the whole span of nature all is tied together into an uninterrupted gradation through the eternal harmony which makes all members related to one another. The perfections of God have clearly revealed themselves on our levels and are not less majestic in the lower classes than in the higher:

> Vast chain of being! which from God began,
> Natures aethereal, human, angel, man,
> Beast, bird, fish, insect! what no eye can see,
> No glass can reach! from Infinite to thee,
> From thee to Nothing!
>
> Pope

We have set forth the foregoing considerations by remaining faithful to the directives of physical relationships, which would keep those considerations on the path of rational credibility. Should we permit ourselves one more escapade from these [197] tracks into the field of phantasy? Who shows us the limits where the well-founded probability ends and arbitrary fiction begins? Who is so bold as to dare an answer to the question whether sin would exercise its dominion also on the other bodies of the world-edifice, or virtue alone has her regime set up there?

> The stars are perhaps abodes of glorious souls,
> As vice rules here, there virtue is the lord.
> von Haller

Does not a certain middle position between wisdom and unreason belong to the unfortunate faculty of being able to sin? Who knows, are not the inhabitants of those faraway celestial bodies too noble and wise to debase themselves to that stupidity which is inherent in sin, but that those who inhabit the lower planets are not linked too fast with matter and endowed with all too weak spiritual faculties to be obligated to carry the responsibility of their actions before the judgment seat of justice? In such a way, only the Earth and perhaps Mars (so that we would not be deprived of the miserable comfort of not having companions in misery) would alone be in the dangerous middle road, where temptations of sensible stirrings against the domination of the spirit would possess a strong potential for seduction. And yet, the spirit cannot deny that it has the faculty by which it [198] is in a position to put up resistance to them provided its sluggishness does not take pleasure in being carried away by them; [but] where the dangerous middle point is between weakness and ability, there precisely those advantages, which put *him* above the lower classes, place *him* at a height from which he may sink again infinitely deeper below them. In fact, both planets, the Earth and Mars, are the middle members of the planetary system, and not without probability an intermediate physical as well as moral constitution between the two extremes may be assumed about their inhabitants; however, I will readily leave these considerations to those who feel they can muster more assurance in the face of undemonstrable considerations and more readiness in providing answer.

Conclusion

It is not really known to us what actually man is today, however self-awareness and reason might instruct us on this point; how much more may we err as to what man is destined to become! Still the human soul's thirst for knowledge reaches out eagerly after these topics so distant from her and strives to find some light in such a dark field.

Shall the immortal soul during the whole infinity of her future life, which the grave itself [199] does not interrupt but merely transforms, remain tied forever to this point of space, to our Earth? Shall she never share in a closer contemplation of the other wonders of creation? Who knows, if it is not her destiny that she should once know at a close range those faraway celestial bodies of the world-edifice and also the excellence of their establishments which excite so much her curiosity from a distance? Perhaps there are in the process of evolving further members of the system of planets, so that after the completed course of time, which is prescribed to our sojourn here, there may be new habitats ready for us in other heavens. Who knows, whether the moons do not orbit around Jupiter to shine finally on us?

It is permitted, it is pleasing to entertain oneself with such speculations; but nobody shall base the hope of future life on such uncertain pictures of the force of imagination. After frailty had exacted its due from human nature, the immortal soul will with a rapid swing raise herself above all that is finite and place her existence with respect to whole nature in a new relationship which derives from a closer connection with the Highest Being. From there on, this more elevated human nature, which has the source of happiness in itself, will not let herself dissipate amidst external objects and search for repose in them. The whole aggregate of creatures, which has a necessary harmony for the pleasure of the Highest [200] Origin, needs the same harmony also for its own pleasure which it will not reach except in the never-ending happiness.

In fact, when man has filled his soul with such considerations and with the foregoing ones, then the spectacle of a starry heaven in a clear night gives a kind of pleasure which only noble souls can absorb. In the universal quiet of nature and in the tranquillity of mind there speaks the hidden insight of the immortal soul in unspeakable tongue and offers undeveloped concepts that can be grasped but not described. If there are among the thinking creatures of this planet lowly beings who, unmindful of the stirrings through which such a great vision can attract them, are in the position of fastening themselves to the servitude of vanity, then how unfortunate that planet is to have been able to generate such miserable creatures! On the other hand, how fortunate is that same planet, since a road is open for it under the most desirable conditions to reach a happiness and nobility which are infinitely far above those advantages which nature's most exceptional dispositions can achieve on all celestial bodies.

End

REFERENCES

1. A. N. Whitehead, *Science and the Modern World* (Macmillan, New York, 1926), p. 199; K. Popper, On the State of Science and of Metaphysics, *Ratio* **1**, 97 (1957); F. S. C. Northrop, Natural Science and the Critical Philosophy of Kant, in *The Heritage of Kant*, ed. by G. T. Whitney and D. F. Bowers (Princeton University Press, 1939), p. 42.

2. E. Kant, *Allgemeine Naturgeschichte und Theorie des Himmels* (Johann Friederich Petersen, Königsberg and Leipzig, 1755).

3. *Kant's Cosmogony as in his Essay on the Retardation of the Rotation of the Earth and his Natural History and Theory of the Heavens*, with introduction, appendices, and a portrait of Thomas Wright of Durham, ed. and transl. by W. Hastie (James Maclehose and Sons, Glasgow, 1900).

4. *Kant's Cosmogony* [etc.], with a new introduction by G. J. Whitrow (Johnson Reprint Corporation, New York, 1970).

5. *Kant's Cosmogony*, with an introduction by W. Ley (Greenwood Publishing Corporation, New York, 1968).

6. C. Wolf, *Les hypothèses cosmogoniques. Examen des théories scientifiques modernes sur l'origine des mondes, suivi de la traduction de la Théorie du Ciel de Kant* (Gauthier-Villars, Paris, 1886).

7. E. Kant, *Allgemeine Naturgeschichte und Theorie das Himmels*, ed. with a postscript by F. Krafft (Kindler, Munich, 1971).

8. Carl Sagan and Frank Drake, A Message from the Earth, *Science* **175**, 881–84 (1972).

9. B. Erdmann, *Martin Knutzen und seine Zeit: Ein Beitrag zur Geschichte der Wolfischen Schule und insbesondere zur Entwicklungsgeschichte Kants* (Verlag von Leopold Voss, Leipzig, 1876; reprinted by H. A. Gerstenberg, Hildesheim, 1973).

10. S. L. Jaki, *Planets and Planetarians: A History of Theories on the Origin of Planetary Systems* (Scottish Academic Press, Edinburgh, 1977), Chapter VI, The Angular Barrier.

11. S. L. Jaki, *The Milky Way: An Elusive Road for Science* (Science History Publications, New York, 1972).

12. C. V. L. Charlier, On the Structure of the Universe, *Publications of the Astronomical Society of the Pacific* **37**, 63 (1925).

13. *Immanuel Kants Werke* (E. Cassirer's edition), *Vorkritische Schriften*, Band I, ed. by A. Buchenau (Bruno Cassirer, Berlin, 1912), pp. 524–525.

14. A. O. Lovejoy, *The Great Chain of Being: A Study of the History of an Idea* (Harvard Univ. Press, Cambridge, Mass., 1936).

15. E. Adickes, *Kant als Naturforscher* (W. de Gruyter, Berlin, 1925), Vol. II, p. 204.

16. E. Kant, *Theorie-Werkausgabe*, Vol. I, *Vorkritische Schriften bis 1768* (Suhrkamp Verlag, Wiesbaden, 1960), pp. 218–396.

Index of Proper Names

Subject Index